MW01250472

Urban and Highway Stormwater Pollution

Concepts and Engineering

Thorkild Hvitved-Jacobsen
Jes Vollertsen
Asbjørn Haaning Nielsen

CRC Press is an imprint of the
Taylor & Francis Group, an **informa** business

CRC Press
Taylor & Francis Group
6000 Broken Sound Parkway NW, Suite 300
Boca Raton, FL 33487-2742

Printed in the United States of America on acid-free paper
10 9 8 7 6 5 4 3 2 1

International Standard Book Number: 978-1-4398-2685-0 (Hardback)

Library of Congress Cataloging-in-Publication Data

Hvitved-Jacobsen, Thorkild.
Urban and highway stormwater pollution : concepts and engineering / authors, Thorkild Hvitved-Jacobsen, Jes Vollertsen, and Asbjørn Haaning Nielsen.
p. cm.
"A CRC title."
Includes bibliographical references and index.
ISBN 978-1-4398-2685-0 (hard back : alk. paper)
1. Urban runoff--Management. 2. Storm sewers. 3. Water quality management. I. Vollertsen, Jes. II. Nielsen, Asbjørn Haaning. III. Title.

TD657.H895 2010
628'.21--dc22 2009051355

Visit the Taylor & Francis Web site at
http://www.taylorandfrancis.com

and the CRC Press Web site at
http://www.crcpress.com

Contents

Preface

It is the objective of this book to give students and professionals within the area of environmental engineering a comprehensive text on wet weather pollution originating from urban drainage systems and road runoff. This text will give the reader theories and fundamental concepts of wet weather pollution as a basis for solving engineering problems in practice. In addition to being a text for engineering students, it is also a reference intended to support professional engineers, consultants, planners, designers, and operators. The book will provide the reader with knowledge on sources, pathways, fate, impacts, and effects of the pollutants associated with the wet weather hydrologic cycle as it occurs in areas with impervious or semi-impervious surfaces. The corresponding mathematical formulations for prediction and modeling, and the methodologies for pollution abatement, control, and management are central subjects within the text. The entire urban wastewater system including drainage catchments and networks, treatment systems, and local receiving waters will also be dealt with in this context. Analysis, design, and management of the drainage systems are addressed as key points.

It is the authors' main objective to give engineering students and engineers working in municipalities, agencies, and consulting firms sound methodologies and tools to solve urban wet weather problems in terms of pollutant reduction, abatement, and control. To observe this goal, theory and empirical knowledge are combined. Theory is needed to explain fundamental characteristics and interactions. Empirical knowledge is needed to give a solid and sound understanding of the numerous types of pollutants, their importance and impacts, and finally realistic solutions for control. The text emphasizes description and quantification of the basic phenomena and processes that are relevant for urban wet weather pollution. It is—whenever possible—an intention to deal with these aspects in a way that can be subsequently applied worldwide, irrespective of different urban infrastructures, climate conditions, cultural constraints, and institutional limitations. Methodologies with an engineering approach are given high priority.

Although the text has emphasis on pollution related to urban storm drainage, it is in terms of both methodology and applicability relevant for road and highway runoff. The basic characteristics and phenomena of the runoff from impervious urban surfaces and highways are identical. Furthermore, the pollutant contribution from traffic is central in both cases. The similarities between urban runoff and road runoff also concern measures for the control of pollution. The text is therefore also central to engineers dealing with highway runoff and the adverse environmental impacts of traffic.

The world's population of about six billion is growing with an increase of 80–90 million people per year. Furthermore, there is a strong trend of population moving from rural areas to urban settlements, particularly in the developing countries. The rapid growth of urban areas and traffic is an extreme challenge to deal with in terms of pollution. The development of sustainable cities must therefore be given high priority. In this respect, the technical, environmental, economical and socio-cultural aspects are important elements to include when formulating the sustainable urban drainage (SUD) concept. Such integrated solutions should also take into account that water in most areas is a key and scarce resource and that clean water is a basis for any improvement in health and economical development of a growing population. Handling of the water resource (whether a scarcity or what temporarily can occur during heavy storm events) is needed for all types of catchments. In cities with an existing drainage system, appropriate solutions to reduce pollutant impacts must take into account the existing system in terms of investments and area available for the drainage structures. In the developing part of the world where an infrastructure for the daily wastewater flow and storm runoff is often extremely defective, if nonexistent, these flows must be managed from the beginning with limited available financial resources. In all cases, both urban drainage and road runoff management requires rethinking to cope with the quality dimension of the wet weather flows, taking into account a large number of conditions and constraints.

It is the authors' opinion that a sound scientific and technical approach to the basic phenomena is needed to find sustainable solutions in the drainage of cities and highways. Two different aspects are considered fundamental in this respect. The first one is related to the flow of water and concerns hydrology and hydraulics of the water cycle in cities and at the roads. The second aspect concerns the pollution related to rain events and comprises pathways, fate, and effects of the numerous substances that originate from diffuse urban sources including traffic. Several textbooks concentrate on the hydrology, hydraulics, and transport phenomena relevant for urban drainage. However, when it comes to the methodologies of the quality aspects in the case of urban drainage and road runoff, the authors have found a profound need for a comprehensive text.

When seeking solutions to urban drainage and road runoff problems, one cannot deal with the quality and effect-related aspects without taking into account hydrology, hydraulics, and pollutant transport phenomena. The opposite approach—to omit the quality aspect and to focus on the physical aspects of wet weather—seems easier and is the traditional way that was followed in sanitary and civil engineering. Over the past three to four decades, however, the quality-related aspects of urban drainage and road runoff in terms of the effects of pollutants and the corresponding requirement for mitigation and management methods have been widely acknowledged. This fact calls on methodologies for the handling of phenomena in a process-oriented, systematic, and sound way. It is an ambition of this text that those methods and procedures complying with such quality aspects can be given a general description, worldwide relevance, as well as be applicable for site-specific purposes.

The authors acknowledge that a physical understanding in terms of rainfall and runoff characteristics is needed to understand and describe the quality-related drainage phenomena. These physical aspects are only dealt with in this text to a level needed for basic understanding, thereby leaving room to address the pollutant-related phenomena. The text includes the water flow and pollutant transport aspects to an extent that enables readers having only a sketchy knowledge of these subjects to understand and apply all they have learned.

The text is organized in a way that leads the reader through the theory and concepts of urban drainage pollution into their application. It is therefore important to keep in mind not only a "how" but also a "why" (i.e., giving insight to conditions and constraints that are associated with management of the urban wet weather flows). A rigid way of selecting structures and control methods will not lead to sustainable solutions. Taking this "why" seriously, the entrance is open for robust engineering solutions via sound scientific methodologies. The reader will therefore start with an overall view of the "nature" and basic characteristics of urban drainage and road runoff with the quality aspects of the water flows and receiving environment being the target. From there the reader will proceed to a description of central physical, chemical, and biological phenomena and methods that are important for urban drainage quality issues. This text will then narrow into an analysis of how the different urban drainage systems generate pollution, how it can be quantified, and how the discharges affect the environment. The reader will continue on to applications in terms of methodologies, tools, and potential solutions for improved quality of the wet weather flows that include an overview of institutional frameworks under which the solutions might be implemented. The book has an overall logical structure and at the same time focuses on specific subjects.

The specific organization of the book follows from its concepts. First, the basic understanding of pollution related to wet weather conditions and runoff in urbanized areas and at roads is dealt with in Chapter 1. Chapter 2 offers an insight into specific concepts and definitions that are basic for quality management of wet weather flows. This chapter therefore outlines the relationship among the runoff event characteristics, the pollutants, and the corresponding impacts and effects. Chapter 3 is a toolbox that concerns basic physical, chemical and biological characteristics, theories and methodologies important for understanding, and quantification of wet weather urban and road pollution. In Chapters 4 and 5, characteristics and pollutant loads for stormwater runoff, road runoff, and combined sewer overflows are dealt with in terms of both general methodologies and site-specific approaches. Chapter 6 concerns the effects of the different wet weather pollutant discharges onto the environment. In this respect, emphasis is on the different types of surface waters: streams, lakes, and coastal waters, but also soil systems (in case of infiltration) are dealt with. Principles and methodologies for sampling, monitoring, and analysis of the runoff quality are the main scope of Chapter 7. In Chapter 8, specific aspects of urban drainage quality management are focused on followed by different types of mitigation methods in Chapter 9 (i.e., best management practices, particularly related to stormwater and road runoff). Chapter 10 concerns the general aspects of predicting tools and modeling of pollutant transport and loads, impacts and effects of the runoff onto both surface waters and soil systems. Chapter 11 rounds out the text by focusing on principles for regulation of wet weather discharges and the corresponding need for sound criteria and site-specific information.

Thorkild Hvitved-Jacobsen
Jes Vollertsen
Asbjørn Haaning Nielsen
Section of Environmental Engineering
Aalborg University, Denmark

Acknowledgments

During the period this text was being developed, several people contributed to its present state. The authors are grateful for the contributions they received from colleagues and students at the Section of Environmental Engineering, Aalborg University, Denmark.

In particular the authors acknowledge several encouraging and fruitful discussions on matters related to urban drainage with Dr. Yousef A. Yousef, former professor at the Department of Civil and Environmental Engineering, University of Central Florida, Orlando. Thorkild Hvitved-Jacobsen spent a sabbatical leave during the period 1982–1983 at the University of Central Florida where he participated in a number of research projects on quality aspects of urban and road runoff headed by Dr. Yousef. The discussions with Dr. Yousef and his research group have influenced this text in several ways. In particular, it is appreciated that Dr. Yousef opened the door for an understanding of how stormwater quality and hydraulic phenomena can be integrated and studied in terms of their environmental impacts. Professor Poul Harremoës, Environment & Resources, Technical University of Denmark is also deeply acknowledged. The close cooperation with Poul during several years on a large number of urban drainage projects was always inspiring. Poul encouraged us to seek the sound and solid relations between theory and practice. It is the authors' intention and hope that this text should also reflect this approach. Poul passed away in 2003—much too early.

Authors

Thorkild Hvitved-Jacobsen is a professor emeritus at Aalborg University, Denmark. In 2008 he retired from his position as professor of Environmental Engineering at the Section of Environmental Engineering, Aalborg University, Denmark. His primary research and professional activities concern environmental process engineering of urban wastewater collection and treatment systems, including process engineering and pollution related to urban drainage and road runoff. His research has resulted in over 270 scientific publications primarily published in international journals and proceedings. He has authored and coauthored a number of books published in the United Kingdom, the United States, and Japan. He has worldwide experience as a consultant for municipalities, institutions, and private firms and has produced a great number of technical reports and statements. He has chaired and cochaired several international conferences and workshops within the area of his research and organized and contributed to international courses on sewer processes and urban drainage topics. He is a former chairman of the IWA/IAHR Sewer Systems & Processes Working Group. He is a member of several national research and governmental committees in relation to his research area. Web sites: www.sewer.dk and www.bio.aau.dk

Jes Vollertsen is a professor of environmental engineering at the Section of Environmental Engineering, Aalborg University, Denmark. His research interests are urban stormwater and wastewater technology where he combines experimental work on a bench scale with pilot scale and field studies. He integrates the gained knowledge on conveyance systems and systems for wastewater and stormwater management by numerical modeling of the processes. His research has resulted in over 130 scientific publications primarily in international journals, conference proceedings, and book chapters. He has authored and coauthored a number of books and reports published in Europe. He is an experienced consultant for private firms and municipalities as well as on litigation support. He has cochaired international conferences, been a member of numerous scientific committees and organized international courses on processes in conveyance systems and stormwater management systems. He is a reviewer for a national research committee in relation to environmental engineering. Web sites: www.sewer.dk and www.bio.aau.dk

Asbjørn Haaning Nielsen is an associate professor of environmental engineering at the Section of Environmental Engineering, Aalborg University, Denmark. His research and teaching have primarily been devoted to wastewater process engineering of sewer systems and process engineering of combined sewer overflows and stormwater runoff from urban areas and highways. He has authored and coauthored more than 40 scientific papers, which have primarily been published in refereed international journals and proceedings from international conferences. He is secretary and cashier of The Water Pollution Committee of The Society of Danish Engineers and he is a committee member of the Danish National Committee for the International Water Association (IWA). Web sites: www.sewer.dk and www.bio.aau.dk

1 Fundamentals of Urban and Highway Stormwater Pollution

This chapter aims toward giving the reader an understanding and overview of the basic phenomena that are relevant when dealing with wet weather pollution in urban areas, from roads and highways. This overview is considered important as a basis before further details can be dealt with.

In natural lands, a water balance including groundwater and surface waters (wetlands, lakes, and streams) has developed over time. This balance, although basically dynamic and dependent on the frequency of extreme events, long dry periods, and seasonal rainfall pattern, is fundamentally changed when the land is subject to urbanization. A major characteristic of urban development is associated with the more or less impervious surfaces that exist in terms of pavement and roofs for example. These urban surfaces cause two major impacts. Firstly, extreme rainfall on these surfaces could create flooding in the city and watercourses located downstream if the surface runoff is not managed properly. Secondly, the runoff will include a large number of pollutants that are mainly the result of human activity (e.g., related to input from the atmosphere, materials used for constructions, and traffic). Typically, the precipitation exists as rain, however, under cold climate conditions, it can be snow followed by runoff in the snowmelt periods.

It is an engineering task to deal with flooding and pollution of the runoff from storm events. A combination of structural "hard" engineering solutions like underground concrete sewers and "soft" solutions that more or less mimics the natural hydrology of the catchment is applied. In all cases, the engineering solutions must observe the local rainfall conditions and fit into the city structure and its surroundings. It is important that engineers who work with such subjects have a sound background in hydrology and hydraulics that can be combined with environmental process engineering originating from chemistry and biology.

Management of stormwater runoff from a "physical" point of view means that collection and transport of the water itself and its associated solid constituents are dealt with. In addition, process engineering in terms of chemical and biological transformations of constituents and their environmental impacts (i.e., the "quality" aspects of the "pollutants"), is an integral part of stormwater management. These physical, chemical, and biological aspects are from an engineering point of view thereby closely related. The traditional drainage management that was practiced by civil engineers and road engineers having a background in hydrology and hydraulics has focused on the quantity aspects. Except for the transport of sewer solids for example, methodologies to analyze and manage stormwater from a collection and transport point of view have markedly developed during the past century and are today well understood and well described (cf. e.g., Mays 2001). However, the chemical and biological aspects that include the pollution of the runoff, have only been in focus for the past three to four decades.

Management of the urban wet weather flows in terms of the quality aspects is being continually applied in a well-structured form based on sound scientific and engineering knowledge and principles. From an engineering point of view and based on the public's perception of water quality in terms of health, recreation, and protection of the environment, there is a need to support this ongoing process. A main objective of this text is to assist engineering students and practicing engineers by focusing on the process engineering concepts of urban drainage.

1.1 THE URBAN HYDROLOGIC CYCLE

The natural hydrologic cycle as it may appear in rural areas includes a great number of processes that must be dealt with for a detailed description of the entire water balance of a catchment. Figure 1.1 shows the main basic phenomena and processes associated with the natural hydrologic cycle. This hydrologic cycle is strongly related to the hydraulics of the watershed and is crucial for both local ecosystems and those located downstream. Furthermore, groundwater and surface waters are the two primary sources for both potable water and water for irrigation. A great concern is therefore devoted to the protection of this resource in terms of quality and quantity. Several textbooks deal with the details of the hydrologic cycle and its importance in the ecosystem and for human needs, for example, Chow (1964), Hammer and MacKickan (1981), and Wanielista and Yousef (1993).

Urbanization fundamentally changes the hydrologic cycle from a "natural" predevelopment state to an "artificial" postdevelopment state. Urban stormwater related pollution and management thereby have their origin in the precipitation that causes runoff from impervious or semi-impervious surfaces like streets, parking lots, and roofs of

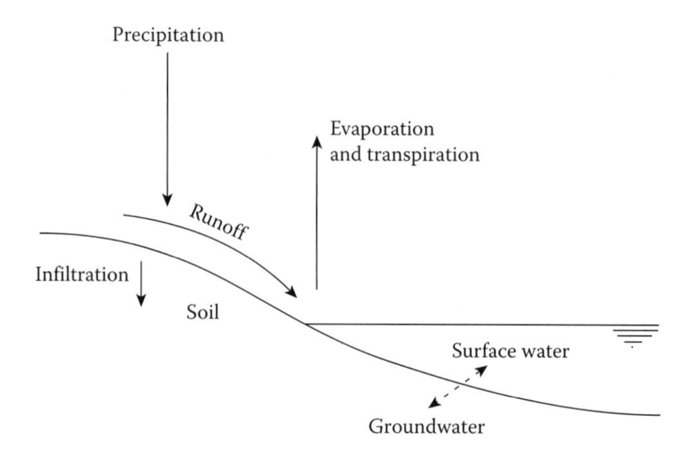

FIGURE 1.1 Principles of the natural rural hydrologic cycle.

buildings. Because of several similarities, road and highway runoff management is also included in an extended understanding of urban stormwater management.

Compared with the natural hydrologic cycle, urbanization in terms of concentration of the population and occurrence of infrastructures like streets, roads, and highways, influence the different processes of the entire water balance. Furthermore, the result of human activities (e.g., in terms of diffuse spills, air-borne pollution, and substances produced by traffic and released from constructions) contribute to the pollution of the runoff water from impervious surfaces. Compared with Figure 1.1, infiltration from these surfaces is considerably reduced if not completely suppressed. The runoff from impervious areas must therefore be managed to avoid flooding that is associated with the fact that the rate of the urban runoff processes is accelerated compared with predevelopment conditions. This time scale aspect is central because it is required that big volumes of polluted runoff water be managed within a short period of time depending on rainfall patterns. The time scale of managing the runoff water is thereby relevant in terms of both detention storage and treatment (i.e., when dealing with both quantity and quality aspects). These aspects are closely related caused by the fact that the runoff water is the vehicle for the pollutants that have been temporarily accumulated on the urban surfaces and roads.

1.2 URBAN DRAINAGE AND POLLUTION: BASIC PHENOMENA AND CHARACTERISTICS

Management of the pollution from storm events in urban areas and from roads is complex compared with the pollution originating from a specific and well-defined facility like a treatment plant or an industry. The fact that the pollution is "event-based" is without comparison the main characteristic of urban storm pollution in contrast to pollution associated with the more or less continuous discharges from most dry weather point sources. In addition to the event-based nature, the structure of the local catchment and the human activities associated with it, the following phenomena and characteristics compose generally formulated central contributions to the specific nature of urban wet weather pollution:

- *Predictability*
  Rainfall is a stochastic phenomenon. Corresponding stochastic characteristics of the pollution from runoff events therefore occur. Specification of pollution related phenomena in terms of a probability, a frequency of occurrence, or a return period therefore becomes both needed and relevant.
- *Local climate conditions*
  The climate at a specific location determines the precipitation pattern (e.g., as an annual rainfall depth, a seasonal rainfall distribution, rainfall event characteristics, and temperature). The local climate thereby affects the pollutant load and its varying impacts during all four seasons of the year.
- *Nonpoint diffuse sources*
  Compared with the continuous point sources, from a pragmatic point of view, it is not possible to identify stationary and mobile sources in

detail. Furthermore, stochastic occurring and nonintended or illegal spills may contribute to pollution of the runoff. It is therefore impossible to make a corresponding detailed description of the input of the pollutants to the urban environment. This aspect will be further dealt with in Section 1.2.

- *Pollutants*
  The identification of what are important pollutants is generally more complex and less well defined than is the case for discharges from continuous point sources like effluents from wastewater treatment plants. The relative loads of the pollutants may vary and different pollutants have different fate and environmental impact depending on the type of receiving environment. Specific pollutant characteristics in terms of their acute or accumulative nature of effects may directly determine those management measures that are feasible and relevant. Runoff water is to a great extent characterized by inorganic substances associated with small particles and typically has a relatively low content of biodegradable organic matter. Such polluted water is therefore not well suited for conventional wastewater treatment.
- *Variability*
  The collection, transport, and load of a pollutant follow the precipitation and runoff pattern and the occurrence of interevent dry periods where pollutants accumulate on the urban and road surfaces. A time dependent variability of the pollutant concentrations and loads occurs within each single event and between the different events at a specific location. As an example—and for a specific location—the standard deviation for pollutant concentrations that originate from a series of runoff events is typically of the same order of magnitude as their mean value.
- *Dilution*
  Even moderate storms will generate large volumes of runoff water during a short period of time and with relatively low (however, not insignificant) pollutant concentrations. When compared with conventional treatment of wastewater, these three facts will create difficulties for treatment of the runoff water. A situation becomes even more difficult because of the relatively long periods of dry weather—the interevent dry periods—where no input of runoff water to a treatment facility takes place. Therefore, the specific characteristics of urban storm pollution call for methods and technologies that are quite different compared with those that are common for conventional treatment of domestic wastewater.

The nature of the origin—the sources—of pollution needs further consideration and fundamental understanding. In principle, this aspect is closely associated with the characteristics of wet weather pollution mentioned above (i.e., the event-based nature and the six formulated phenomena). As a basic characteristic of the runoff from urban and road surfaces, the nature of the origin of pollution is associated with both a time and a space related element:

- *The time element of pollution*
  The two extremes of this approach are the continuous and the stochastic occurring nature of the pollution, respectively. Although the generation of the pollution (i.e., the process of generation at the urban and road surfaces), occurs more or less continuously, both management and impacts related to the pollution from the observer's point of view is stochastic. The reason is that collection and transport of the pollutants to the point where management (treatment) and impacts occur are closely related to the intermittent occurring runoff flow.
- *The space element of pollution*
  Like the time element, the space element also includes two extremes, namely a point and a diffuse-related element. Although the pollution in several cases is associated with a well-defined point element as the source (e.g., a car, a chimney, or an outlet from a factory), the associated area related occurrence of the pollution at the urban and road surfaces often overrules this fact. Therefore, it becomes appropriate to consider the pollution as occurring from a nonpoint diffuse source (cf. one of the six points mentioned above).

It is important to understand this more or less philosophic approach of the nature of pollution in terms of its origin. As an example, source control is closely associated with the understanding of the point element of the pollution. It is well known that the diffuse occurrence of lead is controlled via the input of unleaded petrol to vehicles (i.e., well-defined point sources). However, when dealing with pollutant concentrations—a very important characteristic related to engineering of the polluted wet weather flows—the concept of diffuse pollution becomes central (i.e., a pollution arising from numerous widespread and different sources and therefore typically considered as a surface area related load).

As mentioned above, wet weather pollution is normally associated with a more or less continuously occurring generation of pollutants at the urban surfaces and roads. However, on top of that it is important to note that both illegal and nonappropriate activities (e.g., pollutant spills and leaks from factories, accidents and handling of pollutant flows like car wash) can increase the wet weather loads considerably in a stochastic way. Although such sources are not directly included in management procedures of wet weather pollution, it is crucial to understand that they may occur and that these events are taken into account by appropriate design and selection of robust procedures, constructions, and facilities.

The occurrence and interactions of the phenomena and characteristics dealt with in Section 1.2 make the description and prediction of urban wet weather pollution complex. It is crucial to understand that the well-known methodologies for monitoring, sampling, treatment, computation, and system design must be looked upon and assessed from an angle that takes into account these facts. It is a challenge within this text to give the reader a basic understanding of the relevant phenomena and the knowledge of appropriate tools that can be applied for selecting sound solutions to stormwater pollution problems.

A systematic way to approach and predict the urban wet weather pollution problems from their origin to the final stage in the environment is needed. The following framework is logically following the pathways of the pollutants in this respect:

- Identification of the level of nonpoint diffuse pollutant sources, stationary as well as mobile, and assessment of catchment characteristics that affect the generation of pollutants.
- Identification of the pollutants that are relevant in terms of their impacts and their amount—often in a pragmatic way considered relative to the pollutants from point sources like conventional wastewater treatment plants but also because of their intermittent occurrence.
- Identification of the pathways of the pollutants including both mode of transportation and potential transformations.
- Determination of the loads of the pollutants into the environment.
- Determination of the effect of the pollutants.

Taking this framework as a starting point, it will be possible to select and assess both methods and measures that can be applied to solve stormwater pollution problems. The overall objective in this respect is to protect human health and well-being, a diverse environment, and resources of both material and nonmaterial nature (like financial resources). This objective can be understood as a sustainable goal for stormwater management.

1.3 BASIC POLLUTANT CHARACTERISTICS AND ENVIRONMENTAL EFFECTS

1.3.1 Pollutants and Effects

The pollutants relevant in urban storm drainage can be subdivided in different ways. The following grouping is typically used and relevant for a number of corresponding effects:

- Biodegradable organic matter
- Nutrients
- Heavy metals
- Organic micropollutants (organic priority pollutants)
- Solids (suspended solids)
- Pathogenic microorganisms

These pollutants result in different types of effects. In addition to the effects of these pollutants, the hydraulic conditions also pose an impact on the environment. The integrated effects of both pollutants and flow of water are important and therefore dealt with here. The grouping of the effects shown in Table 1.1 is relevant in most cases.

The effects that are outlined in Table 1.1 will be subject to further detailed description relevant for the receiving environment (cf. Chapter 6).

TABLE 1.1
Overview of Effects Relevant for Urban Wet Weather Pollution

Type of Effect	Subdivision and Comments
Physical habitat changes	1. Flooding in urban and rural areas 2. Erosion caused by overland flow and peak flows in channels and rivers 3. Sediment deposition in the collection and transport systems and in the receiving waters
Dissolved oxygen depletion	Effects on the biological communities
Eutrophication	Effects of both nutrients (N and P) and organic matter as substrates for excessive biological growth and activity
Toxic pollutant impacts	Effects of both heavy metals and organic micropollutants
Public health risks	1. Direct impacts by pathogenic microorganisms and viruses 2. Indirect impacts via contaminated food, both animals (fish) and crops grown on irrigated land
Aesthetic deterioration and public perception	Typically caused by the discharge of gross solids and sediments

In terms of assessment, modeling, and engineering of drainage systems, it is crucial to relate the discharge of the pollutants to their specific adverse effects. Table 1.1 is a first estimate in this respect. Under such conditions and particularly in case of design, it is important to point out pollutants that can be considered appropriate indicators for effects that are caused by the wet weather discharges.

1.3.2 Time and Spatial Scale Effects of the Pollutants Discharged

The event-based characteristics of the wet weather pollution that are briefly outlined in Section 1.2 make it important to introduce the concept of "time scale" related to effects of the pollutants. This time scale is associated with—but not equivalent to—the recovery time of the receiving environment following the intermittent pollutant discharges. The following example will illustrate this: Ammonia that is discharged during a rain event into a river may cause a direct effect in terms of fish kills. However, the recovery time of the river following this harmful event may last a very long time until the river is repopulated by the fish species. When dealing with pollution from urban wet weather discharges, it is typically relevant to focus on the "direct" effect of a pollutant and not the associated "recovery time" of the environment. What is important, however, is to compare the duration of a discharge event with the duration of a discharged pollutant's effect.

In the context of wet weather urban discharges there is no well-defined terminology to clearly distinguish between the word "impact" and "effect." In this text it is considered appropriate to define that urban wet weather discharges cause impacts in the environment and that these impacts can result in adverse effects. From the receiving system's point of view, it is therefore considered appropriate to use the term *effect* as a more specific and precise indication of what is a harmful response of the environment to such discharges. The understanding of *impact* is less precise and often seen from a discharge point of view.

Basically, the effect of a pollutant is acute or accumulative (i.e., cumulative or chronic). An adverse effect of a pollutant at a specific location will either result in a shock effect that decreases relatively fast or the pollutant will accumulate in the system and result in a long-term effect. Such pollutant and receiving water related characteristics are of course also valid for discharges from continuous point sources. However, under such conditions, a pollutant effect reaches a kind of equilibrium in the environment irrespective of its type. For intermittent point discharges like those occurring for urban wet weather discharges, the time scale effect of a pollutant will directly manifest itself because of a nonpolluting, inter-event dry weather period.

If the effect is acute, the impact from single events are important, particularly for those of extreme magnitude and relatively frequent occurrence. If the effect is of the cumulative type, it is important to consider the discharges over a certain period of time, typically a season or a year. In this case the variability of the pollutant load from storm to storm is not important. Figure 1.2 depicts the principles of the time scale concept for pollutants relevant for wet weather discharges into receiving waters.

With reference to Figure 1.2 and as examples, dissolved oxygen depletion that is caused by readily biodegradable organic matter will result in an acute effect whereas nutrients, for example, causing eutrophication are pollutants with a long-term,

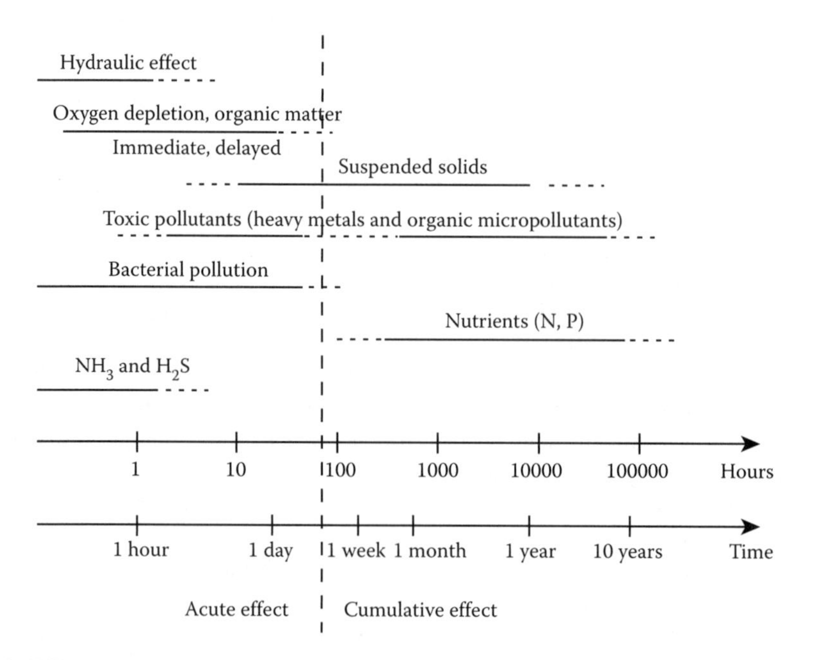

FIGURE 1.2 Time scale for receiving water effects from intermittent pollutant discharges. (Data from Hvitved-Jacobsen, T., Johansen, N. B., and Yousef, Y. A., *Science of the Total Environment,* 146/147, 499–506, 1994.)

cumulative effect. Figure 1.2 only shows the principle and therefore just serves the purpose of a basic understanding and distinction between the two phenomena. Details concerning the distinction between acute and accumulative effects are determined by the kinetics of the relevant processes. Furthermore, the possibility for reuse or recycling of a substance, as is the case for the continuously occurring nutrient cycling in a lake, will also play a role.

Because of the time scale of the effects of the pollutants, corresponding computational and treatment aspects are affected:

- *Acute effects*
  Since a short-term effect is related to a single event, computation of pollutant discharges becomes relevant for each runoff event that causes a discharge, particularly the extreme ones. It is therefore important to deal with the extreme event statistics of a historical rainfall or runoff series. It is also important to deal with the extreme events when focusing on possible methods and measures for reduction of pollutant loads and treatment of the discharges.
- *Accumulative effects*
  In contrast to the acute phenomenon, all discharge events during a season or a year contribute to the observed harmful effect. Therefore, the relevant measure for the effect is basically changed from an actual concentration or a pollutant load for a given event to the total amount of a pollutant that is discharged during a number of events corresponding to the relevant period of time. A variability of the pollutant load from storm to storm is in this case not important and emphasis is on a mean or median pollutant concentration for the site as the basis for determination of the total pollutant load.

These facts are important in terms of which method should be selected for computation and which measures should be used for reduction of the pollutant load.

The influence of the time scale on the nature of the pollutants' effect as depicted in Figure 1.2 can be extended with a spatial scale varying from a small local environment to a large scale receiving water system (see Figure 1.3). This figure shows that different strategies must be implemented when monitoring and calculating the effects that occur under varying time and spatial scale conditions. The extent of the different physical, chemical, and biological processes in the receiving environment (e.g., dilution, sedimentation, accumulation, and degradation) will determine the strategy that is the more relevant.

When dealing with computation for an acute effect, more detailed information on urban runoff characteristics is required compared with what is needed in case of an accumulative effect. It also becomes more difficult to find efficient measures for pollutant reduction in the case of a single extreme event than is the case if the reduction of the pollutant load is important for all events, small ones included. In terms of engineering, we are therefore faced with the most challenging situations in case of acute effects. The fact that the accumulative effects are typically more difficult to assess compared with the acute effects has nothing to do with this statement.

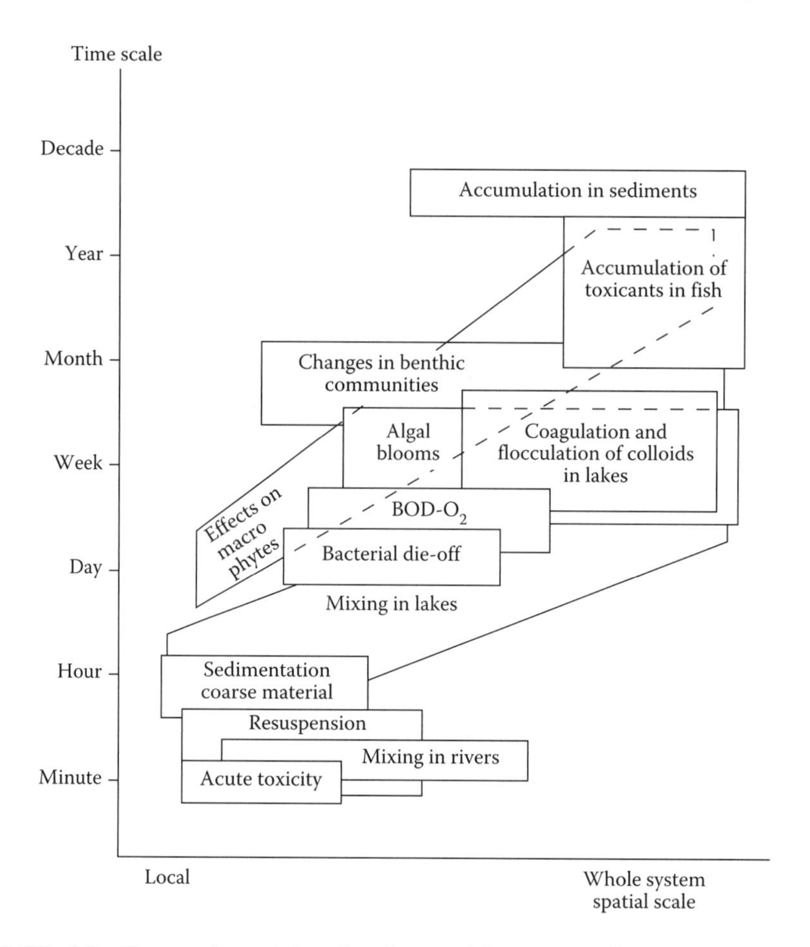

FIGURE 1.3 Time and spatial scales for receiving water effects. (Modified from House, M. A., Ellis, J. B., Herricks, E. E., Hvitved-Jacobsen, T., Seager, J., Lijklema, L., Aalderink, H., and Clifforde, I. T., *Water Science and Technology,* 27, 117–58, 1993.)

1.4 TYPES AND PERFORMANCE OF URBAN DRAINAGE SYSTEMS

Until now in this text, the urban drainage aspects (that includes road runoff) have only been dealt with in general terms, primarily focusing on basic characteristics of the pollutants and their potential adverse effects following a discharge event. The sewer networks and urban drainage systems that are designed to manage the flows of discharges have not been described. It is evident that the design and construction details of the collection and transport systems are very central to the magnitude of the polluted flows and water volumes and the corresponding water quality of the discharges.

1.4.1 Characteristics of Urban Drainage Networks

Each drainage network has its specific characteristics in terms of its construction and performance. In general, the following two main types of sewer systems will result in different performances of the wet weather discharges. This is important for those engineering approaches and environmental effects of the wet weather flows that are relevant:

- Combined sewer networks
- Separate sewer networks

The fundamental differences in performance of the two systems are depicted in Figure 1.4. The daily flow of wastewater and the runoff water from the urban surfaces are transported in the same network of a combined sewer catchment whereas the two types of flows are separated in two independently operating networks in a separate sewer catchment. The two different networks for collection and transport of the municipal wastewater flow and the wet weather runoff in the separate sewer catchment are a sanitary sewer network and a storm sewer network, respectively (cf. Figure 1.4).

Combined sewers and separate sewer networks may include both gravity sewers and force mains (pressure mains). However, force mains are typically not

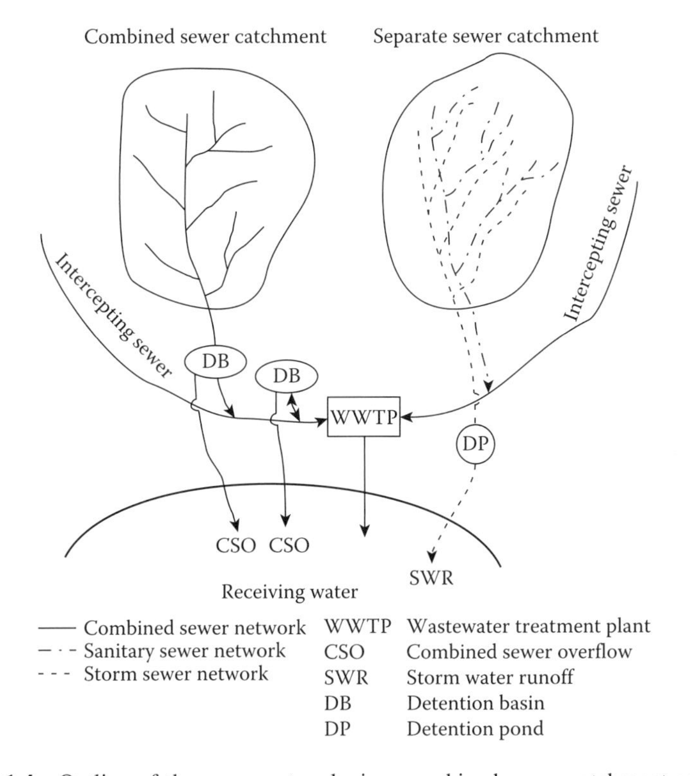

FIGURE 1.4 Outline of the sewer networks in a combined sewer catchment and a separate sewer catchment. (Modified from Hvitved-Jacobsen, T., *Sewer Processes: Microbial and Chemical Process Engineering of Sewer Networks,* Boca Raton, FL, CRC Press, 2002.)

applied in storm sewers. Further characteristics of the two types of networks are as follows:

- *Combined sewer networks*
  Municipal wastewater and runoff from urban impervious surfaces are collected, mixed, and transported in the same network. The surface runoff typically enters the network through inlets located in street gutters and from roof runoff pipes. During dry weather periods, a combined system operates, in principle, like a sanitary sewer network. However, because of the ability to serve runoff purposes, combined sewer networks are designed differently in terms of both capacity and construction details that facilitate their function during runoff events. The pipe diameter size of a combined sewer is therefore larger than for a corresponding sanitary sewer serving an identical catchment. Main construction details are overflow structures and detention basins (see Figure 1.5). Overflow structures exist to protect a network from overloading, a catchment downstream from flooding, and a treatment plant from improper flows during extreme runoff events. The surplus mixture of wastewater and runoff water will typically be temporarily stored in detention basins in the sewer system or discharged untreated into an adjacent watercourse via an overflow structure. This discharge is defined as a combined sewer overflow (CSO). In order to reduce the number and volumes of Combined sewer overflows (CSOs), on-line (in-line) or off-line detention basins are constructed to

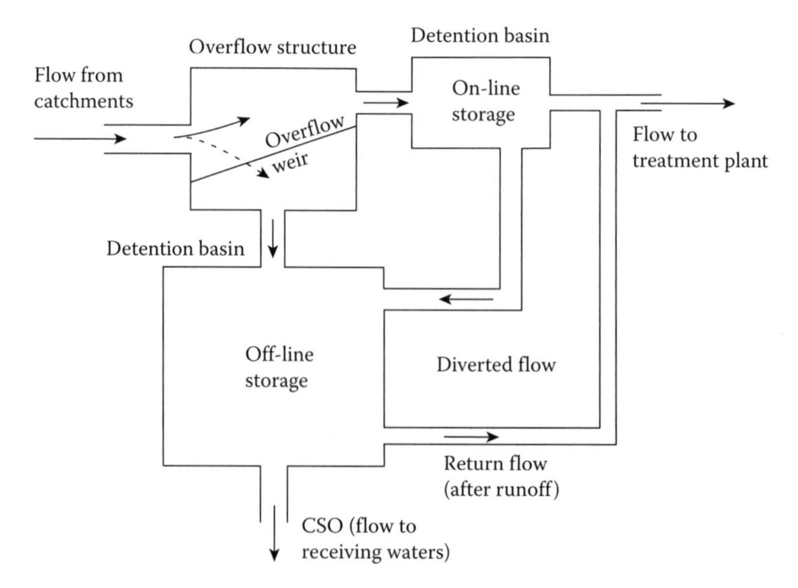

FIGURE 1.5 Outline of a combined sewer network with an overflow structure and corresponding on-line (in-line) and off-line storage facilities. The incoming mixture of wastewater and runoff water either flows in the intercepting sewer to the treatment plant or is discharged as CSO.

temporarily store the diluted wastewater. When the capacity of the network located downstream and treatment facilities allows an increased flow rate, the wastewater stored on-line or off-line can be returned into the intercepting sewer (interceptor; cf. Figures 1.4 and 1.5).

- *Separate sewer networks*
  In the separate sewer network, surface runoff is collected and transported in the storm sewers and discharged into adjacent receiving waters, whereas municipal wastewater is collected and transported in the sanitary sewers to a wastewater treatment plant located downstream, Figure 1.4. A storm sewer—or stormwater sewer—is therefore constructed just for collection and transport of stormwater (runoff water) originating from impervious or semi-impervious surfaces like streets, highways, parking lots, and roofs. The storm sewers only operate during wet weather periods. Traditionally, they divert the runoff water into watercourses with no or limited treatment. In urban areas, the storm sewers are normally underground pipes, however, in suburban or rural areas, the collection of runoff water (e.g., along roads) often take place in open channels like swales and filter strips. The flow of stormwater into the receiving environment is called stormwater runoff (SWR). Stormwater runoff and CSO are thereby terms that define wet weather discharges from separate sewer catchments and combined sewer catchments, respectively (cf. Figure 1.4). A separate sewer network may include a pond for detention, and potentially treatment, of the runoff water. Basically, SWR describes what is discharged from urban catchments but the term may also include runoff water from roads and highways.

The wet weather performance of the two types of networks describes an ideal design situation. In practice, however, a number of mixed systems exist. As an example and because of the absence of an appropriate local watercourse, a storm sewer may be connected to a combined sewer network. Furthermore, collectors for sanitary sewage may have been wrongly connected to a storm sewer leading to an increased pollution of SWR and runoff from wastewater during dry weather periods. Contrarily, if runoff takes place into a sanitary sewer or pipe blockages occur, overflow from such a system—sanitary sewer overflow (SSO)—may take place. Such overflows and discharges will result in unintended loads of concentrated wastewater into the environment and should, of course, be avoided.

When dealing with the wet weather pollution from urban areas, it is crucial to consider the interactions between the three systems: the sewer network, the treatment plant, and the receiving water (cf. Figures 1.4 and 1.5). These interactions are strongly related to the impact of urbanization on the hydrologic cycle and the systems selected to transport and treat wastewater and urban surface runoff. An important point for engineering of these systems is that the design affects the potential pollution. As an example, Figure 1.5 shows that storage in the sewer network of the diluted wastewater during rain may affect both the performance of the treatment plant and the impacts on the receiving water.

1.4.2 Quantity, Frequency of Discharges, and Origin of Pollutants in Drainage Systems

As described in Section 1.4.1, combined sewers and separate sewer networks are designed based on different concepts of performance. This fact affects those engineering methods that should be applied when quantifying the pollution from combined sewer overflows and stormwater runoff.

1.4.2.1 Volume of Water Discharged

In principle, all runoff water volumes collected from a separate sewer catchment will be discharged as SWR. In contrast to this, only a part of the runoff water that enters a combined sewer will (mixed with the daily flow of wastewater) be discharged as CSO. The rest will be diverted to a wastewater treatment plant located downstream. Therefore, and depending on the design and capacity of the interceptor and the treatment plant, the event-based, as well as the yearly amount of collected volumes of water that is discharged, are smaller for a combined system than for a corresponding separate sewer catchment.

The consequence is that although the pollutant concentrations (except for the heavy metals) are typically lower in stormwater runoff than in combined sewage, the yearly load of pollutants discharged per unit area of catchment may easily be the highest for a separate sewer catchment. Runoff from such catchments is definitely not clean as was generally believed among sanitary engineers 30 to 40 years ago. Detention ponds or other facilities to reduce flows and receive water impacts in terms of physical habitat changes and loads of pollutants, for example, are therefore typically needed in separate sewer catchments (cf. Table 1.1 and Figure 1.4).

1.4.2.2 Frequency of a Discharge

All rain events will cause SWR, except for what is temporarily being retained at the surface (i.e., infiltrated or evaporated). In combined systems it is the capacity of the interceptor that is a central parameter for design and that determines when the overflow structure becomes active. This design capacity thereby determines the frequency and extent of the CSOs under given catchment conditions. The yearly number of overflows from a combined system vary considerably in time and place but is often designed to be between 2 and 20, depending on, for example, local practice and requirements, financial resources available, protection level of the receiving water, and capacity of a downstream sewer network and treatment plant.

1.4.2.3 Origin of Pollutants

The pollutants found in CSO and SWR originate from a wide variety of local and regional sources. A pragmatic—and for several applications useful approach—is to subdivide the origin as follows in main groups:

Stormwater runoff (SWR) with pollutants originating from only one group of sources:

- Pollutants originate from the impervious and semi-impervious urban or road areas. Such pollutants may have their primary origin in what is airborne and transported with the precipitation and what is the result of activities and processes that proceed at and close to the urban or road surfaces (e.g., what is generated by degradation of constructions and road materials and what originates from traffic and spills).

CSO with pollutants from three types of sources:

- The same group of sources from the urban surfaces as for the SWR.
- The daily flow of municipal wastewater that is mixed with the runoff water in the sewer network before overflow takes place.
- Pollutants that temporarily accumulate in the network as sewer solids (i.e., sewer sediments and biofilms, slimes). Pollutants that are associated with such deposits may have their origin in the wastewater flow from antecedent dry weather periods. During such periods, sedimentation and biofilm growth can result in accumulation of sewer solids that may be eroded, resuspended, and transported as a part of the mixture of wastewater and runoff water following high flow periods or runoff events.

As already mentioned, this grouping of the pollutant sources is simple, however, in practice very useful. The reason is that it is possible to follow the pollutant pathways and to perform relevant computations to some extent. It is also possible to direct solutions for pollutant abatement to the right type of source to some extent. As can be directly seen, CSO includes, in addition to the pollutants from the urban and road surfaces, two additional groups of sources that are not part of the SWR. Generally, CSO is therefore more polluted than SWR in terms of concentration. In particular, pollutants that originate from erosion of sediments and biofilms in the network itself often contribute considerably to the pollutant concentrations of the CSO. However, as mentioned in Section 1.4.2.1, without further information and analysis it is not possible to conclude that CSOs per unit area of catchment will pollute more than SWR. An important cause is that an essential part of the runoff water in combined sewers is diverted to a treatment plant. The volumes discharged as CSOs and the frequency of these discharges is therefore considerably reduced compared to what is the case from a similar separate sewer catchment.

1.4.3 Drainage Network Developments

Looking some few hundred years back in time, the two different types of sewer networks, the combined system and the separate system, were developed under different conditions. Streams that naturally existed in areas where urban development took place were often used as a transport system for the daily wastewater and solid waste flows as well as for the wet weather flows. Such streams and associated constructions of channels were, in principle, transformed to an open combined sewer system. In areas with climate conditions corresponding to frequent and heavy rainfall, the separate system was often developed for capacity reasons. It became evident that when the dry weather municipal wastewater flow was required to be treated—at least to some extent—in a facility downstream, it was not feasible to use combined systems under such climate conditions. In general, combined systems will therefore operate inappropriately in a tropical climate and in many subtropical areas.

The combined systems of today are mainly found in temperate—and to some extent subtropical—climates. With several exceptions, the combined systems are concentrated in the Northeast and Great Lakes regions of the United States and Canada, Europe, Australia, and to some extent Japan. During the past 50 years, separate systems have, however, dominated the new developments in these regions. The situation today in these countries is that combined sewer networks dominate the central and old parts of the cities whereas separate systems exist in the suburbs.

A fundamental cause for making a change from constructing combined sewer systems to establishing separate systems was in several countries not just to reduce the impact from CSOs but also to develop more efficient sewer network systems and control the load onto treatment plants located downstream. It was, however, the general understanding that stormwater generated at the urban surfaces and roads was clean and could be discharged untreated and with a minimum of adverse effect into receiving waters. As mentioned in Section 1.4.2.3, it was later (i.e., during the past 30 to 40 years) realized that this statement was not necessarily correct.

It must be realized that the polluting performance of the two types of sewer network systems is rather complex and that a comparison in general terms cannot be expressed without exceptions. However, CSOs are for single discharge events typically associated with a higher load of pollutants to adjacent receiving systems than SWR. Although strongly dependent on site-specific conditions and the type of wet weather management, the polluting impact of well designed and operated combined sewer systems need not (on a yearly basis) be more serious than from similar separate sewer systems.

Combined systems are often maintained in areas where they already occur. The financial aspects associated with a change to a separate sewer system often play a dominating role. Combined systems may be further extended to minimize the magnitude of the CSOs and thereby reducing the corresponding adverse effects. From a pragmatic engineering point of view, it becomes relevant to deal with the polluting aspects of both combined sewers and separate sewer systems.

1.5 WET WEATHER QUALITY MODELING AND PREDICTION

From the previous sections, it is clear that a number of basic problems must be considered when predicting pollution of wet weather discharges from urban areas and roads. We are faced with a variability of the dominating phenomena and processes that basically do not exist when dealing with pollution from dry weather continuous outlets. Therefore, it is important to give a description of those basic characteristics and the corresponding tools that can be applied to solve urban drainage problems in the following chapters. In this respect it is crucial that predictive tools in terms of models can be developed for both analysis of performance and design of drainage systems. A typical objective of a simulation is to assess and select appropriate measures for the reduction of impacts from wet weather sources.

Models are not just tools in terms of heavy computer models. When focusing on wet weather pollution, the extent of information that is needed generally exceeds what is required for similar hydraulic computations. Often the data requirements of computer models for simulation of pollution related aspects do not fully comply with what is available. Data that do not have their origin on the site in question must often be extracted and assessed based on general available information. It is therefore important to realize that simple conceptual or empirical models that are in agreement with basic theory and experience in terms of a reliable output often turn out to be superior to complex models. This statement concerning simple and complex quality models for prediction of pollution loads or effects from urban drainage systems is, of course, not general but must be seriously considered in each specific case.

Models can be subdivided in several ways depending on structure and mode of operation. This aspect will be further dealt with in Chapter 10. A distinction between a conceptual and an empirical description or between a deterministic and a stochastic description is fundamental. As can be understood, complexity plays a central role in urban drainage modeling. The degree of model complexity, may also be understood as complexity of computation, can be achieved in different ways. The following is a first approach in this respect:

- *Model complexity in terms of structure*
  A simple empirical model expresses a relation between input and output parameters without taking into consideration the detailed mechanisms involved. Such models in their application are restricted to the range and origin of data on which they were originally based. In contrast, a conceptual model reflects underlying mechanisms in terms of processes for transport and transformation of the pollutants, for example. Because of the general ability to simulate the governing conditions of the system, such conceptual models are likely to have greater generality and ability to simulate site-specific aspects. The structure of a conceptual model is therefore in general more complex than it is for an empirical model.
- *Model complexity in terms of parameter numbers*
  Model complexity also concerns the extent of parameters. For a conceptual model, the degree of details included must be judged based on their importance for the final result and the knowledge that exists for description of a given phenomenon. Lumping of elements or use of empirical expressions as a part of a conceptual description is possible. Furthermore, the number of processes and parameters that are included must be considered depending on the specific purpose of the computation.

The level of computation that is selected will depend primarily on what is the objective and to what extent relevant data are available. In general, it is important that different levels of procedures for computation exist. As an illustration, the yearly amount of a pollutant discharged into an adjacent environment from an overflow structure in a combined sewer catchment can be calculated at the following levels starting with the simplest:

- *Selection of values for yearly pollutant loads per unit area of catchment*
  Such values can be selected among precalculated values based on standard values for pollutant concentrations in wastewater and runoff water, the actual available volume of detention basin, and the capacity of the interceptor.
- *Simple empirical equation*
  A simple empirical equation for pollutant load can be formulated based on the volume discharged originating from computations with a runoff model. A yearly average pollutant concentration in this water volume can be estimated based on pollutant concentrations in wastewater and runoff water and a yearly mean value for the mixing ratio between these two flows of water.
- *Pollutant transport model*
  A rainfall-runoff model that includes a pollutant transport module can be applied. A pollutant transport model formulated at this level is therefore a conceptual model. It requires that detailed information on the system in question is available.

Although this book has emphasis on the pollution that is related to urban runoff, it is clear that such a focus cannot be expressed without its relation to the relevant hydrologic, hydraulic, and transport phenomena. Especially when it comes to computation, the pollutant transport, transformations, and effects must be dealt with in an integral way.

1.6 URBAN DRAINAGE IN THE FUTURE

The urban drainage system of today has its basic structures like pipes and channels as a result of process starting some 4000 years ago. Its present constructions are often 50 to 150 years old, still in several ways performing well. It is, however, clear that both actual requirements and estimated future needs must result in an upgrading of this inherited network. Because future needs are basically unknown, it is important that the drainage system is constructed with a high degree of flexibility. The impact of the potential climate changes is in this respect a driving force. A separate sewer network that allows operation and management of two fundamentally different flows has a higher degree of flexibility and potential to adapt to unknown conditions than a combined sewer system.

Sustainable urban drainage (SUD) has become an issue for the management of the wet weather flows. The concept of SUD basically means that the rainwater in the cities and from the roads is managed in a way that solves the problems of today, take into account future needs, and preserve resources. In its ultimate state it means that we do not obstruct (directly or indirectly) basic conditions for the life of future generations. Nature's way of managing rainwater in rural areas is, in this respect, an approach to managing the wet weather flows in urban areas and from road surfaces. It should be realized that the basic difference between rural and urban areas is just caused by a construction detail (the impermeable surfaces) and the fact that human activities result in a nonintended use and spreading of numerous harmful substances: the pollutants.

An objective of this text is to give the reader a basic and quantitative understanding of the nature of wet weather pollution as it occurs in urban areas and from roads. At the same time it is important that this knowledge be applied in engineering of the drainage networks. In particular, focus will be on the quality related aspects in terms of analytical methods, concepts, and solutions. This text will provide the reader with a number of alternatives.

REFERENCES

Chow, V. T., ed. 1964. *Handbook of applied hydrology*. New York: McGraw-Hill, Inc.

Hammer, M. J., and K. A. MacKickan. 1981. *Hydrology and quality of water resources*. New York: John Wiley & Sons, Inc.

House, M. A., J. B. Ellis, E. E. Herricks, T. Hvitved-Jacobsen, J. Seager, L. Lijklema, H. Aalderink, and I. T. Clifforde. 1993. Urban drainage: Impacts on receiving water quality. *Water Science and Technology* 27 (12): 117–58.

Hvitved-Jacobsen, T. 2002. *Sewer processes: Microbial and chemical process engineering of sewer networks*, 237. Boca Raton, FL: CRC Press.

Hvitved-Jacobsen, T., N. B. Johansen, and Y. A. Yousef. 1994. Treatment systems for urban and highway run-off in Denmark. *Science of the Total Environment* 146/147: 499–506.

Mays, L. W., ed. 2001. *Stormwater collection systems design handbook*. New York: McGraw-Hill.

Wanielista, M. P., and Y. A. Yousef. 1993. *Stormwater management*. New York: John Wiley & Sons, Inc.

2 Pollution from Urban and Highway Wet Weather Flows: Concepts and Definitions

This chapter aims at giving the reader a solid basis on concepts and definitions related to pollution from the urban and highway wet weather flows. A main aspect is to show how the nature of urban runoff pollution relates to the variability in time and place of the relevant phenomena. Last but not least it is important to understand what methods can be used to quantify pollutant related aspects irrespective of the large variability that is associated with these phenomena. The large number of factors that influence the quality of the wet weather discharges and the variability associated with these factors typically requires that a correspondingly large number of site-specific rainfall and runoff data be collected over long periods. Statistical methods for the analysis of such time series therefore have quite a different role to play compared with methods used for analysis of the variability of pollution from continuous point sources.

2.1 RAINFALL CHARACTERISTICS

Although precipitation (rainfall and snowfall) generally decides in which form the transport media for wet weather pollutants is generated, this section will deal with rain as the most frequently occurring phenomenon. Cold climates use snow as the basis for both pollutant accumulation and pollutant transport as a very important substitute for rain. This specific case is the subject of Section 4.2.

In principle, rainfall should not be a core subject in the context of urban drainage. What is interesting is what happens with the rain when it hits the urban surfaces, namely runoff. However, rainfall and rainfall event characteristics are important because it is easy to provide basic and central data on wet weather. Before these phenomena are further dealt with in this section, important issues of runoff in relation to rainfall are focused on.

2.1.1 RAINFALL EVENT AND RUNOFF EVENT CHARACTERISTICS

In normal language usage, the word "event" expresses something that happens intermittently and that randomly takes place over time. The same basic understanding is associated with the hydrologic phenomenon "rain." From the general population's point of view, it may be sufficient to have a rather soft understanding of what a rain event is. However, when it comes to computations and management in urban drainage, further characteristics and definitions are needed.

It is important to understand that the concern in urban drainage pollution in terms of engineering approaches is associated with the runoff from the urban surfaces. Therefore, the rain (precipitation) must in this context be looked upon as the origin but also as the transport medium (vehicle) for the pollutants and what follows in terms of management and impacts. The following two basic aspects related to the rain and the runoff influence how we should consider and define the phenomenon "rain event" and "runoff event":

- *Consecutive rainfall events*
  Consecutive rainfall events must (to use the word "event") be statistically independent of each other. The point is therefore to identify the shortest period of no rainfall between two rainfall events (interevent dry period) that will observe the requirement for a series of rainfalls at a specific site. Statistical methods in terms of an autocorrelation can be applied for analysis of independence between events.
- *Consecutive runoff events*
  When dealing with the management of the runoff in terms of its collection, transport, and treatment, it is correspondingly important to require independence between the runoff events. Independence between runoff events is more complex to define than independence between consecutive rainfall events. The reason is that runoff event independence is related to how far from the catchment area that management or impact of the runoff occurs and also concerns what type of management method or type of impact is actually considered.

The two aspects that are related to independence for runoff events affect the length of the interevent dry period. Generally, independent runoff events require extended interevent dry periods compared with the interevent dry periods for rainfalls. Firstly, it is caused by the fact that there is a time delay between the precipitation and the corresponding appearance of the runoff at a downstream location during that period the runoff gradually undergoes dispersion. During the transport of the runoff, dispersion will affect the runoff pattern of a specific rain event. This transport time includes the drainage transport time at the urban surface itself (time of concentration) and a following transport time in a channel or pipe. Therefore, the runoff pattern not only depends on the rainfall event but also catchment characteristics. Hydrologic processes (e.g., the distribution of the rain at the catchment) as well as hydraulic processes are therefore important. Secondly, and related to the requirement of statistically independent runoff events, the time scale of management and pollution control measures also affect the length of what in a specific situation determines a minimum interevent dry period of runoff. A minimum value of a runoff related interevent dry period must therefore be defined in a way that makes it possible to relate a runoff event not just to one specific rainfall event but also to its downstream management and impact.

As mentioned, it is important to determine a minimum value of an interevent dry period for both rainfall and runoff. The principle of consecutive rainfall events delimited by interevent dry periods is shown in Figure 2.1. The figure also shows the occurrence of a dry weather period within a rainfall event where the minimum requirement of the interevent dry period is not fulfilled. If only the phenomenon rain is considered, a rainfall event is based on an autocorrelation analysis typically defined by an interevent dry period of one to two hours. However, intervals ranging up to four or five hours (or even higher) have also been proposed (Wenzel and Voorhees 1981; Arnell 1982; Schilling 1983; Medina and Jacobs 1994). In principle, each rainfall series should be analyzed to decide when a rainfall event starts and when it stops. However, such a procedure is time consuming and therefore not relevant for practical purposes. To overcome this problem, an automatic procedure can be applied. As an example, a rainfall event definition can be related to a recording by a rain gauge that registers the intensity with a constant sampling frequency of one minute and applying a tipping bucket that corresponds to 0.2 mm of rainfall. A rainfall event is in this case defined as being delimited by two interevent dry periods each a minimum of one hour where no tipping of the rain gauge has been observed. Only one single tipping observed during such an interevent period is accepted and not taken into account.

When dealing with phenomena related to runoff, extended interevent dry periods are required to observe independence between events including their management and potential impacts. The cumulative effects of successive rainfall events must under such conditions be taken into account and the minimum interevent dry period must be adjusted to the ultimate type of control or impact. If, as an example, treatment to a predesignated level in a detention pond requires two days, the definition of a rainfall event (runoff event) delimited by interevent dry periods of a minimum 48 hours is a relevant approach.

As an illustrative example, Figure 2.2 shows rainfall depth/interevent dry period/frequency curves developed from a 33 year rainfall record for the city of Odense, Denmark (Hvitved-Jacobsen and Yousef 1988). The figure depicts rainfall depth for interevent dry periods between 2 and 96 hours and return periods between three months and 3.3 years (i.e., the time that passed until a statistically identical event occurs). As an example, Figure 2.2 shows that a rainfall event with a six-month return period and an interevent dry period of 48 hours corresponds to a rainfall depth of about 27 mm. If such an event is the design requirement for drainage control, a runoff volume of 270 m^3 ha^{-1} of impervious area must be managed between two

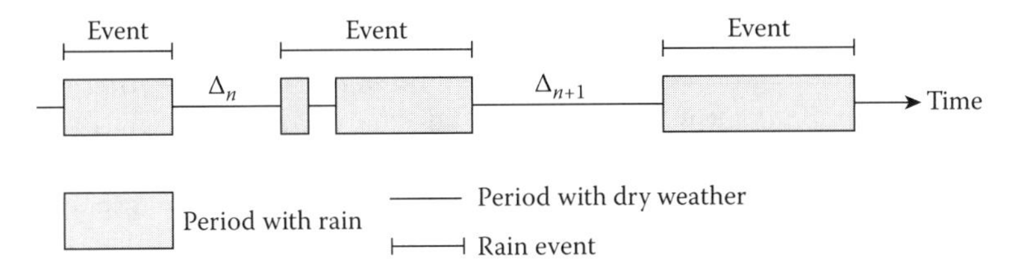

FIGURE 2.1 Principles of consecutive rainfall events and corresponding interevent dry periods (Δ).

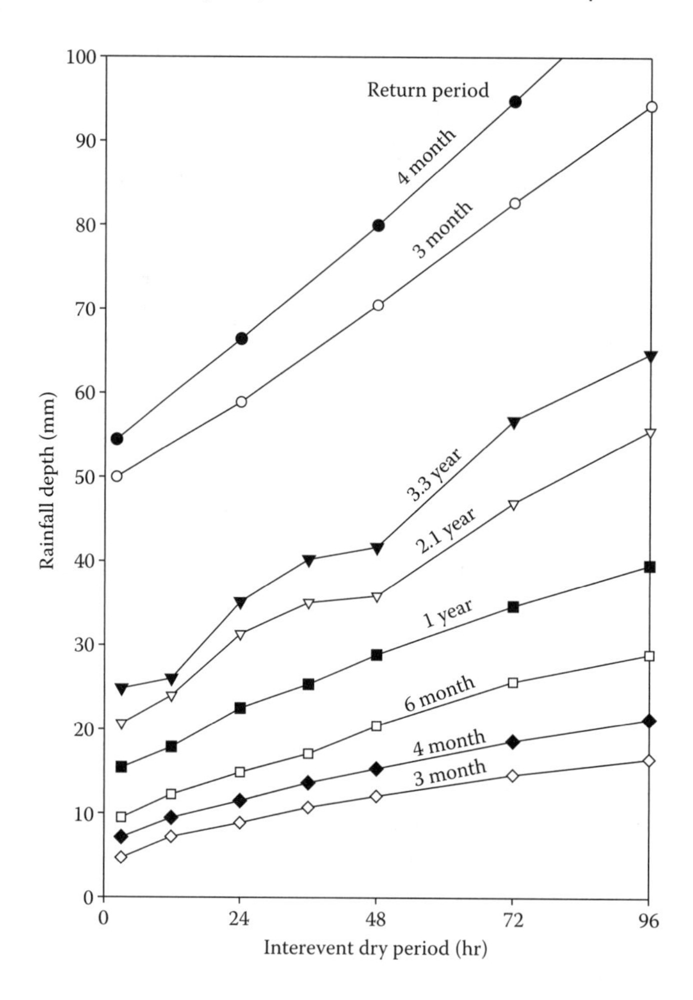

FIGURE 2.2 Rainfall/interevent dry period/frequency curves. The curves originate from a 33-year rainfall record for the city of Odense, Denmark. (Data from Hvitved-Jacobsen, T. and Yousef, Y. A., *Water Research,* 22, 491–96, 1988.)

rainfalls (runoff events). It is important to notice that the rainfall depth of an event will considerably increase with increased interevent dry period and an increased return period of the event (Hvitved-Jacobsen, Yousef and Wanielista 1989).

As mentioned, one of the main characteristics of rainfall in terms of their impact is the event-based nature. It should, however, be understood that urban drainage phenomena might be dealt with without considering rainfall and runoff as events. The capacity of today's computers makes it possible to process data from a historical rainfall series more or less continuously with a resolution that corresponds to the sampling frequency that is defined by the rain gauge (e.g., corresponding to a sampling frequency of one minute). In this way, modeling based on data originating from a local rainfall record can be performed continuously for a series of consecutive wet and dry periods. The impact of a rainfall can thereby be extended into the successive dry period and possibly interact with the impact of a succeeding rainfall.

The state-of-the-art makes it possible to perform such an approach as far as hydraulic modeling is concerned. However, when it comes to pollution, it is still needed to consider the event/interevent concept of urban wet weather pollution. Not just the tradition but also the fact that sampling and corresponding mass balances of pollutants and their effects have been recorded and associated with events still makes it appropriate to follow the event concept. In general, this text will consider pollution related to urban storm drainage as being event-based, however, still bearing in mind that pollution can also be related to continuous time series of precipitation.

2.1.2 Characteristics of Rainfall and Rainfall Series

2.1.2.1 Recording of Rainfall Events

Information on rainfall constitutes the starting point for computation and modeling related to urban wet weather pollution. The validity of rainfall measurements is therefore crucial for the following determination of pollution related to runoff.

Rainfall and other forms of precipitation are measured in terms of their depth, typically expressed in mm or inches in the United States and Canada. A number of different methods and types of equipment can be applied for that purpose. A simple vessel and monitoring of the daily amount of rainfall is still used. However, monitoring that provide detailed information on each rainfall event is generally needed for determination of runoff pollution.

A type of equipment often used for monitoring rainfall intensity is the tipping bucket rain gauge (cf. Figure 2.3). The rain collected in the receiver passes a siphon that allows the water to enter a bucket with a constant momentum. The bucket consists of two separated compartments where the water enters one compartment at a time. When a specific volume of water has been collected (e.g., corresponding to

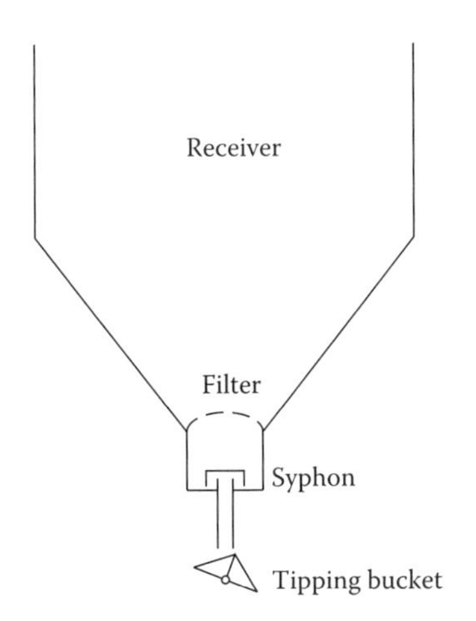

FIGURE 2.3 Principle of a tipping bucket rain gauge.

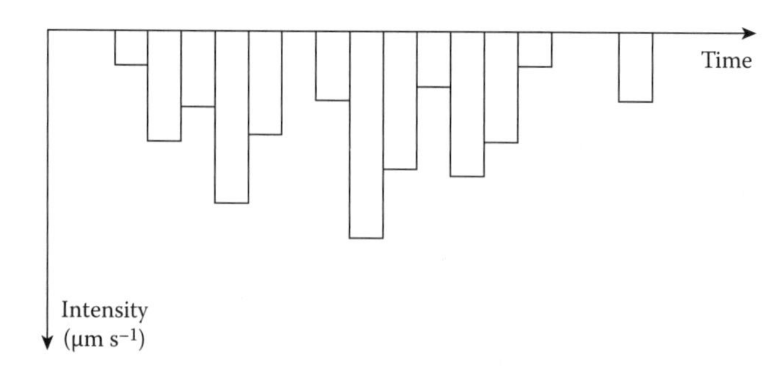

FIGURE 2.4 A rainfall hyetograph, intensity versus time curve, for a rain event that can be achieved by measurements with a tipping bucket rain gauge.

0.2 mm of rainfall) the bucket tips and water will then enter the other compartment. This procedure continues and the number of tips is recorded and transferred into intensity, normally corresponding to a sampling frequency of one minute. The intensity versus time curve is a rainfall hyetograph or just hyetograph (cf. Figure 2.4). The unit of intensity is typically $\mu m\ s^{-1}$ or inch s^{-1}.

In addition to rainfall measurements using the tipping bucket rain gauge or similar techniques, a number of supplementary methods for rainfall information exist. Radar measurement of rainfall is a method that is useful for on-line control and recording of the geographical distribution of the rain. By this method, the radar detects the falling raindrops. When calibrating these signals against measurements from a rain gauge, the spatial variation of the intensity of the rain can be determined, however, typically not with the same accuracy as a well-functioning rain gauge.

2.1.2.2 Rainfall Series

Worldwide, there is an overwhelming amount of rainfall data and other relevant hydrological data available for urban drainage purposes. Of course, these data vary considerably in both quality and in terms of those details that are recorded. The stochastic and site-specific nature of rainfall and associated pollution, however, basically requires that the spatial and temporal scale resolution and the length of the rainfall time series is taken into consideration when using the information in practice.

The need for details concerning rainfall depend on the purpose, but from a pragmatic point of view, also on what is available at the site in question. For several purposes—particularly when focus is on impacts of extreme events—longtime series of rainfall data (historical rainfall series) are required. In order to reduce the uncertainty when considering extreme event statistics, time series covering rainfall data for 30 to 50 years are often required. As an example, it must be noticed that within a historical rainfall series of 30 years, there are statistically only three events with a return period exceeding 10 years. For modeling purposes it can be required that the information on rainfall data geographically is high. Depending on the topography of a catchment, even a few hundred meters between two rainfall gauges may result in different rainfall series. Hydraulic computations may basically require a temporal resolution of minutes but often only hourly or daily precipitation data are available.

Furthermore, computer modeling requires that the rainfall record is available in terms of an electronic data file.

When dealing with the pollution from urban runoff and combined sewer overflows (CSOs), the need for details concerning the rainfall will depend on whether acute or accumulative effects are relevant. For assessment of cumulative effects, it may be sufficient to know the yearly or seasonal amount of water discharged and a corresponding mean or median value of the pollutant concentration. However, in terms of rain events that contribute to the runoff, it becomes important that all runoff events (e.g., rainfall events larger than a rainfall depth of 0.5–2 mm) are recorded and included in the rainfall series. In case of an acute effect that may be associated with extreme events, longtime rainfall series—not necessarily with a high temporal resolution—and a corresponding event-based pollutant concentration may be needed. In terms of pollution control, it may also be relevant to assess the impact of a varying pollutant concentration within an event (e.g., in case only the most polluted part of the runoff volume will be subject to treatment). Therefore, the time scale resolution of each rainfall event becomes relevant. In most cases, however, information on the pollutant variability within an event is not available and impact assessment must be done on a time scale corresponding to an event.

Several years of rainfall monitoring at a site in terms of a rainfall series is the solid basis for estimating any kind of polluting effect of wet weather discharges and corresponding control measures. In several countries, a rather densely located network of rainfall gauges exists and corresponding local rainfall series are produced. Such series can be used directly for modeling purposes or via a statistical analysis for providing data that are central for the calculation of the wet weather load of pollutants.

Historical rainfall series are an important basis for the design of drainage networks. The use of such series is particularly central to achieve an appropriate hydraulic performance, but indirectly quality aspects depend on the reliability of the rainfall series used. As an example, rainfall/interevent dry period/frequency curves as those shown in Figure 2.2 can be used for design of treatment facilities for stormwater. It is therefore crucial that rainfall series can reflect a future rainfall pattern. Possible climate changes at a specific site should therefore be taken into account when using such historical rainfall series for design purposes. Estimated information on the magnitude, frequency, and duration of future rainfall events can be used to modify historical rainfall series that can be applied for prediction and design purposes.

2.1.3 Rainfall Intensity/Duration/Frequency Curves

It requires many years of measurements and extensive resources to establish detailed information in terms of historical rainfall series. In many countries and at specific locations, such information may be scarce or even missing and only rather crude data on the local rainfall pattern may exist.

A first estimate of a local rainfall pattern can be based on recording the total daily rainfall during a limited time period and combining such data with information that originates from regions with similar climate conditions. Following this procedure, a frequently used and still relative simple method is to describe a local rainfall series in a comprehensive form by combining the following characteristics:

- Intensity of a rain event, i ($\mu m\ s^{-1}$)
- Duration of a rain event, t_r (min)
- Frequency of occurrence of a rain event, n (yr^{-1})

Although other units are also applied, the units that are indicated show what is often used in practice.

The term "return period" or "return interval," T, is defined as the inverse of the frequency of occurrence and often substitutes this term:

$$T = n^{-1}. \quad (2.1)$$

The return period is the average recurrence interval that exists between two events having a magnitude of a property that equals or exceeds a certain value. In other words, the return period is the statistically determined period between occurrences of such events.

Intensity/duration/frequency (IDF) curves that combine the three rainfall characteristics represent the basic statistical properties of rainfall data originating from a specific site. Details on IDF curves can be found in any textbook on hydrology (e.g., Wanielista 1979). Simple empirical expressions of the relation between the intensity and the duration of a rain event—at constant frequency of occurrence—can, as exemplified in Equations 2.2 through 2.4, be expressed based on two or three parameters:

$$i = a\ t_r^b, \quad (2.2)$$

$$i = \frac{a}{b + t_r}, \text{ or} \quad (2.3)$$

$$i = a\frac{t_r}{(b + t_r)^c}, \quad (2.4)$$

where

a, b, and c = empirical site specific parameters depending on the return period.

Equations 2.2 through 2.4 can be extended with the frequency of occurrence or the return period (cf. Wanielista 1979).

Equation 2.2 represents a linear curve in a double-logarithmic representation (see Figure 2.5 for comparison). This figure shows how rainfall characteristics in terms of IDF curves perform for three different sites. These curves are particularly

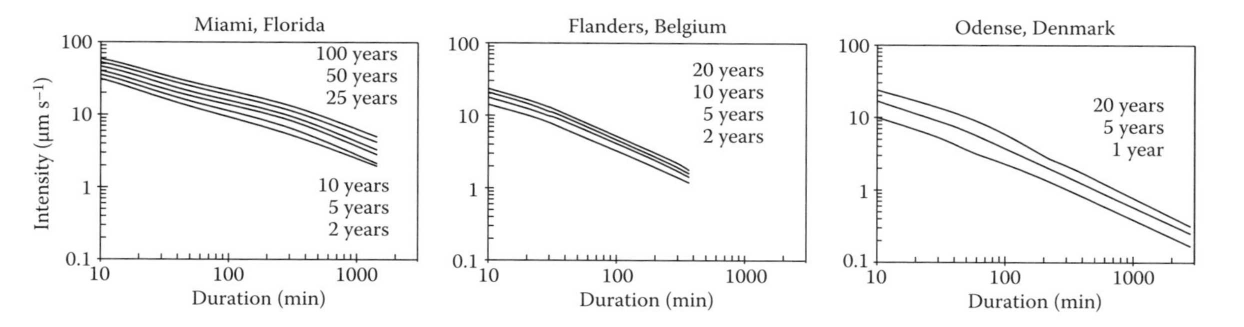

FIGURE 2.5 Rainfall intensity/duration/frequency curves. The figure shows examples from three different locations. (Modified from Wanielista, M. P., *Stormwater Management: Quantity and Quality,* Ann Arbor Science Publishers Inc., Ann Arbor, MI, 1979; Vaes, G., Clemens, F., Willems, P., and Berlamont, J., *Extended Abstracts of the 9th International Conference on Urban Drainage,* Portland, OR, September 8–13, 2002; IDA Spildevandskomitéen, Regional Variation of Extreme Rainfall in Denmark, Technical Report No. 26, Danish Water Pollution Committee, The Danish Society of Engineers, 1999.)

useful for statistical analysis of extreme events. What can be characterized as extreme depends on the local climate conditions and in which context the event is relevant (e.g., corresponding to return periods typically larger than 0.2–0.5 a year). In addition to the total yearly rainfall depth at a specific location, relatively simple produced IDF curves form a sound alternative to a long historical rainfall series.

The IDF curves can be used to construct artificial rainfall events called design storms. A great number of such design storms have been suggested. Well-known examples are the "Chicago design storm" suggested by Keifer and Chu (1957) and the design storm proposed by Sifalda (1973). In existing historical rainfall series, artificially defined storms will typically not be used.

2.2 RUNOFF EVENT CHARACTERISTICS

2.2.1 Rainfall-Runoff Relations

The rainfall on an urban surface or a road creates the runoff. For a completely impervious and plane surface where the entire rainfall is transformed into runoff, the relation between rainfall intensity, runoff flow, and runoff volume is as follows:

$$dV = Q(t)\,dt = i(t)\,A\,dt, \tag{2.5}$$

where

V = runoff volume (m^3)
$Q(t)$ = runoff flow (i.e., runoff flow rate (m^3s^{-1}))
t = time (s)
$i(t)$ = time dependent intensity of the rain event (ms^{-1})
A = contributing catchment area (m^2).

During a rainfall event, the intensity typically varies considerably and a corresponding variation in the runoff flow occurs. The curve that depicts the runoff flow during a rainfall event versus time is designated a runoff hydrograph. Figure 2.6 shows the principle of a runoff hydrograph, which can be measured in a pipe or a

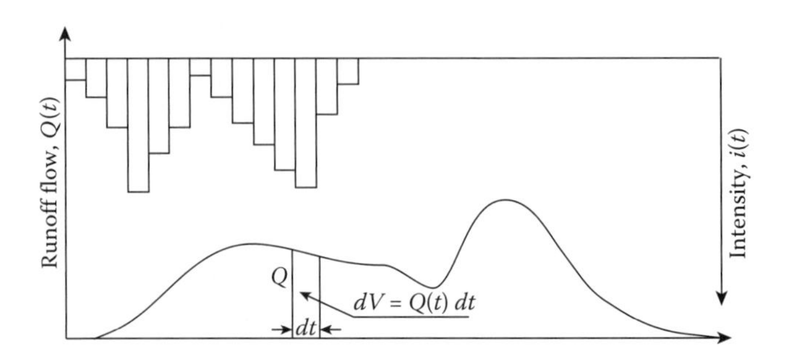

FIGURE 2.6 The principle of a runoff hydrograph for a rainfall event and a corresponding rainfall hyetograph.

channel catchment downstream. Figure 2.6 also shows the corresponding variation in the rainfall intensity at that area. The highest runoff flow rate is called the peak discharge.

As already mentioned, Equation 2.5 corresponds to a situation where the total rainfall volume produced during a storm event equals the runoff volume. Under real conditions, a number of phenomena and processes will reduce the volume that is available for runoff. What is available for runoff is called rainfall excess. The runoff reducing factors include, for example, infiltration into pervious and semi-pervious areas at the catchment connected with the sewer network, outflow from paved areas to adjacent pervious areas, storage in depressions (initial loss of surface runoff), and evaporation. The area under the hydrograph that equals the excess rainfall, V, is therefore less than the total volume of rain during the rainfall event taking place from time 0 to time t:

$$V < A\int_0^t i(t)dt. \tag{2.6}$$

To account for the reduction in the runoff compared with the rainfall, a dimensionless runoff coefficient, φ, is defined:

$$\varphi = \frac{V}{RA}, \tag{2.7}$$

where

R = rainfall depth $= \int_0^t i(t)dt$ (m).

The runoff coefficient can be interpreted as the proportion of the total rainfall of an event that becomes runoff (i.e., the ratio of runoff to rainfall for an event). When taking into consideration the reduction in the runoff volume, the rainfall-runoff relation that is expressed in Equation 2.5 therefore transforms to

$$dV = Q(t)\, dt = \varphi\, i(t)\, A\, dt. \tag{2.8}$$

The runoff coefficient is a pragmatic approach of the relations between rainfall and runoff, however, a useful and simple empirical substitute for a description of the rather complicated rainfall-runoff processes. For a specific catchment, the value of φ basically depends on rainfall intensity, rainfall volume, previous rainfall events, and actual climate conditions. Despite this fact, standard values for different types of urban surfaces are typically used for design purposes (see Table 2.1). Such values can be used to estimate an average runoff volume and flow rate for a given catchment. The runoff coefficient in Equation 2.7 is defined for an event whereas Table 2.1 shows what might be considered typical values for different types of catchments.

Simple empirical models for estimation of the runoff coefficient have been developed (Driscoll, Shelly, and Strecker 1990). The following equation is an

TABLE 2.1
Typical Values of Runoff Coefficients for Selected Urban Surfaces Based on Information from Several Literature Sources

Character of Surface	Runoff Coefficient, φ
Roof	0.9–1.0
Asphalt and concrete pavement	0.70–0.95
Semipermeable pavement (surface blocks)	0.60–0.70
Lawns, heavy soil (dependent on slope)	0.15–0.35
Lawns, sandy soil (dependent on slope)	0.05–0.20

example that describes the runoff coefficient in terms of the imperviousness of the catchment:

$$\varphi = \alpha I + \beta, \tag{2.9}$$

where

I = impervious fraction of the catchment (–)
α and β = coefficients determined by calibration based on locally monitored data (–).

The description of the urban hydrologic and hydraulic phenomena is not a major objective of this book and therefore only dealt with to an extent that is considered a basic prerequisite to understanding urban wet weather pollution. As an example, the description of the relation between rainfall and runoff is therefore held at a rather simple level. When dealing with the engineering of urban runoff, the methods applied in case of rainfall-runoff relations not only depend on the level of computation but also to a great extent on common practice. Although there are similarities, the practice varies from country to country. Further details in this respect can be found in manuals and textbooks relevant for the actual country.

Finally, it is important to mention that the unit "length" in urban drainage practice often should be interpreted as "volume." As an example, a "5 mm detention basin" should be understood as a basin with a volume that can store 5 mm of runoff from the catchment in question (i.e., 0.005 m^3 of runoff per m^2 of impervious area). If the effective (impervious) catchment area is 10 ha, the volume V of a 5 mm detention basin is

$$V = 5\ (\text{mm}) \times 10^{-3}\ (\text{m mm}^{-1}) \times 10\ (\text{ha}) \times 10^{4}\ (\text{m}^2\ \text{ha}^{-1}) = 500\ \text{m}^3.$$

2.2.2 Transport and Loads of Pollutants during Rain Events

The generation of pollutants and their pathways during a rain event will to a great extent depend on the actual urban system, the human activities and the processes that proceed. The main pathways for the pollutants during wet weather in an urban system, and the potential interaction with the daily wastewater flows, are depicted in Figure 2.7.

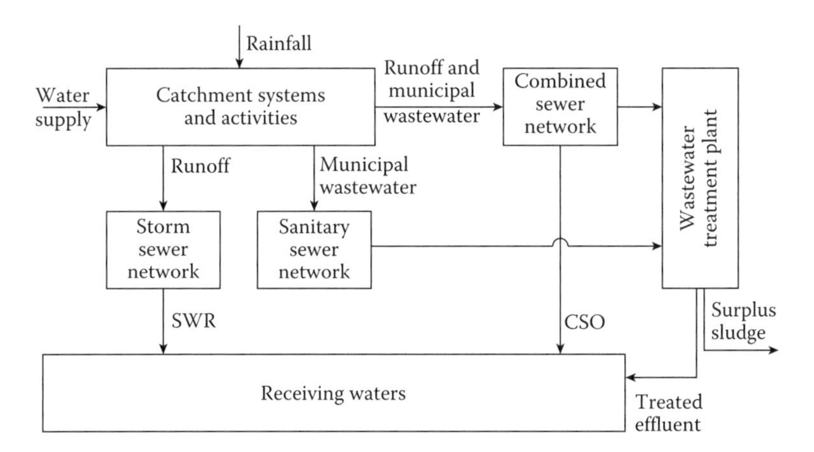

FIGURE 2.7 Overview of the pathways of pollutants during wet weather flow in an urban area.

The flow of pollutants during a runoff event through the urban system depicted in Figure 2.7 is in both separate and combined sewer catchments quantified in terms of a pollutograph. A pollutograph describes the flow of a substance versus time at a specific point of the urban system. Two types of pollutographs can be defined:

- The pollutant concentration (e.g., in units of g m^{-3}) versus time.
- The transport of a pollutant (e.g., in units of g s^{-1}) versus time.

Both types of pollutographs are relevant. The first one when the concentration of a pollutant is important in terms of treatment or an acute effect. The transport pollutograph becomes central in the case of an accumulative effect or when a mass balance should be established, as a load. It can be readily seen that the transport pollutograph can be formed by successive multiplication of the hydrograph with the concentration pollutograph (cf. Figure 2.8). It is also seen that the area under the transport pollutograph equals the amount of pollutant that is transported during the runoff event. The combined information that is given in terms of the hyetograph, the hydrograph, and the pollutographs for a rain event forms an essential basis for any engineering approach related to pollution.

2.3 SOURCES OF POLLUTANTS

A philosophic approach and corresponding basic understanding of the nature of pollutant origin for the wet weather flows are dealt with in Section 1.2. In this section, a more pragmatic approach of the pollutant sources is in focus. Both approaches are important. The first one is not only of academic interest but also important for practical reasons when managing and controlling the flows and when developing new ideas for sustainable solutions. The second pragmatic approach is needed when calculating and modeling the wet weather flows in order to determine a specific pollutant load or impact.

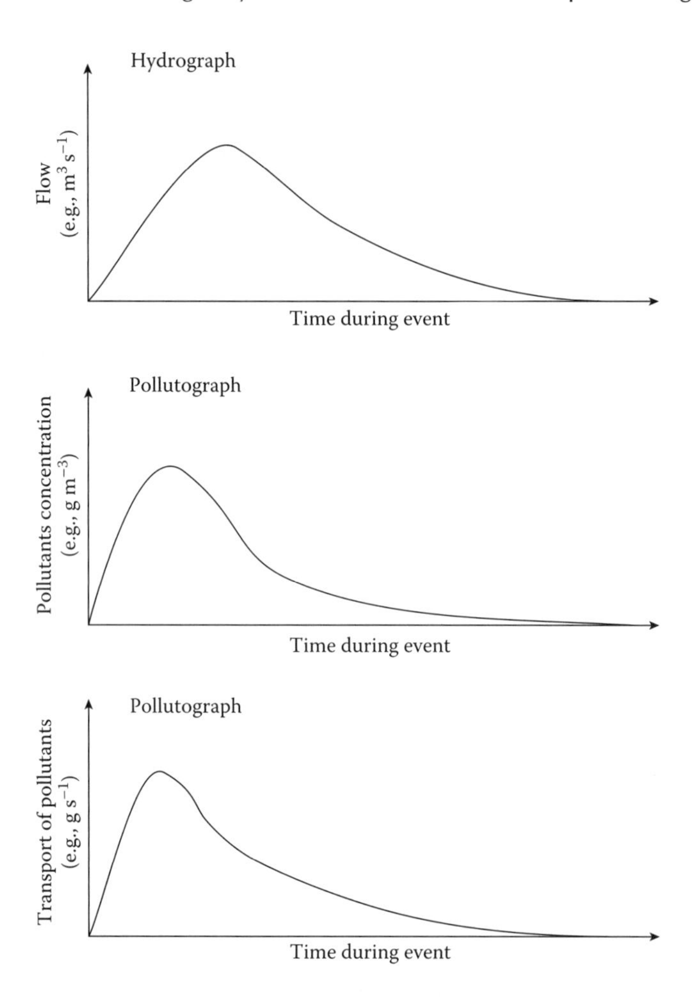

FIGURE 2.8 A hydrograph and the corresponding two types of pollutographs.

Both natural processes and human activities contribute to the pollutant level in urban runoff and both accepted and illegal pollutant contributions exist. The urban system is open to the environment and all types of pollutants may therefore appear in the runoff from urban areas (cf. Section 1.3.1). Contrary to the point sources like those from wastewater treatment plant outlets that often can be identified in detail, a quantification of an urban runoff related pollutant load in terms of a specific and well-defined origin of the pollutants is basically not possible (cf. Section 1.2). The origins of the pollutants in urban runoff are therefore defined as being from nonpoint sources, stationary as well as mobile, irrespective of the fact that the discharge into adjacent receiving waters may occur from well-defined outlets and therefore from this point of view are defined as point outlets.

In this section, the sources for the pollutants will be briefly dealt with. As we are concerned with nonpoint sources, it will not be possible to identify the different subsources in terms of loads and a lumped approach for the determination of such loads is needed. This section primarily concerns the pollutant sources and their qualitative description in general terms. For stormwater runoff (SWR) and CSOs, specific quantification of pollutant loads will be dealt with in Chapters 4 and 5, respectively.

Based on a pragmatic distinction of the origins for the pollutants, it is to some extent possible to quantify from where the pollutants occur. These origins will be identified in this section. The grouping of the pollutant sources follows what has been outlined in Section 1.4.2.3.

Although the pollutant sources cannot be quantified in detail in terms of loads, it is still important to understand where the pollutants have their origin. To some extent it will be possible to identify important and less important sources. It is crucial in terms of source control. As an example, it has never been possible to quantify the sources for lead in SWR in detail. However, because of the knowledge on the previous extensive use of Pb in leaded gasoline and how this metal affects the environment, there is no doubt that leaded gasoline was a major source for urban runoff loads with lead. In countries where source control for Pb was implemented, the concentration of lead in urban runoff has been considerably reduced. Although the pathways of Pb were only qualitatively formulated, this knowledge could obviously be applied with success.

An overview of the origins for the pollutants in SWR and CSOs is given in Figure 2.9. The details of this overview will be dealt with in Sections 2.3.1 through 2.3.3. From this overview, it is readily seen that CSOs include contributions from sources other than those for SWR. It is therefore clear that these two types of urban

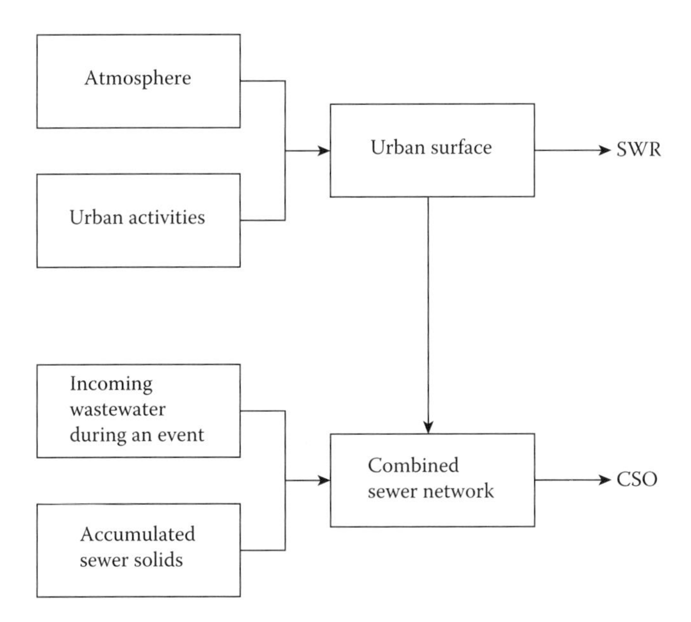

FIGURE 2.9 Overview of the pollutant sources and pathways for stormwater runoff (SWR) and combined sewer overflows (CSO).

runoff are normally different in terms of the pollutants that are present and in which concentrations they are found.

A characteristic feature for the pollutant sources is their different behavior in time and from site to site. It is the case in terms of both system characteristics (structure of the urban area and drainage system) and level of activity (land use). These aspects will, in addition to the stochastic nature of the rain, result in a very large variability of both the quality and the quantity of pollutant contributions. As an example, it is by nature not possible to identify a characteristic level of pollutant concentration in urban runoff without having specified the conditions (e.g., the different characteristics of the site).

The overview of pollutant sources shown in Figure 2.9 is the basis for quantification of pollutant loads dealt with in Chapters 3 and 4. Referring to Figure 2.9, it is in practice not possible to subdivide between the pollutant contributions from the atmosphere and those from the urban activities as the pollutants from these two main sources will be mixed and temporarily accumulate at the urban surfaces. The following subdivisions of the urban pollutant sources are therefore the pragmatic approach:

- Pollutants from the urban surfaces
- Pollutants from the daily wastewater flow
- Pollutants from accumulated sewer solids

In the next three sections, basic characteristics of these pollutant sources will be dealt with further.

2.3.1 Pollutants from Urban Surfaces and Roads

Referring to Figure 2.9, the pollutants at the impervious surfaces decide what may appear in SWR or road runoff and what contributes to the level of pollutants in CSOs.

The pollutants that are temporarily accumulated at the impervious and semi-impervious urban surfaces have their origin in inputs via the atmosphere, inputs from road materials, and nearby stationary objects or inputs from human activities (including traffic) at or close to the urban surfaces or roads. The last two types of inputs are often characterized by the term “land use.” Pollutants from urban constructions can be liberated directly to the rainwater but pollutants can also be airborne ranging from a few centimeters to hundreds of kilometers. In spite of this difference, a pragmatic and often used approach is to relate the pollution to local activities (i.e., pollution caused by the population density and mainly originating from the traffic at the site and the nearby stationary objects). In other words, a pollutant that occurs at an urban surface is in principle not considered associated with transport via the atmosphere. It is, of course, not a clear or correct statement. However, as we are dealing with nonpoint sources—and it is realized that a detailed description of the loads from the different pollutant sources is not possible—for practical purposes, it does not become an obstacle for a following quantification of the mass of pollutants associated with the runoff (cf. Figure 2.9).

In spite of the fact that the atmosphere in practice is just a subsystem for input of pollutants to the urban and road surfaces, it is still important to consider its characteristics and importance. The following three main type of sources for pollutants in SWR will be further discussed:

- *The atmosphere*
  The pollutants that appear in the atmosphere exist as gases (volatile substances), aerosols (liquid particles, i.e., associated with raindrops and fog), and suspended solid particles (dust). The pollutants that are transported to the urban surface during both dry and wet weather periods are identified as atmospheric fallout (cf. Section 4.1). The pollutants that appear in the atmosphere may originate from both stationary and mobile regional sources outside the catchment in question (e.g., from heating, traffic, soil erosion, and industry) and from long-distance transport from similar sources. The sources are typically found on land, but salts (chlorides) may originate from the sea. Often the local sources play a major role and thereby contribute to a higher pollutant level in dense populated areas compared with suburban catchments or rural areas. To some extent, roof runoff will primarily include pollutants that originate from the atmosphere and to a less extent pollutants more directly produced close to the street level. Therefore, roof runoff is typically less polluted compared with the runoff from streets. Further details of atmospheric pollutants are dealt with in Section 4.1.
- *Human activities at and closely related to the urban surface*
  The urban surfaces include in particular roofs, streets, parking lots, roads, and urban highways. The polluting activities and processes associated with these surfaces are often identified by the term "land use." The term should in this context be rather broadly understood. It refers to both urban and industrial catchments and pollutants originating from both mobile sources (automobile traffic) and stationary objects (e.g., soil erosion and corrosion products from houses, road materials, and urban installations).
- *Spills, accidents, and illegal or inappropriate activities*
  Spills, accidents, and illegal or inappropriate activities may result in the input of pollutants to urban runoff. Such specific input occur stochastically and pollutant contributions may result in extreme pollutant loads, in particular occurring as discharges from single events or—in the case of illegal or inappropriate activities like a car wash or industrial discharges—often as an increased and unexpected concentration of a limited number of pollutants in the wet weather flow. Due to high pollutant loads or concentrations, such inputs may result in acute effects in the environment. Since the inputs are not occurring regularly and as a normal pollutant contribution, they are difficult to include in planned procedures for management of the pollutants in the wet weather flows. However, it must be realized that they are potentially occurring. The choice of robust structures and management methods is considered the pragmatic way of handling such specific extreme and stochastically occurring polluting flows. Possible structural means are holding tanks and basins as a part of the collection and transport system for the wet weather flow. Methods

for source tracking of illegal discharges of heavy metals in the biofilms of sewer networks are other means to reduce illegal and inappropriate inputs.

The automobile traffic (moving vehicles) is an important example of a source for pollutants that may accumulate at the urban surfaces. The following subsources for pollutants that are traffic related are

- Degradation products from automobile tires
- Degradation products from brake pad wear
- Corrosion products from the automobile body
- Transported load losses
- Fuel combustion products
- Contaminants from road surface materials (asphalt)
- Use of de-icing agents (road salt)

Depending on which materials are in use, a great number of different heavy metals and organic micropollutants appear as a result of automobile traffic. Cu, Pb, Zn, and Cd—sometimes extended with Ni and Cr—are considered the most important heavy metals associated with both mobile and stationary sources. Examples of organic micropollutants are the Polycyclic Aromatic Hydrocarbons (PAHs) that are a result of a partial combustion of gasoline and diesel fuels.

A number of external conditions influence the extent of traffic-related pollution. Such conditions are traffic volume; the extent of vehicle control performed by the authorities; the general level of vehicle and road maintenance; and the materials used for vehicles, roads, and highways. A number of site-specific factors that influence the pollution from the traffic are not directly considered integral. Examples of such factors are the geographic characteristics (i.e., climate parameters and particularly the rainfall pattern). In addition, the method and effectiveness of street cleaning may also affect the pollution related to street and highway runoff.

A great number of investigations have been performed worldwide to assess the level of pollutants originating from the nonpoint sources dealt with in this section. As previously mentioned, the variability of these data is very high. Further details in this respect are the subjects of Chapters 4, 5, and 9. An overview with a number of examples of the origins of pollutants originating from urban surfaces is found in Ashley and others (2003). Very detailed and comprehensive information on the pollutant contribution from different urban sources originate from the Nationwide Urban Runoff Program (NURP) performed in the United States during the period 1979–1982 (U.S. EPA 1983). The data presented in Table 2.2 originate from these investigations and serves the purpose of giving an impression of the concentration level for central pollutants and their variability in SWR depending on the land use. Further examples and details are found in Section 4.6.2.

Although the data shown in Table 2.2 originates from the period 1979 to 1982, they are still today (2009) considered an important contribution to assess the pollutant concentration level in SWR. The exception is the concentration level of lead that was caused by the shift from leaded to unleaded gasoline for cars and is

TABLE 2.2
Characteristic Pollutant Concentrations in Stormwater Runoff Depending on Land Use

Pollutant (Unit)	Residential		Open Space	
	C_m	COV	C_m	COV
BOD_5 (g m^{-3})	10	0.41	–	–
COD (g m^{-3})	73	0.55	40	0.78
TSS (g m^{-3})	101	0.96	70	2.92
Total Kjeldahl Nitrogen, TKN (g m^{-3})	1.9	0.73	0.97	1.00
Total P (g m^{-3})	0.38	0.69	0.12	1.66
Total Pb (mg m^{-3})	144	0.75	30	1.52
Total Cu (mg m^{-3})	33	0.99	–	–
Total Zn (mg m^{-3})	135	0.84	195	0.66

Source: Modified from U.S. EPA (Environmental Protection Agency). 1983. Results of the nationwide urban runoff program, Volume I: Final report. Water Planning Division, U.S. EPA, NTIS No. PB 84-185552, Washington, DC.

Note: Data show median concentrations (C_m) and corresponding coefficients of variation (COV = standard deviation divided by mean value).

considerably reduced to approximately 10–20% of what Table 2.2 shows for residential areas.

The COV values shown in Table 2.2 are, of course, subject to a significant variability. Driscoll and colleagues (1990) found that COV = 0.75 was a good estimate for highway sites and any pollutant. Furthermore, they propose site-specific refinements be made using COV = 0.71 for urban highways and COV = 0.84 for rural locations.

2.3.2 Pollutants Originating from the Dry Weather Flow of Wastewater

As seen in Figure 2.9, wastewater in combined sewers will contribute to the contents of pollutants in CSOs. It does so in two different ways: directly by mixing the inflow of runoff water to the sewer with the wastewater flow and indirectly by erosion and resuspension of sewer solids that have been temporarily accumulated during preceding inter-event dry periods. Both aspects are important for the level of pollutants in the CSOs. In terms of calculation of pollutant loads from CSOs, the two phenomena are normally dealt with separately. Only the first phenomenon caused by the mixture of runoff water and wastewater will be dealt with in this section. The contribution caused by erosion and resuspension of accumulated sewer solids will be the subject of Section 2.3.3.

During a runoff event, wastewater in combined sewers will be mixed with the inflow of the runoff from the urban surfaces. In other words, the wastewater is diluted with the runoff water. Figure 2.10 shows the inflow hydrograph, $Q_i(t)$, for a simple overflow structure without storage capacity. When $Q_i(t)$ exceeds the flow capacity,

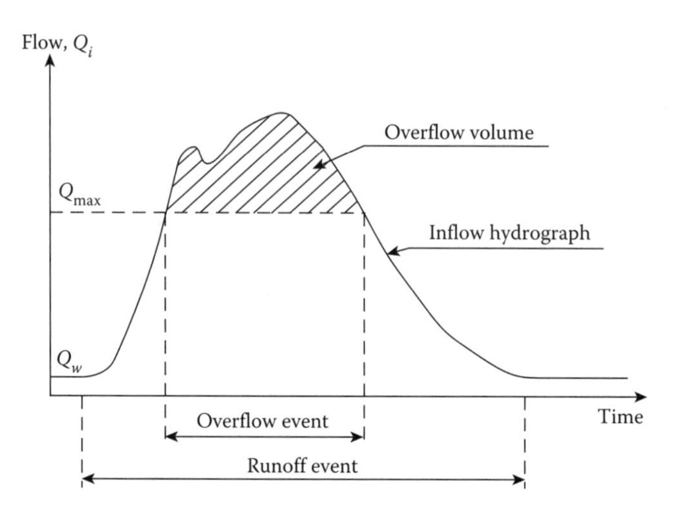

FIGURE 2.10 Runoff hydrograph, $Q_i(t)$, in an overflow structure of a combined sewer during an event where overflow takes place. Q_w is the (dry weather) wastewater flow and Q_{max} the capacity of the interceptor.

Q_{max}, of the interceptor that is located downstream of the overflow structure, overflow will take place. It is readily seen that if Q_w is the actual flow of municipal wastewater, the mixing ratio between runoff water and incoming wastewater when the overflow event starts can be determined as

$$m_s = \frac{Q_{max} - Q_w}{Q_w}, \tag{2.10}$$

where

m_s = mixing ratio between runoff water and wastewater in a combined sewer overflow structure at the start of the overflow event (–)
Q_{max} = flow capacity of the interceptor (m^3 s^{-1})
Q_w = flow of municipal wastewater at the start of the overflow event (m^3 s^{-1}).

As seen from Equation 2.10, the mixing ratio at the start of an overflow event is determined by the actual flow of wastewater and the flow capacity of the intercepting sewer. The mixing ratio is a simple and important (however, not sufficient) design parameter for an overflow structure in terms of what can be accepted quantity of discharge and quality of the overflow. It should be noted that by using the mixing ratio, the pollutants originating from eroded sewer solids are not considered (cf. Section 2.3.3). This fact may have a significant impact on the load of pollutants discharged into an adjacent receiving water body. Further details related to the mixing ratio will be dealt with in Section 5.1.2.1.

The dry weather composition of the municipal wastewater in a sewer network is one of the factors that determine the quality of the discharge during a CSO event. Worldwide, there is considerable variability in the composition of wastewater. At a specific site the composition of the wastewater is furthermore subject to daily and seasonal variations and to a number of conditions in terms of sewer processes, infiltration, and exfiltration. Keeping these aspects in mind, Table 2.3 shows typical levels of wastewater composition for a few selected parameters.

In addition to the variability of the composition of wastewater, the dilution in terms of the mixing ratio, m, at the start of an overflow event also varies considerable depending on the design of the sewer and the level of dry weather flow during the overflow event. In some countries, m equals typically 2–3. In other countries sewer networks are normally designed for values of m equal to 10–15 corresponding to a rather high dilution of the wastewater in the CSOs.

2.3.3 Pollutants from Accumulated Sewer Solids

As mentioned in Section 2.3.2, CSOs will, in addition to the pollutants that are directly associated with the incoming wastewater, also include those originating from erosion and resuspension of temporarily accumulated sewer solids. Two types of accumulated sewer solids occur:

- Sewer sediments (i.e., deposits and near-bed materials)
- Biofilms

According to Crabtree (1989), the sediments in sewer pipes include a coarse granular fraction (type A sediment) with an overlying mobile fraction (type C sediment). Typical characteristics for these types of sediments and biofilms (type D) are shown in Table 2.4.

The origin of the pollutants in the accumulated sediments in combined sewers is mainly two sources: It is wastewater constituents that settle during the antecedent dry weather periods and those constituents that originate from previous storm inputs.

TABLE 2.3
Typical Composition of Domestic Wastewater in Sewer Networks in Terms of Central Pollutant Parameters

Pollutant (Unit)	Diluted	Moderate Strength	Concentrated
BOD_5 (g m^{-3})	110–150	190–250	350
COD (g m^{-3})	250–320	430–530	740–800
TSS (g m^{-3})	120–190	210–300	400–450
TKN (g m^{-3})	20–30	40–50	70–80
Total P (g m^{-3})	4–10	7–16	12–23
Total Pb (mg m^{-3})	30	65	80
Total Cu (mg m^{-3})	40	70	100
Total Zn (mg m^{-3})	130	200	300

Source: Modified from Henze, M., Harremoës, P., Jansen, J.la C., and Arvin E., *Wastewater treatment: Biological and chemical processes*. Berlin: Springer-Verlag. 2002; Tchobanoglous, G., Burton, F. L., and Stensel. H. D., *Wastewater engineering: Treatment and reuse*. New York: McGraw-Hill. 2003.

TABLE 2.4
Basic Characteristics for Accumulated Sewer Solids (Sediments and Biofilms)

Type of Sewer Solids	Description	Typical Wet Density (10^3 kg m^{-3})	Typical Organic Content (%)
A	Coarse, granular bed material	1.72	7
C	Mobile, fine grained overlying type A	1.17	50
D	Biofilm (wall slime)	1.21	60

Source: Modified from Crabtree, R. W., *Journal of the Institution of Water and Environmental Management,* 3, 569–78, 1989.

Substances may thereby originate from the urban surfaces washed into the network during particularly low-intensity rainfalls where accumulation of course sand particles, for example, occurs. Furthermore, growth of biofilms at the sewer walls and to some extent also at the sediment surface during dry periods will take place. In addition, wastewater constituents may be adsorbed to such slime layers. Table 2.4 only focuses on characteristic values for the organic content in sewer solids but a wide range of pollutants from wastewater and urban runoff are, of course, also present in sewer solids (Ashley and others 2003).

Sewer solids are always found in all combined sewers, however, in varying amounts. Low gradient sewer pipes with low dry weather flow velocities favor conditions for sedimentation and development of bulky biofilms. The potential for a significant contribution to suspended solids in the water phase originating from eroded sediments and detached biofilms is therefore relatively high during wet weather flow conditions. The duration of an antecedent dry period and the wet weather runoff pattern also affect the amount of sewer solids that is subject to erosion and resuspension during wet weather flows. Accumulated sewer solids typically contribute considerably to the pollutant level in the CSOs (cf. Example 2.1 and Chapter 5). Compared with values as shown in Table 2.2, the pollutant level of SWR that has entered a combined sewer is, in general, considerably increased with resuspended sewer solids. In addition to other aspects like the flow capacity of a sewer network, it is therefore crucial to reduce the accumulation of sewer solids that might occur during dry weather periods.

Until now, it has not (with sufficient accuracy) been possible to describe in conceptually based model terms, the dry weather accumulation of sewer solids and the subsequent wet weather erosion and resuspension. The modeling of the pollutant contribution from the sewer solids during wet weather flow therefore applies an empirical and rather pragmatic approach, an approach that is based on measured constituents for the runoff water in the combined sewer (cf. Chapter 5).

2.4 POLLUTANT VARIABILITY

From the description of the sources for pollutants dealt with in Section 2.3, it is clear that the transport of urban wet weather pollutants is subject to variability in time and place for several reasons. That is not to the same extent the case for the dry weather continuous sources. This fact results in a corresponding impact of both SWR and CSO.

In terms of both environmental effects and the choice of an appropriate control methodology for pollutant reduction, it is important to note how the variability of the pollutant load occurs. For this reason it becomes important to distinguish between

- Pollutant variability within an event
- Pollutant variability between events at a specific site
- Pollutant variability between sites

These three types of pollutant variability will be described and discussed in the following sections.

2.4.1 Pollutant Variability within an Event

Pollutants in SWR as well as in combined sewage including CSOs will typically appear in varying concentrations during a storm runoff event. It is often (but not always) observed that the highest concentrations or the largest mass of pollutants transported will occur during the initial period of a runoff event compared with the later stages of the same event. This phenomenon is defined as "first flush" or sometimes "first foul flush."

As an example, Figure 2.11 depicts the variations of flow and concentration of COD of the mixed runoff water and wastewater measured during a runoff event in a combined sewer pipe. It is readily seen that the concentrations are relatively high in

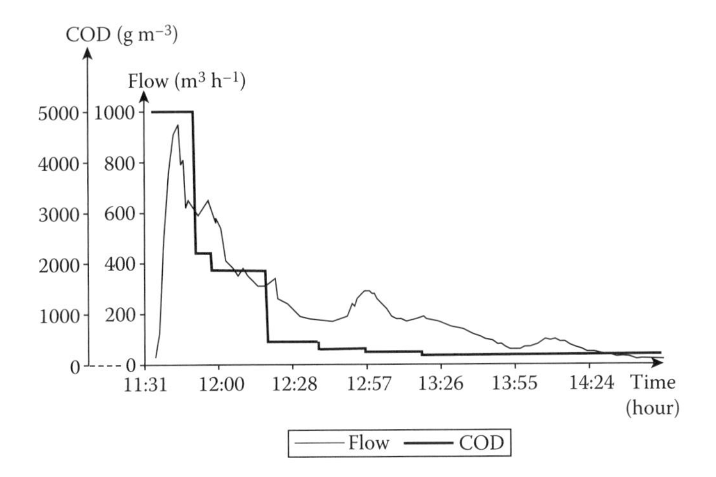

FIGURE 2.11 Variations of flow and COD concentration during a runoff event in a combined sewer pipe. The measurements are performed in a combined sewer pipe. (Modified from Larsen, T., Broch, K., and Andersen, M. R., *Water Science and Technology,* 37, 251–57, 1998.)

the beginning of the event in contrast to what follows. The immediate impression is therefore an occurrence of a first flush phenomenon.

The first flush phenomenon can be quantified based on the development of the cumulated relative flow and the cumulated relative pollutant mass transport during a runoff event. Referring to the hydrograph and the pollutograph, respectively, shown in Figure 2.12, the following two terms are defined (cf. Equations 2.11 and 2.12).

Cumulative relative runoff volume during a runoff event:

$$f_{\text{flow}} = \frac{\sum_{j=1}^{i} v_j}{\sum_{j=1}^{n} v_j}, \tag{2.11}$$

where

$j = 1,\ldots, i$; interval number (–)
$i = 1,\ldots, n$; interval number (–)
n = total number of intervals—often equal to the number of samples analyzed (–)
v_j = volume of runoff water in interval number j (m^3).

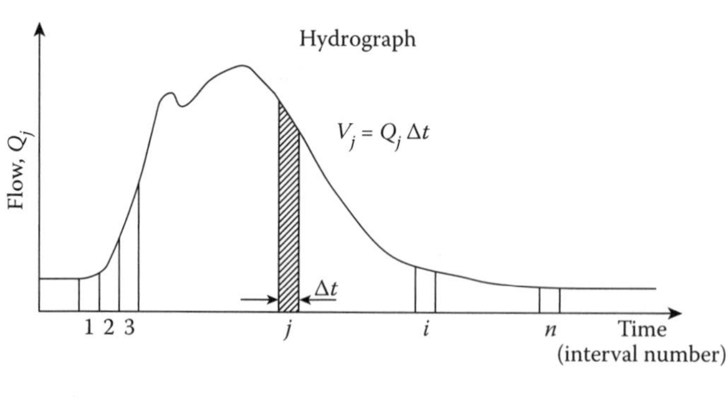

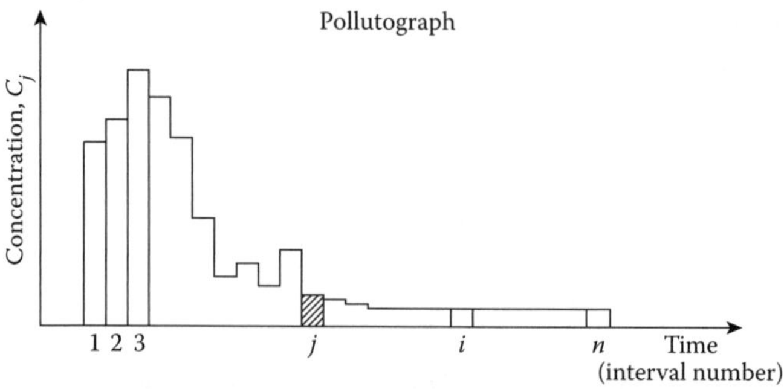

FIGURE 2.12 Hydrograph and corresponding pollutograph for an event as a basis for assessment of the first flush phenomenon (cf. Equations 2.11 and 2.12).

Cumulative relative pollutant mass transport during a runoff event:

$$f_{\text{mass}} = \frac{\sum_{j=1}^{i} c_j v_j}{\sum_{j=1}^{n} c_j v_j}, \tag{2.12}$$

where

c_j = pollutant concentration of the runoff water at interval number j (g m^{-3}).

The denominators in Equations 2.11 and 2.12 express the total runoff volume and the total pollutant mass load for the entire runoff event, respectively. The nominators are the corresponding values from the start of the event to interval number i (cf. Figure 2.12).

Based on the results shown in Figure 2.11, Figure 2.13 depicts the corresponding cumulative relative mass of pollutant, f_{mass}, transported during the event versus the cumulative relative volume, f_{flow}. Such dimensionless mass (M) versus volume (V) curves are often referred to as $M(V)$ curves. If the pollutant (in the example COD) concentration was constant through the event, a straight line from point (0.0) to point (1.1) would be the result. The deviation (to the left side) between the actual curve and this straight line illustrates the potential occurrence of a first flush phenomenon.

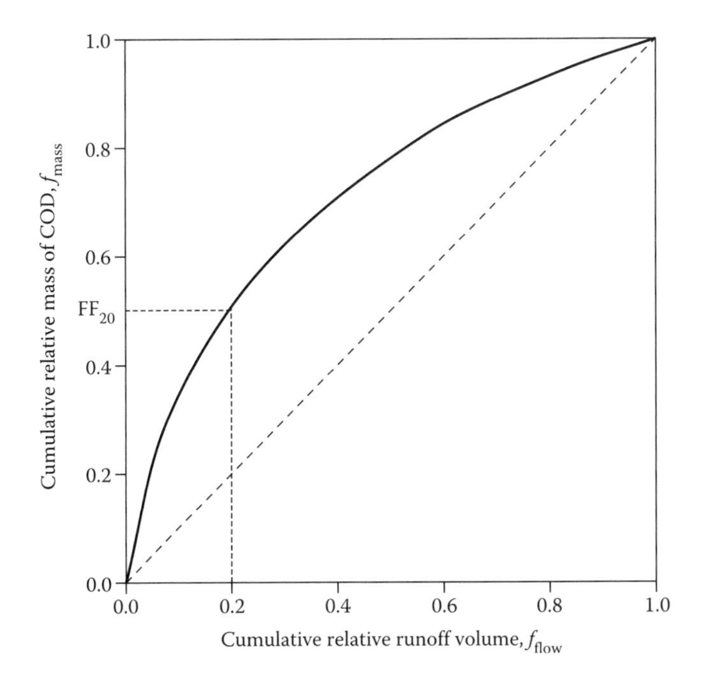

FIGURE 2.13 $M(V)$ curve for illustration of the first flush phenomenon corresponding to the event characteristics depicted in Figure 2.11.

A simple empirical description of a $M(V)$ curve as shown in Figure 2.13 is

$$f_{\text{mass}} = (f_{\text{flow}})^m, \tag{2.13}$$

where

f_{mass} = cumulative relative mass, cf. Equation 2.12 (–)
f_{flow} = cumulative relative Volume, cf. Equation 2.11 (–)
m = empirical first flush coefficient (–).

Based on Equation 2.13, a potential first flush phenomenon exists in case $m < 1$ whereas $m > 1$ corresponds to a situation where the mass transport is concentrated in the last part of the runoff event. If $m = 1$, the mass concentration is evenly distributed during the runoff event. Sometimes this type of distribution is referred to as being neutral.

Typically, dimensionless $M(V)$ curves like the one depicted in Figure 2.13 are used to assess the occurrence of a first flush. Based on such curves, numerous criteria can in principle be used to define a first flush. The following three criteria for occurrence of first flush have been proposed:

1. The slope of the $M(V)$ curve
   A first flush phenomenon can be defined if the initial slope of the curve is larger than 45°. The quantification of the first flush can thereby be defined as its maximum divergence of the curve from the 45° slope line (Geiger 1987).
2. A first flush ratio, $\text{FFR}_{a,b}$
   In Han and others (2006) a first flush ratio is defined as follows:

$$\text{FFR}_{a,b} = \frac{\sum_{j=1}^{k} c_j v_j / M_{\text{tot}}}{\sum_{j=1}^{k} v_j / V_{\text{tot}}}, \tag{2.14}$$

where

$\text{FFR}_{a,b}$ = first flush ratio defined as the relative ratio for $b\%$ pollutant mass transported in $a\%$ of the runoff volume (–)
k = interval number corresponding to transport of $a\%$ of the runoff volume
M_{tot} = total pollutant mass transport for the runoff event (g)
V_{tot} = total runoff volume for the event (m^3).

Saget, Chebbo, and Bertrand-Krajewski (1996) propose that first flush occurs when at least 80% of the pollutant load is transported in the first 30% of the runoff volume. Based on the analysis of a large number of $M(V)$ curves, Bertrand-Krajewski, Chebbo, and Saget (1998) consider a criterion of this magnitude being crucial for a significant first flush. The criterion can be defined by $FFR_{30,80} = 80/30 = 2.67$. The choice of the values $a = 30$ and $b = 80$ is, of course, arbitrary and other pairs may in principle be selected to define the first flush phenomenon.

3. The FF_{20} criterion
   A third—and less restrictive—first flush criterion can be defined by calculating a "first flush value," FF_{20}, as follows (Deletic 1998):
   FF_{20} = the relative load of a pollutant transported by the first 20% of the runoff volume.

The criterion for first flush being present in an event can thereby be defined as a "significantly higher value" than 20% (Deletic 1998). As an example, Figure 2.13 shows that the first 20% of the mixed runoff and wastewater transports about 50% of the COD mass corresponding to $FF_{20} = 0.50$ or a first flush ratio $FFR_{20,50} = 50/20 = 2.50$. Although only about 61% and not 80% of the mass is transported in the first 30% of the runoff volume, there should be no doubt that this event shows first flush.

It is clear that the occurrence of first flush will be judged differently depending on which criterion, 1, 2, or 3, is used to assess the phenomenon.

In addition to the method for analysis of a first flush phenomenon based on M(V) curves, other methods can be applied for assessment of its occurrence. Figure 2.14 exemplifies a method where the absolute amount of accumulated pollutant is depicted versus the accumulated rainfall during an event. For those events where the slope of the line is larger in the beginning compared with the end of the event, a first flush phenomenon can be judged to exist. The figure also shows that some of the events have no (clear) first flush and that relatively large

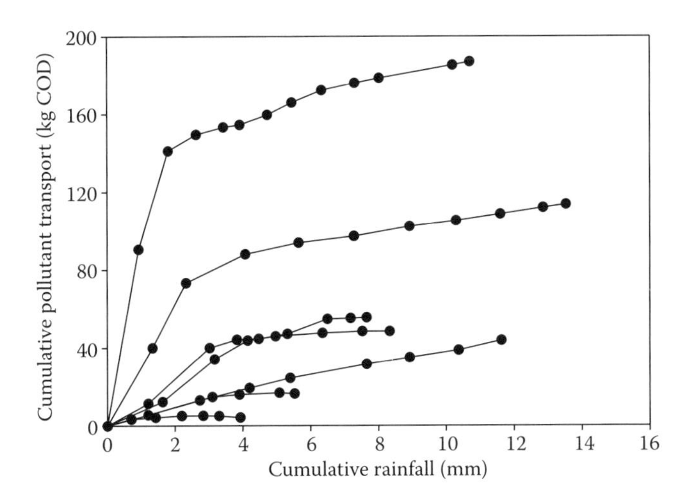

FIGURE 2.14 Accumulated amount of COD versus accumulated rainfall for seven runoff events in a combined sewer catchment. Events that correspond to transport of relatively large amounts of COD show a first flush phenomenon.

amounts of a pollutant transported during an event also seem to favor a first flush. At the site dealt with in Figure 2.14, the first flush seems to take place within the first 2–3 mm of rainfall. This interval depends on a number of factors (e.g., the size and the sewer network characteristics of the catchment and the rainfall pattern). For other catchments, this value may be found considerably larger (e.g., about 10 mm).

There are several reasons why a first flush phenomenon can occur. Basically it occurs because accumulated pollutants associated with particles are subject to erosion and resuspension followed by their transport in the water phase. A relatively high flow at the beginning of an event is therefore important for the phenomenon. First flush is therefore a potential phenomenon for pollutants accumulated at both urban surfaces and in combined sewer networks. First flush is therefore relevant for both SWR and CSO discharges. In combined sewers, first flush may be seen in the case of the occurrence of loose urban dust and type C and D sewer solids. The pollutants at the urban surfaces will in case of first flush concentrate in the runoff water washed into a sewer in the beginning of an event after which the surfaces gradually become more and more clean. In combined sewer systems, biofilm that grows on the sewer walls and sediments that are deposited in the network during the antecedent interevent dry periods can subsequently be eroded and resuspended caused by the increased wet weather flow. In general, the most pronounced first flush phenomena seem to occur in combined sewers. The relative impervious catchment area, the length of the antecedent dry weather period, the duration of the rain event, and the relative distribution of the intensity during the event are major factors that affect the magnitude of the first flush. The fact that heavy rain events are often most intensive in the beginning will significantly contribute to the first flush phenomenon.

A second phenomenon that contributes to first flush in combined sewers is associated with the fact that the incoming runoff water in the beginning of an event "pushes" the wastewater in the sewer in a forward direction to either a trunk sever, an interceptor, or an overflow structure. This type of first flush therefore typically appears at the downstream sections of a drainage system (Deletic 1998). Although the nature of this phenomenon fundamentally is not a real first flush, it is in practice normally not possible to distinguish it from the impact of erosion and resuspension. A detailed monitoring and data analysis would typically be required.

As previously described and in addition to how it is defined, the first flush is a highly variable phenomenon strongly dependent on local conditions. It is therefore not surprising that there is a standing dispute among both scientists and practitioners on its occurrence and relevance. Measurements in sewer networks, however, show that first flush can occur. Particularly it is expected to occur in low gradient sewer pipes in combined sewer catchments where sediments deposit during dry weather flow and in catchments that are not very large where distributed rainfall and time of transport to the point in question will not level out its importance.

Lee and Bang (2000) studied these first flush phenomena. In a comprehensive study of the runoff from nine totally different urban catchments, they observed a first flush phenomenon in catchments with areas smaller than about 100 ha where the impervious area was larger than 80%. It was furthermore observed that the pollutant concentration peaks occurred later than the flow peaks in areas larger than 100 ha where the impervious area was less than 50%.

In addition to the problems that occur because of different definitions of the first flush criterion and of a number of contradictory observations, it is clear that the phenomenon should be considered both complex, site specific, and subject to discussion. Due to the multiplicity of influencing factors and parameters, no general relationship can be established for prediction of the first flush phenomenon (Bertrand-Krajewski, Chebbo, and Saget 1998). The phenomenon has, however, been reported to exist worldwide from observations in particularly combined sewer networks where it is regularly considered to occur (Thornton and Saul 1986; Geiger 1986; Ashley and Crabtree 1992; Gupta and Saul 1996; Larsen, Broch, and Andersen 1998). In an investigation on mass loads from major highways in southern California, Kim and others (2005) observed that high or medium first flush occurred in approximately 80% of the storm events. Other investigations, however, question the importance of the phenomenon (Saget, Chebbo and Bertrand-Krajewski 1996).

Where first flush is considered to be of importance, it should be taken into account in the design process of the drainage system. In order to reduce pollutant loads into the adjacent receiving water system, it might be relevant to catch or treat the first flush part of the flow rather than the "tail" of the runoff. In terms of treatment of both CSOs and SWR it is crucial that the treatment efficiency is generally higher in waters of relatively high concentrations than for those of low concentrations (Strecker et al. 2001). A potential occurrence of first flush may therefore influence selection of an appropriate technology when designing new systems and renovating existing drainage networks (cf. Section 4.5 and Chapter 9).

2.4.2 Pollutant Variability between Events

The stochastic nature of runoff events including pollutant buildup in drainage systems that affects the amount of pollutants that is available for transport, implies that both pollutant loads and pollutant concentrations may vary between the runoff events at a specific site. As an example, Figure 2.14 directly shows that the pollutant loads varies from event to event. According to the varying slopes of the curves, the figure also indicates that the concentrations vary from event to event.

Due to the varying concentrations of the flows from runoff events at a specific site, it becomes important to figure out what a characteristic pollutant concentration for such a site would be. Basically, the load of a pollutant is just calculated as the product between such a concentration and a relevant volume of runoff water. Details in this respect will be dealt with in Section 2.6. In this section, only the qualitative aspects will be discussed.

As an example of the variability of pollutant concentrations, Figure 2.15 shows COD concentrations originating from six combined sewer catchments. However, the figure also shows the variability between the events for each of these six sites. Each concentration measured refers to a mean value for an event. This value includes the contributions from the urban surfaces and the accumulated sewer solids but not the contribution from the incoming wastewater during the event as this part was

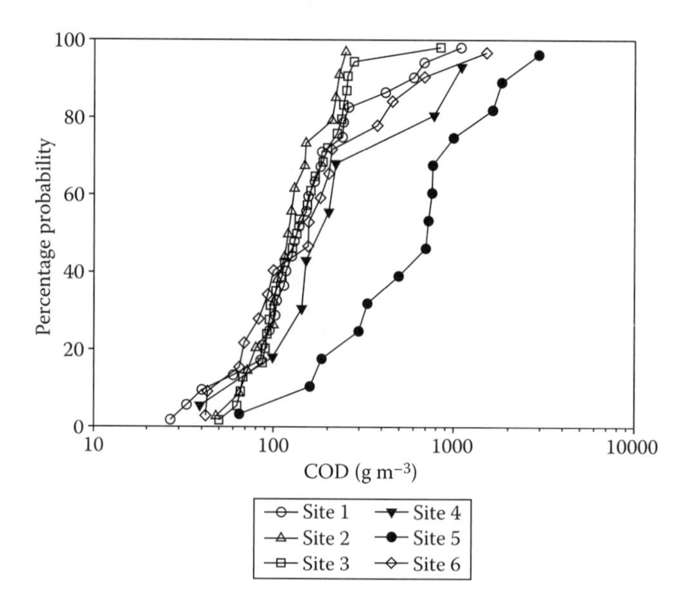

FIGURE 2.15 Variability for COD concentrations originating from runoff events for six combined sewer catchments in Denmark.

deducted (cf. Sections 2.3.1 through 2.3.3 and further explanation in Section 2.6). For each of the six sites, Figure 2.15 shows that the COD concentrations vary over more than a decade. Similar variations from event to event at a specific site are typical and exist worldwide and for all types of pollutants relevant for SWR and CSOs.

2.4.3 Pollutant Variability between Sites

Each site has specific characteristics in terms of urban runoff that concerns climate conditions including rainfall pattern, type of infrastructure, and technology applied. In addition to these aspects, culture (human behavior and activities) will, to a great extent, determine those pollutants that are dominating and to what extent they will turn up in the urban runoff. For such reasons, pollutant variability (both qualitatively and quantitatively) must exist from site to site.

It is difficult to illustrate the site variability. When collecting data to compare, a basic problem is that not just the drainage systems but also the methods applied often vary from country to country. Monitoring, sampling, handling of samples, and analysis of samples are often differently implemented in different countries, and often insufficiently reported. Any attempt to compare such data therefore becomes problematic. However, as shown in Figure 2.15, significant pollutant variability from one site to another exists even within the same country and when applying the same procedures for monitoring, sampling, and analysis.

Within limited regions (e.g., within the United States and to some extent within the European Union where the catchments and network systems are rather comparable), it has been possible to assess the site variability. Figure 2.16 shows pollutant concentrations in urban runoff from European cities. As readily seen, a rather

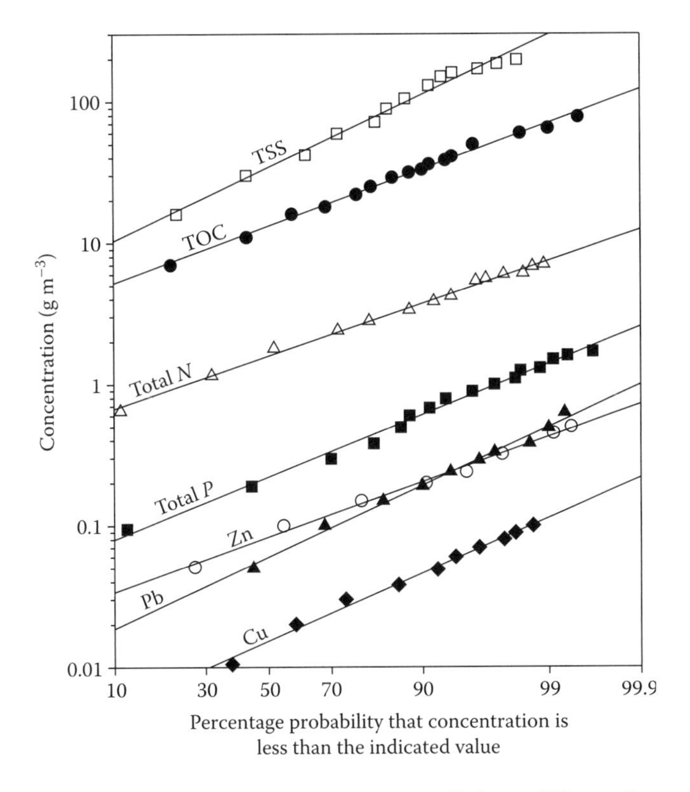

FIGURE 2.16 Pollutant concentrations in urban runoff from different European cities. Each point corresponds to an urban site-characteristic concentration that is, a SMC value, cf. Section 2.6.2. (Modified from EWPCA, Status Report 1987. Report from the Technical Subcommittee on Urban Runoff Quality Data, the European Water Pollution Control Association, 1987.)

significant variability exists. This variability is of the same order of magnitude as the variability between events at a specific site.

It is clear that the variability of urban wet weather phenomena is a challenge when finding sound solutions to any specific problem that might occur. It also concerns the question: is it possible to produce a text that is worldwide relevant and useful? The authors' answer to this question is that at the conceptual level the problems and the solutions to these problems are fundamentally identical irrespective of the context in which the problem exists. What is important is in the description of given phenomena to extract the essential and general part and just to use the site-specific descriptions as illustrative examples.

2.5 STATISTICAL CONCEPTS

The variability of the pollutant concentrations and mass loadings related to urban storm drainage is dealt with in Sections 2.2 through 2.4. The nature of the phenomena described in these sections is the basis for quantification in stochastic terms.

The selection of an appropriate probability distribution is in this respect the first step. In this context, the following characteristics of data are important:

- *Positive random variables*
  The data that are relevant concerns basically pollutant concentrations and mass loadings. In terms of statistics, these data—the random variables—are therefore positive numbers.
- *Extreme event statistics*
  The interval of variability for a random variable is large and can, for the same set of data, include both very low and very high numbers. The point is that both the nature of the rainfalls and the pollution related to urban storm runoff contribute to the extreme event phenomena in terms of a potential risk and fatal effects related to large pollutant concentrations and mass loadings.

The data depicted in Figure 2.15 demonstrate the relevance of the two characteristics mentioned above. As an example, it is readily seen from the figure that 2–3% of the COD concentrations exceeds 1000 g m^{-3}. Compared with the characteristic COD concentrations for domestic wastewater shown in Table 2.3, and taking into account the "dilution" with the incoming runoff water, it is quite clear that pollution caused by extreme events is possible. In accordance with the nature of urban runoff data, statistical analysis has shown that the underlying distribution typically is log-normal (U.S. EPA 1983; Di Torro 1984). The positively skewed log-normally distribution accounts for the fact that extreme values of the stochastic variables might occur.

The characteristic feature of a stochastic variable, x, that is log-normally distributed, is that ln (x), or log (x), follows a normal distribution. Assuming that ln (x) is normally distributed with mean u and variance s^2, then $x = \exp(\ln(x))$ is log-normally distributed with the following mean, α, and variance, β^2:

$$\alpha = \exp\left(u + \frac{s^2}{2}\right), \tag{2.15}$$

$$\beta^2 = \alpha^2\left(\exp s^2 - 1\right). \tag{2.16}$$

A central statistical parameter for comparison of the variation between different sets of data is the coefficient of variation, COV. The coefficient of variation is a measure of the relative variation of the data and defined as follows:

$$\mathrm{COV} = \frac{\beta}{\alpha}. \tag{2.17}$$

The coefficient of variation can be interpreted as a relative standard deviation (i.e., a standard deviation expressed relative to the mean value).

Compared with a simple mean, the median value, x_m, of a log-normal distribution of urban runoff data is considered a robust measure of a central tendency because it is less influenced by the small numbers of large (extreme) values that might occur. The median value, x_m, and the coefficient of variation, COV, are the two very useful statistical parameters for characterization of log-normally distributed data. In Section 2.6.2 it will be described how this median value is determined based on flow-weighted mean values from a number of runoff events at the site in question. The two statistical parameters are often used and also normally sufficient for characterization of urban runoff data originating from a specific site:

$$\frac{\alpha}{x_m} = \sqrt{1+\mathrm{COV}^2} \quad \alpha > x_m. \tag{2.18}$$

In Figure 2.17, an example of a log-normally distributed stochastic variable, x, is depicted. The figure shows both the density function (probability function or relative frequency function), $p(x)$, and the distribution function (cumulative probability or cumulative frequency function), $P(x)$:

$$P(x) = \int_0^x p(t)dt = \int_0^x \frac{1}{t \times s \times \sqrt{2\pi}} \exp\left(-\frac{1}{2}\left(\frac{\ln t - u}{s}\right)^2\right) dt. \tag{2.19}$$

It is often appropriate to depict the distribution function $P(x)$ versus the normally distributed stochastic variable ln (x) (i.e., using a logarithmic scale on the abscissa). For practical reasons, the ln (x) function is typically substituted by log (x). Furthermore, in order to transform the S-shaped distribution function into a linear curve, the ordinate is transferred into a nonlinear function scale corresponding to the normal distribution function shown in Equation 2.20 (c.f. Figure 2.18).

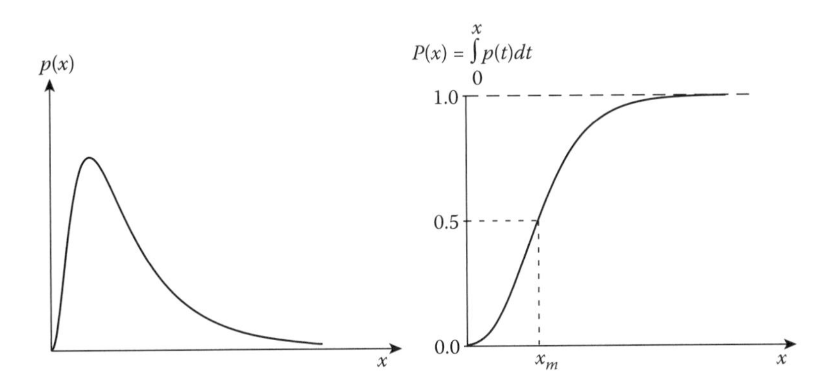

FIGURE 2.17 Illustration of a density function, $p(x)$, and a distribution function, $P(x)$, of a log-normally distributed stochastic variable, x. The t denotes an independent variable of integration.

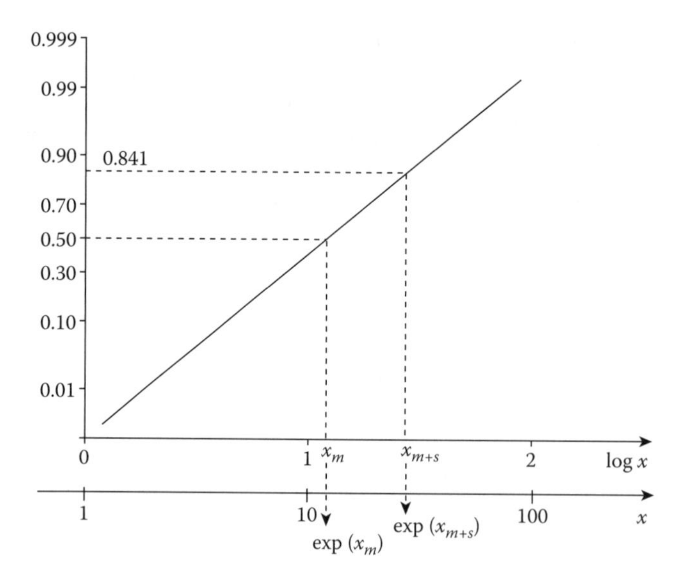

FIGURE 2.18 The distribution function, $P(x)$, depicted in Figure 2.17 shown as transformed from an S-shaped curve into a straight line by plotting the data in a diagram with a normally distributed probability function (normal probability) scale at the ordinate and a log-scale at the abscissa.

$$F(y) = \int_{-\infty}^{y} e^{-\frac{1}{2}t^2} dt. \tag{2.20}$$

Figure 2.18 shows how the distribution function in Figure 2.17 can be transformed.

A first step in analysis of data is often to prepare such probability plots. The observations, x, in a data set are therefore appropriately ranked in ascending order. Probability values, P_i, associated with each observation are calculated as follows:

$$P_i = \sum_{j=1}^{i} p_i(x_j \leq x_i) = \frac{(i-0.5)}{n}, \tag{2.21}$$

where

P_i = probability that an observation, x_j, in a ranked data set observes the criterion $x_j \leq x_i$ (–)
$j = 1, \ldots, i$; ranked observation number (–)
$i = 1, \ldots, n$; ranked observation number (–)
n = total number of observations in the data set (–)
x_j = observation number j.

The advantage of the transformation of the *S*-shaped distribution function into a linear curve is that the median and the standard deviation of the log-normally distribution can be readily illustrated and determined. Furthermore, a regression analysis of data originating from a measurement program will in the case of assessment of the linear approximation indicate if the data can be considered log-normally distributed or not.

A more thorough analysis of monitored data applying statistical tests can be used to assess whether the data follow a log-normal distribution. For this assessment, a number of appropriate tests exist (e.g., the Kolmogorov–Smirnoff test and the Shapiro–Wilk test). These two statistical tests can be found as a computer software package in *NIST/SEMATECH e-Handbook of Statistical Methods* (the Web site address can be found after the reference section).

When testing for normality of measured data where the mean value and variance has to be estimated from the sample, the latter of the two tests is the most powerful. These tests will, however, not indicate the type of nonnormality in case the hypothesis is rejected. For detection of violations of normality test assumptions, the normal probability plot is a valuable graphical aid (e.g., for detection of outliers).

2.6 POLLUTANT CHARACTERIZATION

The determination of concentrations that are characteristic for urban wet weather pollution is a prerequisite for determination of mass loadings and therefore crucial for assessment of environmental effects and design of control measures. The pollutants dealt with in urban drainage in terms of their stochastic and event-based nature, their source characteristics, and their variability require a different approach of characterization compared with what is known for the continuous fluxes of pollutants. The quantification of pollutant matters within urban drainage in terms of mass balances has shown that two different definitions of concentrations are suitable:

- *Event mean concentration (EMC)*
  The EMC is a concentration that characterizes an event in terms of a flow-weighted mean value.
- *Site mean concentration (SMC)*
  The SMC is a concentration that characterizes a site in terms of a median value based on a number of EMCs at that site.

In the following, these definitions will be further dealt with. Although SMC is the acronym of site *mean* concentration, its value is typically determined statistically as a *median* value (cf. Section 2.6.2).

2.6.1 Event Mean Concentration

The event mean concentration (EMC) is defined as a flow-averaged concentration of a constituent determined during a runoff event irrespective of it relating to a separate or a combined sewer catchment. The EMC is therefore defined as the total transport of mass during an event divided by the corresponding total volume of runoff water:

$$\mathrm{EMC} = \frac{M_{\mathrm{tot}}}{V_{\mathrm{tot}}}. \tag{2.22}$$

where

M_{tot} = total mass of a constituent for the event (g)
V_{tot} = total volume of runoff water for the event (m^3).

Referring to Figure 2.12, the EMC can be computed as follows:

$$\mathrm{EMC} = \frac{M_{\mathrm{tot}}}{V_{\mathrm{tot}}} = \frac{\sum_{j=1}^{n} c_j v_j}{\sum_{j=1}^{n} v_j} = \frac{\sum_{j=1}^{n} c_j v_j}{V_{\mathrm{tot}}}. \tag{2.23}$$

An EMC value refers only to the entire event and does not take into account intra-event phenomena like the first flush. The determination of an EMC is directly related to a monitoring or sampling procedure and Equation 2.23 is therefore not shown in an integrated form. The concrete implementation of the procedure in this respect therefore affects the value and the interpretation of the EMC determined.

In a separate sewer catchment, a sampling or monitoring procedure takes place during the runoff event (e.g., in a pipe, in an open channel, or at the corresponding outlets). Typically, the flow is monitored continuously during the event parallel with regular sampling from the flows. The direct relation to the EMC follows from Equation 2.23. The EMC is thereby a characteristic and mass balance related value for the pollutant concentration of the flow into the receiving water or into a treatment facility from that event.

In a combined sewer catchment, the determination of an EMC needs further explanation. The following two approaches can be used for its determination:

- Sampling or monitoring at the overflow takes place during an overflow event.
- Sampling or monitoring at the inflow to a CSO structure takes place during a runoff event.

The fundamental difference between these two approaches follows from Figure 1.5. The first approach (i.e., sampling and monitoring in the CSO during the overflow event) will produce data that results in an EMC for the actual overflow event (i.e., an EMC that is connected with the overflow volume). The EMC is therefore a parameter that is directly related to mass loadings from a specific event and the site in question. This EMC value will depend on the actual design of the overflow structure determined by Q_{max}, for example, and a number of construction details (e.g., the height of the overflow weir). An EMC determined in this way is therefore in principle not suitable for neither the assessment of measures to reduce pollutant mass loads from the structure nor useful for a comparison with CSO characteristics from other sites.

In contrast to this approach, determination of an EMC value can be based on sampling and monitoring at the inflow to an overflow structure (see Figure 1.5). In this way it is possible to include flow and mass transport for the entire runoff event irrespective of the actual design of the overflow structure. The EMC value is therefore independent on how the overflow event occurs and the value refers to the event-based total mass of pollutants originating from the urban surfaces, the daily wastewater flow, and the eroded and resuspended sewer solids. The following mass balance based on transported water volumes and corresponding concentrations of a constituent covering the entire runoff event can therefore be established:

$$V_{tot}\,C_{tot} = V_{ww}\,C_{ww} + (V_{tot} - V_{ww})\,C_{runoff}, \tag{2.24}$$

where

C_{tot} = EMC of a constituent measured for the total runoff event that is, in the inflow to the overflow structure (g m^{-3})

V_{ww} = volume of municipal wastewater inflow during the runoff event (m^3)

C_{ww} = estimated average concentration of a constituent in the dry weather wastewater flow (g m^{-3})

C_{runoff} = EMC of a constituent in the runoff water of the combined sewer (g m^{-3}).

The mass balance established in Equation 2.24 is in accordance with the subdivision of the pollutant sources in three main types for combined sewers (cf. Section 1.4.2.3), and in further details also dealt with in Section 2.3. The first term at the right side of Equation 2.24 is equal to the mass of a constituent originating from the incoming wastewater whereas the second term accounts for the mass originating from two sources: the urban surfaces and the sewer solids. A graphical interpretation of the contents related to the mass balance, Equation 2.24, is shown in Figure 2.19.

Equation 2.24 is from an application point of view useful because all parameters except for C_{runoff} are experimentally accessible under the assumption that it is possible to estimate the volume and concentration of the incoming wastewater during the runoff event based on measurements during dry weather periods:

$$V_{ww} = Q_{ww}\,\Delta t, \tag{2.25}$$

where

Q_{ww} = average estimated wastewater flow relevant for the period where runoff takes place (m^3 s^{-1})

Δt = duration of the runoff event (s).

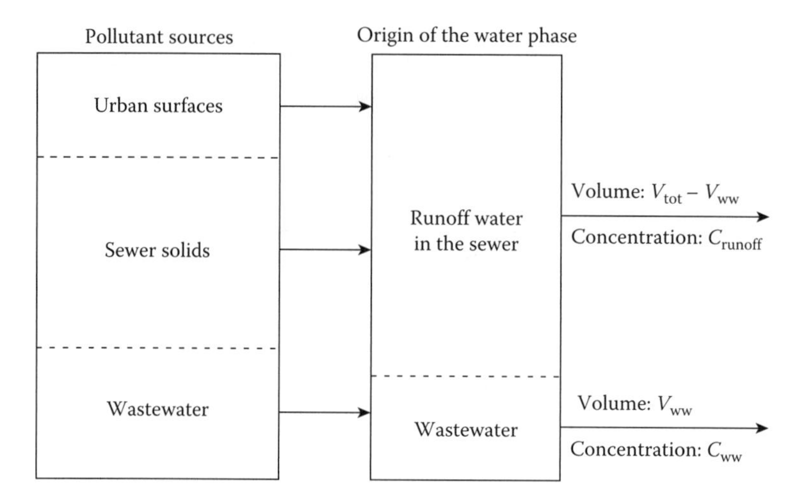

FIGURE 2.19 Outline of the origin of pollutants and water and the definitions of concentrations related to a runoff event in a combined sewer network.

Based on experimental data and estimated values, it is therefore possible to determine C_{runoff} from Equation 2.24. In Equation 2.24, C_{runoff} can be interpreted as an EMC that characterize the runoff event in a combined sewer system. In terms of direct measurement, C_{runoff} is of course fictitious because the runoff water from the urban surfaces in the real network does not exist as a separate flow but is mixed with the wastewater. However, as a characteristic concentration for the wet weather flow in a combined system, C_{runoff} becomes a central parameter. It is important to notice that C_{runoff} can be interpreted as the concentration of the "wet weather pollutants" originating from both the urban surfaces and the sewer solids. These two contributions are associated with the volume of runoff water ($V_{tot} - V_{ww}$) from the urban surfaces. As an example, the data presented in Figure 2.15 are all EMC values for COD expressed in terms of C_{runoff}.

The two different approaches of calculating mass transport in a combined sewer network during a runoff event in this section particularly seen from a concentration point of view (i.e., referring to determination of an EMC. In Example 2.1, it will be further demonstrated. The corresponding mass balance approach is a subject of Section 5.4.1.

Example 2.1: COD concentrations in the inflow to an overflow structure in a combined sewer network during a runoff event

A residential area served by a combined sewer network comprises housing for about 2,100 inhabitants. The total paved area is 85 ha with an average runoff coefficient of 0.32. During a day in June, a rainfall results in a runoff of 8.3 mm from about 1 p.m. to 2 p.m. Based on previous measurements, the dry weather inflow to the intercepting sewer is approximately 4.5×10^{-3} m^3 s^{-1} during a summer period and at this time of the day. The COD concentration of the wastewater is under corresponding conditions estimated to be 650 g m^{-3}. Flow-weighted sampling at the inflow to the overflow structure during 73 minutes, covering the entire runoff event, results after a chemical analysis of the samples in an EMC value for the COD concentration, C_{tot}, equal to 167 g m^{-3}.

With this information, calculate the COD concentration (EMC value) of the runoff water at the inflow to the overflow structure.

When applying Equation 2.25, the total volume of wastewater during the runoff period is approximately

$$V_{ww} = Q_{ww}\,\Delta t = 4.5 \times 10^{-3}\ (\mathrm{m^3\ s^{-1}}) \times 73\ (\mathrm{min}) \times 60\ (s\ \mathrm{min^{-1}}) = 19.7\ \mathrm{m^3}.$$

The volume of runoff water is

$$V_{tot} - V_{ww} = 8.3\ (\mathrm{mm}) \times 10^{-3}\ (\mathrm{m\ mm^{-1}}) \times 85\ (\mathrm{ha}) \times 10^{4}\ (\mathrm{m^2\ ha^{-1}}) \times 0.32$$
$$= 2257.6\ \mathrm{m^3}.$$

Equation 2.24 results in

$$C_{runoff} = \frac{V_{tot}C_{tot} - V_{ww}C_{ww}}{V_{tot} - V_{ww}} = \frac{(2257.6 + 19.7)167 - 19.7 \times 650}{2257.6} = 163\ \mathrm{g\ m^{-3}}.$$

The COD concentration of the runoff water in the combined system, $C_{runoff} = 163$ g m^{-3}, is rather close to the total concentration, $C_{tot} = 167$ g m^{-3}, which includes the contribution from the dry weather wastewater flow.

The reason for this fact is caused by the construction and functioning of the sewer network. Although the COD concentration of the wastewater is relatively high, the corresponding (dry weather) volume is rather small compared with the volume of runoff water. The mass of COD originating from municipal wastewater is, during the relatively short runoff period, rather low compared with the contribution from the runoff that includes materials eroded from the sewer. This phenomenon is typical for networks where the capacity in the sewer is large compared with the dry weather flow. In this example, the volume of municipal wastewater accounts for less than 1% of the runoff volume. In several regions, the dry weather flow may, however, amount to 20–50% of the flow capacity in the sewer, and C_{tot} is in this case considerable larger than C_{runoff}. The example thereby shows that the design of the system, to a great extent, will influence the quality of the CSO discharge.

The numbers used in this example originate from a real case where the COD concentration of the runoff from the urban surfaces is approximately 40 g m^{-3}. The difference between $C_{runoff} = 163$ g m^{-3} and this value shows that a large part of the COD originates from eroded sewer solids, a case that is often observed.

2.6.2 Site Mean Concentration

As previously mentioned, a site mean concentration (SMC), is a concentration that is characteristic for the degree of wet weather pollution at a given site. For both separate and combined systems, a number of EMCs originating from the same location is therefore required for its determination. A number of approaches can be applied for its estimation (Mourad, Bertrand-Krajewski, and Chebbo 2005). In the following, three methods will be shown and discussed.

1. The arithmetic mean method
   According to this method, the SMC value can be determined as follows:

$$\mathrm{SMC} = \frac{\sum_{j=1}^{n} \mathrm{EMC}_J}{n}, \tag{2.26}$$

where
   $j = 1, \ldots, n$; integer (–)
   n = total number of events (–)

   Depending on the number of events, n, a SMC value that is determined by the arithmetic mean method might be strongly influenced by events with high EMCs and small runoff volumes, for example. To overcome such a problem, the following method can be applied.
2. The weighted mean method
   In contrast to Equation 2.26, the weighted mean method takes into account the runoff volumes. It is a method that directly refers to a mass balance. It is therefore appropriate when using a SMC value to estimate a pollutant load (cf. Section 4.5.4).

$$\mathrm{SMC} = \frac{\sum_{j=1}^{n} \mathrm{EMC}_J V_j}{\sum_{j=1}^{n} V_J}, \tag{2.27}$$

where
   V_j = runoff volume of event number j (m^3).
   A third method for estimation of the SMC is the median value.

3. The median value method
   The median value, the 50 percentile, for determination of the SMC is like the weighted mean method not sensitive to high values of EMCs.

Investigations have shown that the uncertainty in the estimation of a SMC value may vary significantly and it is not possible to determine the number of EMCs to be measured to estimate a SMC value with a given uncertainty (Mourad, Bertrand-Krajewski, and Chebbo 2005). Typically, not less than 10–15 EMC values are required, however, this number is rather pragmatically recommended.

The C_m values that are shown in Table 2.2 are SMCs ("characteristic" SMC values depending on the land use) and so are the values depicted in Figure 2.16. Figure 2.15 shows cumulative probability curves that at the 50 percentile determine the SMC values for each of the six catchments.

According to the log-normal probability plot shown in Figure 2.18, Figure 2.20 illustrates the distribution of EMCs and the corresponding SMC values according to the median value method for total phosphorus, zinc, and cadmium in runoff water from a residential catchment in Orange County, Florida (Yousef et al. 1990). The slope of the three straight lines are approximately equal indicating that the standard deviations for the concentrations of the three pollutants in the runoff water are also approximately equal when compared on a log scale (cf. comments to Table 2.2 in Section 2.3.1).

2.7 FINAL REMARKS

Drainage of urban areas has a long history, going back at least 4000 years in time and probably 9000 years to the days of the "urban revolution" of our civilization when the first urban settlements were developed. However, the clear conception that urban drainage is potentially associated with human health problems and pollution of the environment is relatively new. The understanding that drainage of urban areas should also be seen in the context of human health basically started 150 years ago caused by several cases of fatal cholera outbreaks in European cities. It is only recently that concern with urban runoff as a contributor to the pollution of our

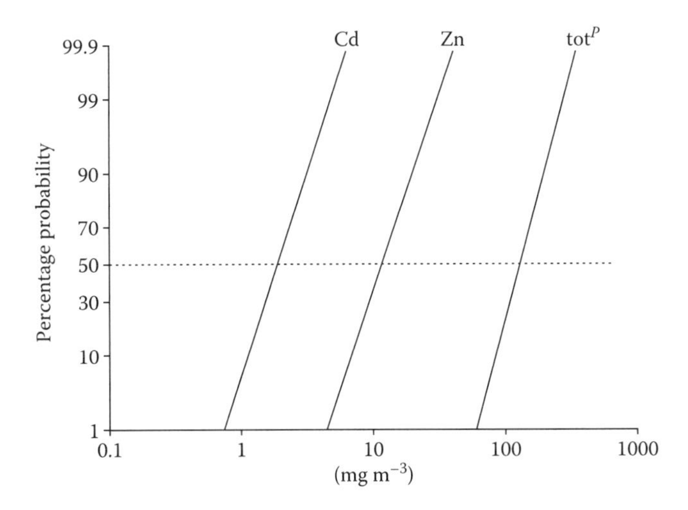

FIGURE 2.20 A log-normal distribution plot of the concentrations for total phosphorus, zinc, and cadmium in the runoff water from a 7.8 ha residential catchment in Orange County, Florida. (Modified from Yousef, Y. A., Wanielista, M. P., Dietz, J. D., Lin, L. Y., and Brabham, M., *Final Report on Efficiency Optimization of Wet Detention Ponds for Urban Stormwater Management,* Department of Civil and Environmental Engineering, University of Central Florida, Orlando, FL, 1990.) Approximately 20 events (i.e., 20 EMC values) determine, in each case, the SMC value.

environment has been addressed. The federal and national laws implemented in this respect are not more than 30–50 years old. The combined theoretically and empirically based understanding of the nature of urban wet weather pollution and corresponding concepts that are the focal point of this chapter were developed during this period. The concepts and definitions related to urban runoff quality are therefore still open for discussion and changes. However, in the authors' opinion, Chapter 2 provides (based on the contents of Chapter 1) the reader with a fundamental and sound understanding directed to the engineering of the sewer networks that we have inherited as well as to those we will develop in the future. The authors consider the understanding of the concepts and definitions as crucial for further insight into the engineering of sewer networks and thereby to cope with "water in the city" in a sustainable way.

A central concept of sustainable urban drainage (SUD) is to minimize the deteriorating effects and at the same time to support solutions for the beneficial uses of the rainfall in urban areas. In particular, the focal point is the quality of life for the coming generations and the preservation of resources. These aspects are rather recently accepted, particularly when compared with the age of the urban drainage networks of today. The acceptance of urban drainage as a potential course for adverse impacts onto our ecosystems cleared the way for this understanding.

New challenges appear in the horizon (e.g., possible climate changes), globally causing different local precipitation patterns than experienced until now. The growing population of the cities, particularly in the developing countries, also requires new thinking. By focusing on the quality aspects, this text will provide the reader with sound concepts and tools for solving both today's problems and those of the future.

REFERENCES

Arnell, V. 1982. Rainfall data for the design of sewer pipe systems. Department of Hydraulics, Chalmers University of Technology, Report Series A:8, Gothenburg, Sweden.

Ashley, R. M., J.-L. Bertrand-Krajewski, T. Hvitved-Jacobsen, and M. Verbanck, eds. 2003. Solids in Sewers. IWA (International Water Association) Scientific & Technical Report no. 14.

Ashley, R. M., and R. W. Crabtree. 1992. Sediment origins: Deposition and build-up in combined sewer systems. *Water Science and Technology* 25 (8): 1–12.

Bertrand-Krajewski, J.-L., G. Chebbo, and A. Saget. 1998. Distribution of pollutant mass vs volume in stormwater discharges and the first flush phenomenon. *Water Research* 32 (8): 2341–56.

Crabtree, R. W. 1989. Sediments in sewers. *Journal of the Institution of Water and Environmental Management* 3 (December): 569–78.

Deletic, A. 1998. The first flush load of urban surface runoff. *Water Research* 32 (8): 2462–70.

Di Torro, D. M. 1984. Probability model of stream quality due to runoff. *ASCE Journal of Environmental Engineering* 110: 607–28.

Driscoll, E. D., P. E. Shelly, and E. W. Strecker. 1990. Pollutant loadings and impacts from stormwater runoff, Vol. III: Analytical investigation and research report. Federal Highway Administration, FHWA-RD-88-008, Washington, DC.

EWPCA. 1987. Status Report 1987. Report from the Technical Subcommittee on Urban Runoff Quality Data, EWPCA (the European Water Pollution Control Association), limited circulation of the report.

Geiger, W. F. 1986. Variation of combined runoff quality and resulting pollutant retention strategies. Proceedings from an International Conference on Urban Storm Water Quality and Effects upon Receiving Waters, TNO Committee on Hydrological Research, Proceedings and Information No. 36, Wageningen, The Netherlands, 71–91.

Geiger, W. F. 1987. Flushing effects in combined sewer systems. Proceedings of the 4th International Conference on Urban Storm Drainage, Lausanne, Switzerland, 40–46.

Gupta, K., and J. A. Saul. 1996. Specific relationships for the first flush load in combined sewer flows. *Water Research* 30: 1244–52.

Han, Y., S.-L. Lau, M. Kayhanian, and M. K. Stenstrom. 2006. Characteristics of highway stormwater runoff. *Water Environment Research* 78 (12): 2377–88.

Henze, M., P. Harremoës, J.la C. Jansen, and E. Arvin. 2002. *Wastewater treatment: Biological and chemical processes.* Berlin: Springer-Verlag.

Hvitved-Jacobsen, T., and Y. A. Yousef. 1988. Analysis of rainfall series in the design of urban drainage control systems. *Water Research* 22 (4): 491–96.

Hvitved-Jacobsen, T., Y. A. Yousef, and M. P. Wanielista. 1989. Rainfall analysis for efficient detention ponds. In *Design of urban runoff quality controls*, eds. L. A. Roesner, B. Urbonas, and M. B. Sonnen, 214–22. ASCE (American Society of Civil Engineers) publication. New York, 10017–2398.

IDA Spildevandskomitéen. 1999. Regional variation af ekstremregn i Danmark [Regional variation of extreme rainfall in Denmark]. Skrift nr. 26 (Technical report No. 26), Report from the Danish Water Pollution Committee, The Danish Society of Engineers (in Danish).

Keifer, C. J., and H. H. Chu. 1957. Synthetic storm pattern for drainage design. *ASCE J. Hydraulic Division* 83 (4): 1332/1–1332/25.

Kim, L.-H., M. Kayhanian, S.-L. Lau, and M. K. Stenstrom. 2005. A new modeling approach for estimating first flush metal mass loading. *Water Science and Technology* 51 (3–4): 159–67.

Larsen, T., K. Broch, and M. R. Andersen. 1998. First flush effects in an urban catchment area in Aalborg. *Water Science and Technology* 37 (1): 251–57.

Lee, J. H., and K. W. Bang. 2000. Characterization of urban stormwater runoff. *Water Research* 34 (6): 1773–80.

Medina, M. A., and T. L. Jacobs. 1994. Integration of probabilistic and physically-based modelling with optimization methods for stormwater infrastructure rehabilitation. In *Modelling the Management of Stormwater Impacts*, ed. W. James, 221–41. Boca Raton, FL: Lewis Publishers/CRC Press.

Mourad, M., J.-L. Bertrand-Krajewski, and G. Chebbo. 2005. Sensitivity to experimental data of pollutant site mean concentration in stormwater runoff. *Water Science and Technology* 51 (2): 155–62.

Saget, A., G. Chebbo, and J. L. Bertrand-Krajewski. 1996. The first flush in sewer systems. *Water Science and Technology* 33 (9): 177–84.

Schilling, W. 1983. Univariate versus multivariate rainfall statistics: Problems and potentials. Proceedings of a Specialized Seminar on Rainfall as the Basis for Urban Run-off Design and Analysis, Copenhagen, Denmark, 129–37, August 24–26.

Sifalda, V. (1973), Entwicklung eines Berechnungsregens für die Bemessung von Kanalnetzen [Development of synthetic storms for design of sewer networks]. *GWF: Wasser/ Abwasser* 114:435–40.

Strecker, E. W., M. M. Quigley, B. R. Urbonas, J. E. Jones, and J. K. Clary. 2001. Determining urban storm water BMP effectiveness. *Journal of Water Resources Planning and Management, American Society of Civil Engineers (ASCE)* 127 (3): 144–49.

Tchobanoglous, G., F. L. Burton, and H. D. Stensel. 2003. *Wastewater engineering: Treatment and reuse*. New York: McGraw-Hill.

Thornton, R. C., and A. J. Saul. 1986. Some quality characteristics of combined sewer flows. *J. Inst. Public Health Engineers* 14 (3): 35–39.

U.S. EPA (Environmental Protection Agency). 1983. Results of the nationwide urban runoff program, Volume I: Final report. Water Planning Division, U.S. EPA, NTIS No. PB 84-185552, Washington, DC.

Vaes, G., F. Clemens, P. Willems, and J. Berlamont. 2002. Design rainfall for combined sewer system calculations: Comparison between Flanders and the Netherlands. In *Extended abstracts of the 9th international conference on urban drainage*, eds. E. W. Strecker and W. C. Huber, 179–180, September 8–13. Portland, Oregon.

Wanielista, M. P. 1979. *Stormwater management: Quantity and quality*. Ann Arbor, MI: Ann Arbor Science Publishers Inc.

Wenzel, H. G., and L. M. Voorhees. 1981. An evaluation of the urban design storm concept. Water Resources Center, University of Illinois at Urbana-Champaign, Research Report UILU-WRC-81-0164.

Yousef, Y. A., M. P. Wanielista, J. D. Dietz, L. Y. Lin, and M. Brabham. 1990. *Final report on efficiency optimization of wet detention ponds for urban stormwater management.* Orlando, FL: Department of Civil and Environmental Engineering, University of Central Florida.

WEB SITES

NIST/SEMATECH e-Handbook of Statistical Methods, http://www.itl.nist.gov/div898/handbook/, February 2007.

3 Transport and Transformations: A Toolbox for Quality Assessment within Urban Drainage

The title of this chapter is selected as a reminder for the reader that knowledge on and relations between transport and transformations of pollutants has a central role in environmental process engineering. It is also the case for urban drainage pollution. In Chapter 1, it was an objective to present the overall approach of urban wet weather pollution and Chapter 2 focused on a deeper understanding of its nature. Following this track, Chapter 3 is a toolbox that includes a number of fundamental physical, chemical, and biological aspects needed for engineering in terms of quality assessment and prediction within urban drainage. This chapter has therefore focused on subjects associated with both equilibrium and dynamic conditions that are relevant within physics, physicochemistry, chemistry, microbiology, and biology. It is important to realize that the quality related aspects typically take place in multiphase systems. The partitioning in terms of exchange of substances between water, species in solution, a solid phase, and a gas phase is therefore crucial.

It is expected that the reader is familiar with the very basic aspects of mathematics, physics, chemistry, and biology. Chapter 3 is, however, not a deep scientific introduction into transport and transformations, but it will highlight concepts and tools needed for some of the most important applications in the text. The authors are aware that the reader, for specific purposes, may need details that exceed what is included in this chapter. Such details will either be included in the following chapters or they can be found in the literature referred to, particularly in the textbooks. Readers with a deep insight in environmental process engineering may find at least part of the contents of this chapter well known. Others with a background in civil engineering or environmental hydraulics may find at least part of the toolbox very helpful when reading the following chapters.

3.1 TOOLBOX FUNDAMENTALS

When discussing the importance of the physical, chemical, and biological phenomena and processes that occur during wet weather conditions in urban areas, it is important to understand that urban drainage pollution is not just a discipline that focuses on the sewer (drainage) network. It also concerns the urban and road environment and therefore widens to include the catchment, the treatment facilities, and the adjacent environment (e.g., the local receiving waters; Figure 3.1). Due to that, the relevant toolbox fundamentals are found in all types of natural sciences, particularly in physics (hydrology and hydraulics), physicochemistry, chemistry, microbiology, and biology. One may say that wet weather—the rainfall and the runoff—in urban areas imply that the physical processes bring the pollutants from one place into another. During this transport and in the system where they temporarily or permanently appear, they may undergo transformations under changed conditions. As an example, substances that originally were accumulated as part of the street dust can be transported through drainage networks where they temporarily accumulate and finally, after erosion and resuspension, end up in a lake where they might exert adverse effects on the ecosystem.

Physical, chemical, and biological processes related to transport and transformations of constituents proceed in the sewer network systems under both dry and wet weather conditions. In both cases, a number of physical processes (e.g., advection, dispersion, diffusion, and exchange processes across the air–water interface) play an important role. In combined sewers, the microbial processes are dominating under dry weather conditions (Hvitved-Jacobsen 2002). Under wet weather conditions, however, the microbial processes often play a minor role caused by a relatively short residence time in the sewer and drainage network and often because of reduced substrate availability. However, the microbial processes may play an important role in the adjacent environment. In contrast, the physical processes often become dominating in the wet weather flows. As an example, the impact of a relatively high wet weather flow may result in erosion and resuspension of deposited sewer solids followed by transport of such materials. Under dry weather conditions, however, these materials will typically be either temporarily retained or transported with a low velocity (e.g., as near-bed materials), whereby the importance of the transforming processes is increased.

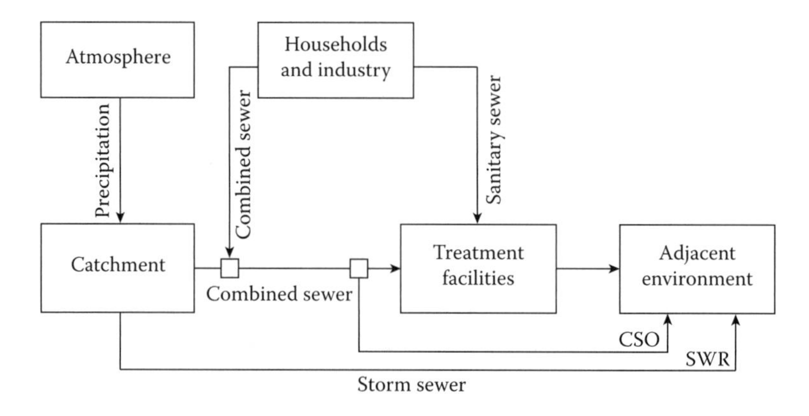

FIGURE 3.1 Principles of wet weather flows of pollutants in an urban environment.

This toolbox has therefore wide objectives to fulfill. The following main subjects are included in this chapter:

- Characteristics of the constituents
- Mass balances
- Equilibrium characteristics
- Physical processes
- Physicochemical processes
- Chemical processes
- Microbiological processes
- Biological processes

In general and as already mentioned, only a brief, however sufficient, overview will be given for these topics. More detailed information is found in other textbooks. The following titles are just mentioned as examples and as help for those readers that need further insight into this chapter: Snoeyink and Jenkins (1980), Pankow (1991), Stumm and Morgan (1996), Trapp and Matthies (1998), Nazaroff and Alvarez-Cohen (2001), Tchobanoglous, Burton, and Stensel (2003), and Duke and Williams (2008).

3.2 CHARACTERISTICS OF CONSTITUENTS

Urban wet weather pollution often concerns the impact of constituents at low concentrations that is, when concentrations are in the order of ppm (parts per million), ppb (parts per billion), or even ppt (parts per trillion). The relevant constituents appear in different phases (i.e., in the gas phase), in (aqueous) solution, and in the solid phase (e.g., associated with particles of different size). The occurrence, the processes, and the associated impacts of the constituents depend on numerous complex conditions.

A quality criterion for a constituent in storm-generated flows in terms of its concentration must relate to its potential harmful effect and a given level of protection against the adverse effects. When it comes to environmental standards set by authorities, this quality criterion is, of course, a starting point but a number of economic and political factors are crucial for the final result. This is also the case when it comes to monitoring and control. A number of agencies are involved in the formulation of both criteria and standards. The World Health Organization (WHO), the U.S. Environmental Protection Agency (U.S. EPA), and the European Commission (EC) are all examples of organizations that have published such information. In this section, however, such aspects will not be the issues. Further details in this respect are dealt with in Chapter 11. It is the major basic characteristics of the constituents (pollutants) that will be the focus in this chapter.

3.2.1 ORGANIC AND INORGANIC POLLUTANTS

The following six main groups of organic and inorganic constituents are typically dealt with because they frequently occur in the wet weather flows and are related to major adverse effects (cf. Section 1.3.1). This grouping is a pragmatic approach that might be differently defined, however, in the context of urban drainage, the list is widely referred to. It is therefore important to realize that similarities among constituents from different groups might easily be more marked than is the case for components within a specific group.

The six groups are

- Biodegradable organic matter
- Nutrients
- Heavy metals
- Organic micropollutants
- Solids (suspended solids)
- Pathogenic microorganisms

The first four groups of the list concern a number of organic and inorganic substances. The fifth group deals basically with the solubility characteristics of the pollutants in water and their association with a solid phase (particles). The sixth group that includes the pathogenic microorganisms accounts for living organisms with a potential harmful effect on human health.

These six groups of constituents only give a very rough first classification of the pollutants. A number of details are typically needed. The following sections give further details in this respect.

In addition to the six groups mentioned, a number of specific constituents and parameters are often needed to give a complete characterization of the runoff. The following list includes commonly selected parameters:

- Dissolved oxygen (DO): a measure of aerobic conditions and the electron acceptor for aerobic processes
- Redox potential: a measure of both equilibrium and process-related conditions
- Conductivity: an overall estimate of ionic substances
- pH
- Alkalinity: particularly associated with the carbonate system
- Chloride: a deicing agent applied on roads

Due to the solute–solid interactions, it is important to mention that the occurrence of pollutants is of interest not just in a water phase (suspension) but also in sediments, soil, and plants.

3.2.1.1 Biodegradable Organic Matter

Biodegradable organic matter is a group of urban drainage pollutants particularly relevant for two reasons:

- Biodegradable organic matter is a DO consuming group of substances. DO depletion in both the drainage systems and the receiving waters can therefore affect the performance of the drainage system and the activity and survival of aerobic microorganisms, plants, and animals.
- Biodegradable organic matter is food (carbon source) for heterotrophic organisms—including higher animals. This group of substances can, when

available in excess, speed up the growth rate for the fastest growing organisms. Consequently, such organisms can become dominating species and thereby result in a reduced diversity of the system.

The most common parameters used for measurement of biodegradable organic matter are BOD (biochemical oxygen demand) and COD (chemical oxygen demand), although COD does not observe the criterion as a measure of "biodegradability." Sometimes DOC (dissolved organic carbon) is used. Standard methods for the determination of these parameters are well defined (APHA-AWWA-WEF 1995). It is, however, important to realize that these measures for organic matter are bulk parameters.

The BOD/COD ratio for organic matter is a measure of its biodegradability. A ratio > 0.5 indicates biodegradability in the order of magnitude as for domestic wastewater. Further details on the biodegradability of organic matter can be achieved by measurement of the oxygen uptake rate (OUR; Vollertsen and Hvitved-Jacobsen 2001; Hvitved-Jacobsen 2002).

The biodegradable organic matter found at urban and road surfaces is typically dominated by plant material. Consequently, it is compared with both domestic wastewater and organics in the combined sewer overflows (CSOs) less degradable in terms of the kinetic characteristics. A COD value measured in stormwater runoff should therefore be judged less problematic for the environment compared with a similar value found for CSOs.

Due to the different transport mode including accumulation as deposits in urban drainage systems, both soluble COD (COD_{sol}) and particulate bound COD ($COD_{part} = COD_{tot} - COD_{sol}$) can be relevant to determine. Furthermore, some kind of simple fractionation between more and less biodegradable COD can thereby be made.

An important characteristic of the COD parameter is that it observes the requirements for establishing a mass balance because it includes the total amount of organic matter. It is not the case for the BOD parameter.

Further details concerning biodegradable organic matter and biodegradation are found in Sections 3.7 and 6.3.1.

3.2.1.2 Nutrients

The group of nutrients consists of substances with biological available and reactive nitrogen, N, and phosphorus, P. These nutrients are key parameters for the assessment of eutrophication. Although a number of species of the nutrients exist, and also show different impacts, only the total concentrations are routinely measured in case of urban drainage. There are, however, exceptions for that statement:

- Ammonia (NH_3 and NH_4^+) because of its potential toxic effects (cf. Sections 3.2.4 and 6.5.3.2)
- To some extent nitrate and nitrite because of their availability for plant and algal growth
- A distinction between soluble (orthophosphate) and particulate forms of P because of transport characteristics, potential for accumulation in solids, and uptake in plants (cf. Section 6.1.1)

3.2.1.3 Heavy Metals

A large number of heavy metals is potentially relevant because of the toxic effects they might exert. The following four heavy metals are generally considered the main group focused on within urban drainage:

- Copper (Cu)
- Lead (Pb)
- Zinc (Zn)
- Cadmium (Cd)

Furthermore, the following two heavy metals are also frequently used for characterization:

- Nickel (Ni)
- Chromium (Cr)

These six heavy metals are generally focused on because they are typically available and potentially toxic with acute or cumulative (chronic) effects at those concentrations found in urban drainage flows. Other metals may, of course, be of interest in specific cases. Examples of such heavy metals are arsenic (As) and mercury (Hg).

Iron (Fe) is a heavy metal, however, in case of urban drainage not generally considered a pollutant. As a constituent that typically occurs in relatively high concentrations, it may play a role by being a substance—in the precipitated and solid state—that adsorb a number of pollutants (cf. Sections 3.2.4.1.1, 3.2.5, and 9.3.2.1.3).

Although heavy metals routinely typically are measured and assessed in terms of the total concentrations, the adverse (toxic) effects are basically associated with their speciation (i.e., the individual chemical species of the different metals). The availability and the toxicity of a heavy metal is to a great extent related to its solubility, the association of the metal to particles, and its potential for formation of complexes with both inorganic and organic substances. These aspects are dealt with in Section 3.2.5.

3.2.1.4 Organic Micropollutants

The group of organic micropollutants, also called xenobiotic organic compounds (XOCs), includes a large number of different organic substances that are typically discharged into the environment in trace amounts. However, often such rather small amounts exert potential toxic effects into the environment including humans. These organic micropollutants originate from a great number of sources and are the result of widespread use of different materials and chemicals in household and industry and in the transport sector (Thevenot 2008).

The following main groups of substances are included under the term organic micropollutants or XOCs:

- Pesticides
- Aromatic hydrocarbons
- Phenols (e.g., nonylphenols)
- Halogenated aliphatic and aromatic organics

- PCBs (polychlorinated biphenyls)
- PAHs (polycyclic aromatic hydrocarbons)
- Softeners: BBP (butyl benzyl phthalate), DEHP (di(2-ethylhexyl) phthalate), and DIDP (diisodecyl phthalate)
- Anionic detergents: LAS (linear alcylbenzenesulphonates)
- Ethers: MTBE (methyl *tert*-butyl ether)
- Dioxins and furans
- Endocrine disrupting chemicals (estrogens)

Numerous organic micropollutants have been identified in urban runoff and combined sewer overflows. As an example, in 150 relevant publications from the period 1980 to 2001, they registered 313 different organic micropollutants (Arnbjerg-Nielsen et al. 2002). A literature survey revealed that at least 656 organic micropollutants could be present in stormwater runoff (Eriksson et al. 2005). The number of relevant substances in the group of XOCs may change because new chemicals are being introduced and others are being banned.

It is generally neither feasible nor relevant to sample and analyze for all potential organic micropollutants. In case of runoff water, selected substances may therefore be particularly focused on and used as potential indicators for the group. As an example, analysis can be performed for selected groups of the pollutants (e.g., total PAHs, specific PAH compounds, and total hydrocarbons). A number of bulk parameters that include organic micropollutants also exist and are to some extent used for screening purposes:

- AOX (adsorbable organo-halogen compounds)
  This parameter has been suggested as an indicator for organic micropollutants (Hahn, Hoffmann, and Schäfer 1999).
- EOX (extractable organohalogen compounds)

Tests of the effect on the bacterial DNA activity have also been proposed as an indicator for the occurrence of organic micropollutants (Ono et al. 1996).

The number of substances that are included under the term organic micropollutants will vary because of different definitions and observations in terms of adverse effects on plants, animals, and humans. The term *priority pollutant* is used to select organic micropollutants that may exert hazardous effects. A priority pollutant can be defined as a micropollutant having a ratio > 1 between a predicted or measured concentration in the environment and the corresponding concentration where no (acute) effect has been observed. Lists of priority pollutants with numbers between 15–20 and more than 100 can be found.

3.2.1.5 Solids

When dealing with urban drainage, the term "solids" covers a wide range of constituents that exert a number of impacts because of their solid form. Solids within urban drainage have different origin, follow different pathways, and have different impacts. Important examples of solids in this respect are "urban dust" from the catchment surfaces and "sewer solids" from combined sewers. To some extent solids relevant for urban drainage are dealt with in Section 2.3.

Sewer solids in this context are defined as follows:

- Suspended solids and colloids (cf. Section 3.2.4.3)
- Sewer sediments (i.e., deposits and near-bed materials)
- Biofilm

The characteristics and processes associated with each of these types of solids are different and will be dealt with in several later chapters. The suspended solids are often focused on when dealing with the characterization of the urban drainage flows. The term total suspended solids (TSS) analytically defined in APHA-AWWA-WEF (1995) is a parameter that is associated with several types of effects:

- Suspended solids may settle and accumulate—perform blockages—in sewer networks and thereby reduce the flow capacity and cause risk of flooding.
- A number of pollutants are associated with suspended solids.
- Sewer solids and dust on urban and road surfaces may be eroded and transported by the runoff. They deteriorate receiving waters in different ways: suspended solids reduce light penetration, they are esthetically unpleasant in bathing waters, and they change the quality of the bottom sediments when deposited.

The particle size of the suspended solids is a factor that is crucial for a number of these effects. The solubility and particle size aspects will be focused on in Section 3.2.4.3.

3.2.1.6 Pathogens

The potential harmful effect of pathogens (pathogenic microorganisms) on humans is primarily related to infections. Infection means that a harmful organism penetrates the human body where it reproduces and develops a disease. Pathogens originating from infected persons may spread via wastewater, and thereby occur in combined sewer overflows (CSOs), as the disease carrier. Several microorganisms are pathogenic and each of these is responsible for a specific course of disease.

Urban wet weather flows are potential sources for contamination with pathogenic microorganisms. In particular, these organisms might originate from the daily wastewater flow and eroded sewer solids and may therefore be present in CSOs (Ashley and Dabrowski 1995; Ellis and Yu 1995). To a minor extent such organisms may, however, also appear in Stormwater runoff (SWR) (Pitt and Bozeman 1982). Although the CSOs in general may result in the loading of receiving waters with these organisms, SWR might contribute because of wrong connections in the sewer network and because of contributions from animals (e.g., cats, dogs, and birds).

The following four main groups of pathogens are potentially present in wastewater flows. Each group includes a big number of subgroups. Only a few examples in terms of their disease-causing impact will be mentioned.

- *Bacteria*

  Examples are bacteria that cause salmonellosisa, cholera, dysentery, typhoid fever, and diarrhea.

- *Viruses*
  Examples are viruses that cause meningitis, polio, and hepatitis.
- *Protozoan parasites*
  Protozoa is a complex group of microorganisms that in some cases are unicellular animals (e.g., amebas).
- *Helminths (eggs of worm parasites)*
  Helminths are multicellular organisms that can physically damage another organism, cause toxic effects, or deprivation of food.

A virus is an infectious agent that replicates within cells of living hosts. A virus is therefore in contrast to the other three groups of pathogens not in itself a living organism, however, often considered in parallel to these groups.

When dealing with human health hazards caused by discharges of pathogens into the environment, indicators for such risks are common within the area of environmental engineering. It is also believed that indicator organism counts are representative of the presence of possible enteric pathogens in storm-generated flows (Field 1993). The following wastewater-related bacteria are typically used as indicators:

- Total coliforms (TC)
- Fecal (thermotolerant) coliforms (FC), often determined as E. coli
- Fecal streptococci (FS)
- Intestinal enterococci (IE)
- Escherichia coli (EC)

It is difficult and time consuming to identify viruses and this parameter is therefore omitted in most investigations. Since viruses are generally more resistant to environmental factors (e.g., pH, temperature, and sunlight) than enteric bacteria, particularly bacteria that survive in the environment for longer periods (e.g., *Enterococci*, may serve as indicators in this respect). As an example, the die-off of fecal organisms in surface waters during a summer period is relatively rapid, 1 to 2 days. In sediments, however, the survival is often extended for weeks or months.

Further details on pathogenic microorganisms related to stormwater may be found in Burton and Pitt (2002).

3.2.2 Microorganisms

Pathogens (pathogenic microorganisms) referred to in Section 3.2.1.6 was only included as a group of constituents because of their potential harmful effect on humans. Microorganisms have a much wider perspective because they are responsible for the course of several biological processes (i.e., the microbial reactions). Depending on the objective, microorganisms are, in addition to a possible pathogenic effect, thereby classified in different ways:

- In terms of evolutionary origin and genetic differences
- As a carbon source for microbial growth

- As an energy source for survival and growth of organisms
- Due to specific redox characteristics

The last three mentioned aspects are relevant when focus is on the microbial process characteristics. In the following (Section 3.2.2.1–3.2.2.3), the first three mentioned types of classification for microorganisms are further described. The redox characteristics are dealt with in Section 3.6.1.

3.2.2.1 Microbial Classification

The two main groups of microorganisms in terms of evolutionary origin and genetic differences are prokaryotes and eukaryotes. These two groups have established different internal structure of the cells:

- *Prokaryotes*
  Compared with the eukaryotes, this group of microorganisms has a relatively simple cell structure with no cell nucleus. Most bacteria belong to this group. The group also includes the archaeas, a subgroup of the methanogenic microorganisms.
- *Eukaryotes*
  This group of organisms has developed a cell nucleus where the genetic material is accumulated. Algae, fungi, and protozoa are important subgroups that belong to the eukaryotes. Plants and animals also have eukaryotic cell structure.

3.2.2.2 Carbon Source

In terms of the carbon source for development of new cell material, the microorganisms are divided in two groups:

- *Heterotrophic microorganisms*
  This group includes organisms that require organic compounds (organic carbon) for the growth process to produce new cell material.
- *Autotrophic microorganisms*
  These organisms use inorganic carbon (CO_2 or HCO_3^-) for the production of new cells. Inorganic carbon, at neutral pH conditions, is HCO_3^- for aquatic bacteria (e.g., nitrifying bacteria) and algae.

The two terms heterotrophic and autotrophic is also valid for multicell organisms. Animals are heterotrophic whereas plants are autotrophic. Plants exposed to the atmosphere use inorganic carbon in the form of CO_2 whereas submersed plants use HCO_3^- (cf. Section 3.2.4.2). The consumption of HCO_3^- (alkalinity) will affect the pH value.

3.2.2.3 Energy Source

There are three main types of energy sources for growth of microorganisms and for maintenance of their basic life processes:

- *Organic compounds*
  Organisms that use organic matter as an energy source are defined as chemotrophic (or chemoorganotrophic). The energy is released for the benefit of the organism by oxidation of the organic matter to less energetic substances (e.g., CO_2 and H_2O). If organic matter is substrate for growth of the organisms by producing new cell materials, the organisms are heterotrophic (cf. Section 3.2.2.2).
- *Inorganic compounds*
  Similar to organic substances, inorganic (reduced) compounds may also serve as an energy source for an organism. As an example, hydrogen sulfide (H_2S) is by the sulfide oxidizing microorganisms transformed (oxidized) to elemental sulfur or sulfate whereby energy is released. Another example is the nitrifying bacteria that oxidize ammonia to nitrite and nitrate. Such organisms are autotrophic (chemoautotrophic).
- *Sunlight*
  Phototrophic (photoautotrophic) organisms, algae and macrophytes (higher plants), utilize the energy from the sun for maintaining fundamental life processes and for the production of new cells.

3.2.3 Transformation or Equilibrium: A Thermodynamic Approach

The state of a potential energy transformation that is associated with physical, chemical, or biochemical processes determines if a change of state can take place. Such energy-related aspects for processes are subjects of thermodynamics. It is not the objective of this text to go into details with the thermodynamics of chemical and biological processes. Such subjects are dealt with in a number of books on physicochemistry (e.g., Atkins and de Paula 2002). Texts particularly dealing with water chemistry are Snoeyink and Jenkins (1980) and Stumm and Morgan (1996). However, the basic relation between energy changes and the potential transformations relevant in the case of urban drainage will be outlined.

It is important to state the basic requirement for a chemical or biochemical process: has it a potential to proceed or not? It is quite a different question than to ask: will it proceed and under what conditions will it take place? The latter question concerns the process kinetics. The answer to the first question is found by determining the potential work that a system can do onto its surroundings. The numerical value of this work is thermodynamically given by Gibbs' free energy (cf. Section 3.6). An illustration of the energy changes that can take place is for a chemical or biochemical process outlined in Figure 3.2 and Table 3.1.

A chemical equilibrium corresponding to $\Delta G = 0$ can be established between substances in different forms and phases. Equilibrium can be established within a one-phase system (i.e., as a homogenous equilibrium). Chemical equilibrium may

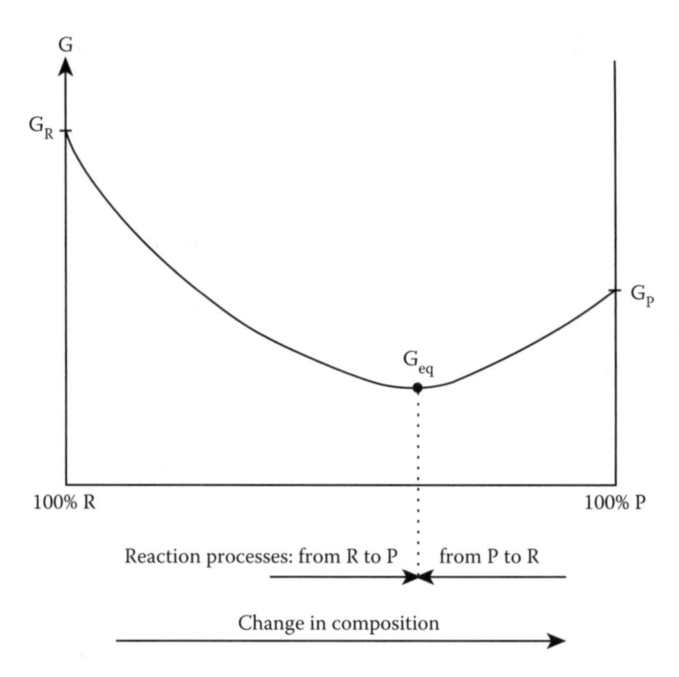

FIGURE 3.2 Illustration of the changes in Gibbs's free energy, G, for a chemical or biological process that might take place for reacting components (R) to form products (P) or the opposite: for P to form R (see also Table 3.1).

TABLE 3.1
Thermodynamic Conditions for a Chemical or a Biochemical Process That Proceeds in the Direction from Reacting Components (R) to Products (P; i.e., R → P)

Thermodynamic Condition for the Process R → P	Comments
$\Delta G < 0$	The process R → P has a potential to proceed. The potential for work (G) of the process is reduced when the starting point is to the left of the point corresponding to G_{eq}. The process is defined as being spontaneous.
$\Delta G = 0$	Equilibrium exists (i.e., no net transformations will take place).
$\Delta G > 0$	The process R → P will not proceed when the mixture of R and P corresponds to a composition to the right of the point for G_{eq}. The reverse process P → R is spontaneous.

also exist as a heterogenous equilibrium between substances that occur in multiphase systems (e.g., across interfaces).

Chemical equilibrium is crucial for a great number of phenomena related to water pollution and therefore also concerns what is relevant within urban drainage. In principle, all processes tend to proceed in the direction toward equilibrium. A chemical equilibrium is, however, basically dynamic because it concerns a situation where the formation of products (P) by a reversible process equals a corresponding backward reaction in terms of the formation of reactants (R). Chemical equilibrium thereby describes a situation where the rate of the forward reaction in principle equals the reverse reaction rate. The following exemplifies chemical equilibrium of a simple, reversible chemical reaction that takes place within a homogenous system (i.e., a one-phase system exemplified by a water phase):

$$r_1R_1 + r_2R_2 \leftrightarrows p_1P_1 + p_2P_2 \quad (3.1)$$

where
R_i = reactants, i.e. reacting components
P_i = products
r_i and p_i = stoichiometric coefficients (–)
i = 1,2; component number (–).

Chemical equilibrium in a homogenous system corresponding to Equation 3.1 is at constant temperature equivalent to the fulfillment of the following equation:

$$\frac{[P_1]^{p_1}[P_2]^{p_2}}{[R_1]^{r_1}[R_2]^{r_2}} = K, \quad (3.2)$$

where
$[R_i]$ and $[P_i]$ = concentrations (equilibrium concentrations) of reactants and products, respectively (mole L^{-1})
K = equilibrium constant (unit depends on the values of the stoichiometric coefficients).

Equation 3.2 is valid for a wide range of equilibrium descriptions in both natural and technical systems. In case of acid–base reactions, and depending on the concrete description, K is referred to as the acid or base dissociation constant, K_a or K_b, respectively, and often in tables given as $pK_a = -\log K_a$ or $pK_b = -\log K_b$. In case a component has established equilibrium between two phases (across an interface), Equation 3.2 will develop into a different form. In Sections 3.2.4 and 3.2.5, examples of such interfacial equilibrium partitioning of substances will be given.

3.2.4 Pollutants in Different Phases: Partitioning

Basic characteristics of pollutants in terms of their transport, transformations, and effects are highly influenced by their relative distribution among the solid, the liquid, and the gaseous phases. The equilibrium conditions as well as the dynamic behavior of the pollutant transport between phases (i.e., the exchange across an interface) are central in understanding the quality aspect of phenomena related to the wet weather flows.

The dynamic behavior of a substance—in this context a pollutant—in terms of its transfer between two phases is defined as partitioning. The ultimate state of a partitioning reaction is the equilibrium where no net transfer takes place. Partitioning includes a number of transfer processes like air–water mass transfer, sorption, and ion exchange, subjects that are dealt with in this section. Partitioning is a complex phenomenon affected by, for example, speciation of the substances (cf. Section 3.2.5), and the physicochemical and chemical processes and phenomena that are subjects of Sections 3.5 and 3.6, respectively.

Partitioning within a suspension (i.e., the transfer of a substance taking place between the water phase and a particle surface) is relevant for all types of wet weather flows. The following are just examples that refer to characteristics of a substance that can be transferred between a soluble and a solid (particulate) state:

- The solubility of a substance determines its potential occurrence in solution. The solubility is therefore closely related to the chemical reactivity of a component (i.e., its potential transformation) and the availability of the substance as a substrate in a biological system.
- Particles can be transported or accumulated (e.g., by sorption or sedimentation) depending on particle characteristics and environmental conditions.
- A number of pollutants are associated with particles and are also subject to changes depending on the transport pattern and the transformation of the particles.

Interactions between soluble and particulate compounds occur in almost all aqueous systems relevant to urban drainage. As an example, such interactions are often dominating phenomena when dealing with infiltration and filtration of the runoff in both natural (soils) and artificial (filter media) systems (e.g., in terms of pollutant retention in filters for treatment of the runoff).

A number of basic characteristics for both soluble substances and particles will be focused on in the following sections. Further details on partitioning can be found in Connell (2006) or the texts referred to in Section 3.1. Specific information related to the behavior of particles in water can be found in Gregory (2006).

3.2.4.1 Solubility and Chemical Equilibrium

The solute–solid interactions proceed in different ways. Both chemical and physicochemical reactions that tend to establish equilibrium are important to note. In both cases, it is basically a question of establishing "bonds" in terms of chemical bonds (electron-pair bonds) or bonds established by electric dipole moments.

3.2.4.1.1 Precipitation, Water–Solid Equilibrium

Dissolution and precipitation are processes that are particularly relevant for components with a low solubility. A great number of salts that exist in urban drainage flow and following in receiving systems (surface waters and soil systems) are components with a low solubility. As an example, the aqueous equilibrium for iron (III) hydroxide is

$$Fe(OH)_3 \leftrightarrows Fe^{3+} + 3OH^- \tag{3.3}$$

Following this example under conditions where equilibrium exists between precipitated iron (III) hydroxide and ionic species, the equilibrium expression shown in Equation 3.2 is expressed as

$$C_{Fe^{3+}}(C_{OH^-})^3 = L \tag{3.4}$$

where

$C_{Fe^{3+}}$ = equilibrium concentration of Fe^{3+} (g m^{-3} or mole L^{-1})
C_{OH^-} = equilibrium concentration of OH^- (g m^{-3} or mole L^{-1})
L = equilibrium constant (e.g., $mole^4$ L^{-4}).

The equilibrium constant (L) is in this case named the solubility product for iron (III) hydroxide.

Aspects related to water-solid equilibrium conditions will, in a wider context related to speciation of substances—particularly exemplified in terms of heavy metals—be dealt with in Section 3.2.5.

3.2.4.1.2 Sorption

The term sorption includes two different phenomena:

- *Adsorption*
  This phenomenon concerns the adherence of a chemical substance from a liquid or gas phase to a solid interface (e.g., onto the surface of a particle). The basic characteristics of the interactions between soluble constituents and particles are dealt with in Section 3.2.4.3.2.
- *Absorption*
  Contrary to adsorption, absorption concerns a case where a substance passes an interface and penetrates into a different phase. Within urban drainage, absorption is as an example related to the transfer of substances from the atmosphere into a raindrop. The word absorption can be understood as "uptake" (e.g., as absorption of a substrate into an organism), typically corresponding to a transport across a cell membrane.

Equilibrium in a two-phase system is described in terms of an adsorption isotherm. An adsorption isotherm is an empirical description of the partitioning of a substance between a liquid or gas phase and the surface of a solid. The word *isotherm* that basically refers to *uniform temperature* is in this case somewhat misleading. However, it can be understood in a way that the constant included in the equilibrium equation is temperature dependent. The following two types of adsorption isotherms, Langmuir and Freundlich, which are formulated for a water–solid partitioning, are commonly applied.

Langmuir adsorption isotherm:

$$C_s = C_{s,max} K_L \frac{C_w}{1 + K_L C_w}, \tag{3.5}$$

where

C_s = equilibrium concentration of a substance in the solid phase (g g^{-1})
$C_{s,max}$ = maximum concentration of a substance in the solid phase (g g^{-1})
C_w = equilibrium concentration in the water phase (g m^{-3})
K_L = equilibrium constant (Langmuir adsorption constant; m^3 g^{-1}).

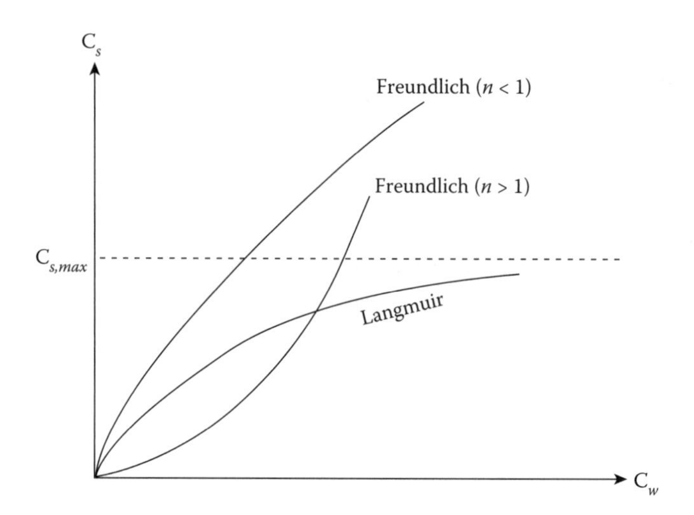

FIGURE 3.3 The principle of adsorption described by the Langmuir and the Freundlich isotherms.

The Langmuir adsorption isotherm describes a phenomenon where a maximum concentration, $C_{s,\max}$, in the solid phase exists irrespective of the fact that the concentration in the water phase increases. It is therefore a case where a limited number of sites for adsorption at the solid surface exists (cf. Figure 3.3). This limitation will typically occur in case a chemical reaction takes place between a substance in the water phase and the solid surface. This type of adsorption is therefore often named chemosorption.

Equation 3.5 can be rewritten in a linear form in a coordinate system with C_w/C_s versus C_w resulting in $1/(C_{s,\max}\ K_L)$ as the intercept and $1/C_{s,\max}$ as the slope:

$$\frac{C_w}{C_s} = \frac{1}{C_{s,\max}K_L} + \frac{C_w}{C_{s,\max}} \tag{3.6}$$

Freundlich isotherm:

$$C_s = K_F C_w^n, \tag{3.7}$$

where

K_F = equilibrium constant (Freundlich adsorption constant; unit depends on n)
n = exponent (–).

Compared with the Langmuir isotherm, the Freundlich isotherm is by nature just an empirical approach and depending on the value of n, the Freundlich isotherm may show different courses (cf. Figure 3.3). It is therefore in general easier to apply the Freundlich isotherm than the Langmuir isotherm in terms of curve fitting. The Freundlich isotherm is often applied in a logarithmic, linear form:

$$\ln C_s = \ln K_F + n \ln(C_w) \tag{3.8}$$

3.2.4.1.3 Air–Water Equilibrium

Equilibrium for volatile substances at the air–water interface is normally described applying Henry's law:

$$p = H\,C_w \tag{3.9}$$

where

p = partial pressure of a volatile component in the gas phase (atm)
H = Henry's law constant (atm m^3 g^{-1} when unit of C_w is g m^{-3}, atm m^3 $mole^{-1}$ when unit of C_w is mole m^{-3}, and atm (mole fraction)$^{-1}$ when units of C_w is mole fraction)

Henry's law constant is a measure of the volatility of a component in an aqueous solution. A low volatility corresponds to a value of H between 10^{-7} and 10^{-5} atm m^3 $mole^{-1}$ whereas a value between 10^{-5} and 10^{-3} atm m^3 $mole^{-1}$ indicates a relatively high volatility.

Henry's law constants for a number of volatile substances that can be found in wastewater of sewer networks are stated in Hvitved-Jacobsen (2002).

3.2.4.2 Carbonate System

The term *carbonate system* refers to a description of the interactions between inorganic carbon components and what is ultimately established in terms of homogeneous and heterogeneous equilibria between the following components:

- Gas phase: $CO_2(g)$
- Water phase: $CO_2(aq)$, H_2CO_3, HCO_3^-, CO_3^{2-} (and corresponding cations)
- Solid phase: carbonates, for example, limestone, $CaCO_3$ and dolomite, $CaMg(CO_3)_2$

The carbonate system is of importance for a number of environmental systems. The following characteristics of the carbonate system are particularly relevant for environmental engineers:

- The carbonate system is a pH-buffer in all types of water systems.
- The carbonate system provides inorganic carbon for the photosynthesis and other autotrophic processes.
- The carbonate system plays a central role for the global climate.
- The carbonate system determines the exchange of CO_2 across the air-water interface and precipitation/dissolution of the solid forms of carbonate.

Figure 3.4 illustrates the carbonate system under conditions where limestone ($CaCO_3$) exists in equilibrium with the water phase, where the water phase is open to the atmosphere and CO_2-equilibrium across the air–water interface is established.

The basic description of the carbonate system corresponding to Figure 3.4 follows from Henry's law (air–water equilibrium), the acid–base equilibria within the water phase and the description of the precipitation-dissolution equilibrium (i.e., the solid–water phase equilibrium). The equations for these equilibria will be outlined in the following.

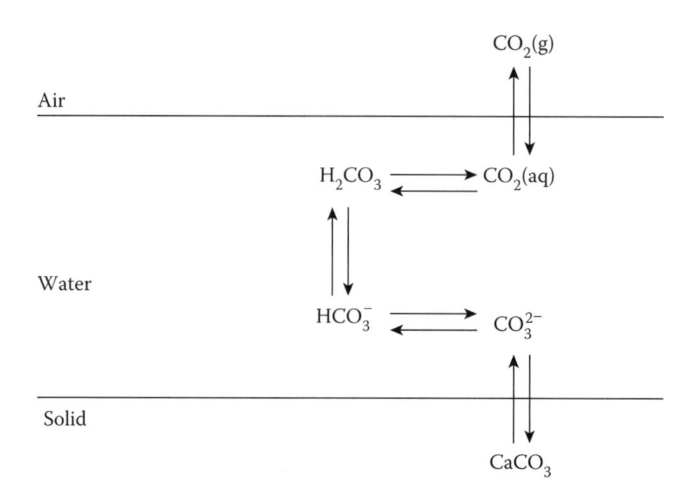

FIGURE 3.4 Illustration of the carbonate system with both homogenous and heterogeneous equilibria.

The values of the equilibrium parameters in the equations for the carbonate system in the following are exemplified at 25°C and 1 atm total pressure corresponding to a partial pressure for CO_2 (g) about 0.00037 atm. The concentrations in the water phase are given in units of mole L^{-1}. Further details concerning the carbonate system are found in Snoeyink and Jenkins (1980), Stumm and Morgan (1996), and Nazaroff and Alvarez-Cohen (2001).

The nomenclature used in the following quantitative description of the carbonate system follows in principle what is used in this section and will therefore not be explained further.

3.2.4.2.1 The Air–Water Equilibrium

The equilibrium between CO_2(g) and CO_2(aq) follows Henry's law (cf. Equation 3.9):

$$p_{CO_2(g)} = H_{CO_2} C_{CO_2(aq)}, \quad (3.10)$$

where $H_{CO_2} = 29.4$ atm L mole^{-1}

3.2.4.2.2 The Water Phase Equilibria

The equilibrium between CO_2(aq) and H_2CO_3 follows the general expression for equilibrium (cf. Equation 3.2):

$$\frac{C_{H_2CO_3}}{C_{CO_{2(aq)}}} = K_m = 1.58 \times 10^{-3}, \quad (3.11)$$

where the total concentration of the nondissociated carbonic acid in the water phase is defined as

$$C_{H_2CO_3*} = C_{CO_2(aq)} + C_{H_2CO_3} \quad (3.12)$$

As shown by the numerical value of K_m, the main part of the nondissociated carbonic acid exists as $CO_2(aq)$.

The acid–base equilibria in the water phase (cf. Figure 3.4) are described in terms of Equation 3.2 and result in the following two equilibrium expressions for dissociation:

$$\frac{C_{H^+}C_{HCO_3^-}}{C_{H_2CO_3*}} = K_{S1} = 4.47 \times 10^{-7}, \tag{3.13}$$

$$\frac{C_{H^+}C_{CO_3^{2-}}}{C_{HCO_3^-}} = K_{S2} = 4.68 \times 10^{-11}. \tag{3.14}$$

3.2.4.2.3 The Water–Solid Equilibrium

The equilibrium between carbonate in the solid state (e.g., limestone) and carbonate in the water phase follows from an expression similar to what is exemplified in Equation 3.4:

$$C_{Ca^{2+}}C_{CO_3^{2-}} = L_{CaCO_3} = 5 \times 10^{-9}. \tag{3.15}$$

3.2.4.2.4 General Physicochemical Requirements

In addition to the Equations 3.10 through 3.15 for equilibrium in the carbonate system, the following two Equations 3.16 and 3.17 are generally essential for an aquatic system.

The dissociation of water:

$$C_{H^+}C_{OH^-} = K_w = 10^{-14}. \tag{3.16}$$

The requirement of electroneutrality (considering a case with an aqueous solution of components that only originate from the carbonate system) implies that the following balance between anions and cations is observed:

$$C_{H^+} = C_{OH^-} + C_{HCO_3^-} + 2C_{CO_3^{2-}}. \tag{3.17}$$

If other ions than those originating from the carbonate system are present, these must be included in an extended version of Equation 3.17.

3.2.4.2.5 Alkalinity

The definition of the alkalinity, A, as being the capacity of an aqueous solution to neutralize acids follows from Equation 3.17:

$$A = C_{OH^-} + C_{HCO_3^-} + 2C_{CO_3^{2-}} - C_{H^+}. \tag{3.18}$$

Other ions than those included in Equation 3.18 may influence the alkalinity, however, typically the components from the carbonate system are abundant in most types of waters. In water systems with a pH between 7 and 10, HCO_3^- normally plays a dominating role.

3.2.4.2.6 Total Inorganic Carbon

Furthermore, the total concentration of inorganic carbon in the water phase is

$$C_T = C_{H_2CO_3*} + C_{HCO_3^-} + C_{CO_3^{2-}}. \tag{3.19}$$

The following example will illustrate the importance of the carbonate system within the area of urban drainage.

Example 3.1: pH and Alkalinity of Rainwater

The example concerns calculation of pH and alkalinity of rainwater before the rain hits the ground. The water in the raindrops is considered saturated with carbon dioxide (CO_2(aq)) corresponding to a CO_2(g) concentration in the atmosphere of 370 ppm (i.e., a pressure of 0.00037 atm). It is assumed that no other components than those originating from the water itself and the carbonate system occur in the rainwater. The temperature is 25°C (i.e., the parameters given in this text can be directly used).

The concentration of CO_2(aq) is under equilibrium conditions calculated from Equation 3.10:

$$C_{CO_2(aq)} = \frac{p_{CO_2(g)}}{H_{CO_2}} = \frac{0.00037}{29.4} = 1.26 \times 10^{-5} \text{ mole L}^{-1}.$$

From Equation 3.12 it follows approximately that

$$C_{H_2CO_3^*} = C_{CO_2(aq)} + C_{H_2CO_3} \approx C_{CO_2(aq)} = 1.26 \times 10^{-5} \text{ mole L}^{-1}.$$

Equation 3.13 is a first estimate of the acid–base equilibrium. In most cases it is a correct assumption, assuming that it is the dominating equilibrium compared with Equation 3.14. It follows therefore from Equation 3.17 that

$$C_{H^+} = C_{OH^-} + C_{HCO_3^-} + 2C_{CO_3^{2-}} = C_{HCO_3^-}$$

Equation 3.13 following results in

$$(C_{H^+})^2 = K_{S1}\, C_{H_2CO_3^*} = 4.47 \times 10^{-7} \times 1.26 \times 10^{-5} = 5.63 \times 10^{-12}$$

$$pH = -\log C_{H^+} = 5.62.$$

The assumption that Equation 3.13 is dominating compared with Equation 3.14 is therefore correct.

According to Equations 3.16 and 3.18, the alkalinity of the rainwater is approximately

$$A = C_{OH^-} + C_{HCO_3^-} + 2C_{CO_3^{2-}} - C_{H^+} = C_{HCO_3^-} - C_{H^+} = 0.$$

As previously stated, these calculations exemplify a case where only CO_2(g) is absorbed in the rainwater and the electroneutral CO_2(aq) is the dominating component in the water phase. Under real conditions, a number of other substances that occur in the atmosphere (e.g., ionic components like sulfates, chlorides, and nitrates coupled with H+ and other cations) can be absorbed and thereby affect both pH and alkalinity. The well-known example is acid rain where such components originating from the combustion of fossil fuels can reduce the pH value of the rainwater considerably (e.g., to a value of pH between 4.0 and 4.5; cf. Section 4.1).

3.2.4.3 Particle Characteristics

TSS, determined according to APHA-AWWA-WEF (1995), is generally used as a bulk parameter to determine the amount of solids (particles) in the suspended water phase. Determined in this way, the distinction between a dissolved fraction and a particulate fraction is defined based on filtration through a 0.45 μm membrane filter. This definition is also widely used within the area of urban drainage. The parameter is useful and easy to determine, however, also rather crude because it gives no information on the particle size distribution and the small particles having diameters < 0.5 μm are not included.

The interaction between and characteristics of the soluble and the particulate forms of the pollutants dealt with in urban drainage are as previously mentioned important for transport, transformation, and effects of the components. The following major characteristics of the particulate form are in this respect basic and will be further discussed:

- The size distribution (the physical characteristics) is important for transport and accumulation.
- Interactions between soluble species and particles (chemical and physico-chemical interactions).

The second point concerning the interaction between different phases is often referred to as partitioning. This term is used for both equilibrium and dynamic conditions and covers, in principle, all types of interactions between a gas phase, a liquid phase, and a solid state (i.e., in multiphase systems).

3.2.4.3.1 Particle Size Distribution

Figure 3.5 outlines how the particle size classification is typically applied within urban drainage. What can settle (i.e., undergo a sedimentation process) and thereby be removed by gravity from an aqueous phase will, in addition to particle size and specific density, depend on a number of external conditions (i.e., water phase characteristics and hydraulic conditions, cf. Section 3.4.2.1). The definitions of particle

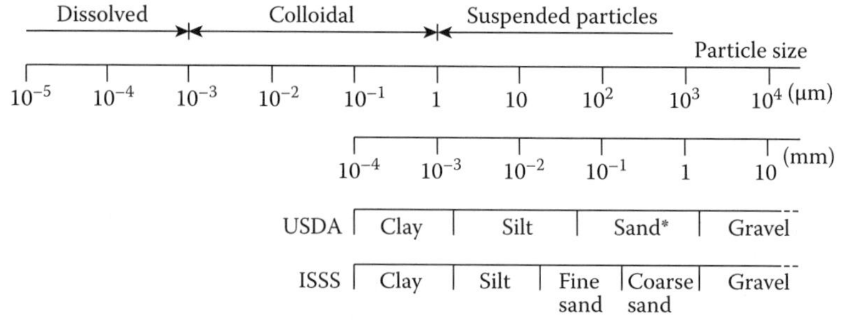

**Divided into very fine, fine, medium, coarse, and very coarse sand*

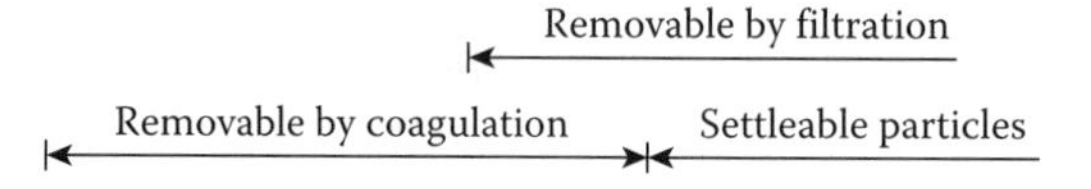

FIGURE 3.5 Size classification of particles within the area of urban drainage. Definitions of the inorganic particle sizes are given according to USDA (U.S. Department of Agriculture) and ISSS (International Society of Soil Science).

size referred to in Figure 3.5 are relevant when dealing with particles in suspension, in soils, and for dust on urban surfaces. Particles that occur in suspension in the air phase are called aerosols.

The density of particles plays a basic role for their behavior, particularly the settling rate. Inorganic particles like clay, silt, and sand have typically a specific density about 2.6 g cm^{-3} whereas the density of organic rich particles generally is in the order of 1.05–1.10 g cm^{-3}.

Particles that potentially settle are typically within the range 10–500 μm. Particles with a diameter < 10 μm will in principle not settle. Particles > 500 μm settle very fast. The small particles (10–100 μm) have even under ideal (quiescent) conditions very low settling rates (cf. Section 3.4.2.1).

3.2.4.3.2 Interactions Between Particles and Species in Solution

This section concerns details in terms of interactions between species in suspensions (i.e., an aqueous phase with soluble substances and particles). The basic characteristics of sorption, particularly in terms of equilibrium formulations, are dealt with in Section 3.2.4.1.2.

The interactions between substances in an aqueous suspension are based on forces that are established across a water–solid interface. Partitioning takes place between the surface of a particle and the constituents in the surrounding water phase. The intermolecular forces occur by the movement and relative distribution of electrons in the solid phase and the soluble species. The internal distribution of electrons results in the formation of either electron-pair bonds or permanent and mutually induced dipoles that more or less can tie together the substances (i.e., keep substances originating from the water phase associated with particles).

It is not the objective in this context to describe detailed theoretical aspects of surface chemistry. Details in this respect can be found in a number of textbooks on physicochemistry (e.g., Atkins and de Paula 2002). When dealing with urban wet weather pollution it is, however, important to establish an understanding of these phenomena for the following reasons:

- Aggregates of small particles including colloids in suspension can be established
- Soluble substances having dipoles may adhere to the surfaces of suspended particles also having dipoles

In addition to the chemical characteristics of the substances, these phenomena are strongly related to the particle size distribution (cf. Section 3.2.4.3.1). The different intermolecular interactions occur in urban drainage flows and are therefore important for the settling characteristics of particles and for the pollutant distribution between the water phase and the particulate phase. Adsorption that was dealt with in Section 3.2.4.1.2 is a phenomenon that can also be caused by the occurrence of such intermolecular forces. In this section, however, adsorption was formulated in macromolecular terms.

Due to a high specific surface defined in terms of $m^2\ g^{-1}$, the small, suspended particles may exert a relatively high potential for surface interaction. The colloidal particles like clay (inorganic) and humic substances (organic) belong to this group of very fine particles (cf. Figure 3.5).

Colloidal particles in stormwater have a significant binding and transport capacity for pollutants. The relatively large specific surface area and a general high affinity to soluble species in the surrounding aqueous environment are the main reason for it. The hydroxyl and acid groups at the surfaces of these small particles will, with neutral pH values, tend to donate protons to the water phase leaving the surface with an overall negative charge that has a tendency to react with cations in the solution. The following illustrates such surface reaction where a hydrous oxide reacts with a bivalent cation:

$$\text{R–OH} + \text{Me}^{2+} \rightarrow \text{R–O}^- + \text{H}^+ + \text{Me}^{2+} \rightarrow \text{R–OMe}^+ + \text{H}^+. \tag{3.20}$$

A number of ions (cations, e.g., heavy metals), can therefore—caused by exchange processes and electrostatic forces—be associated with these surfaces and follow the pathways of particles. Furthermore, cations like Na^+ and K^+ can be exchanged with bivalent heavy metal ions that form stronger bonds, for example. The cation exchange capacity (CEC) used for quantification of this phenomenon is dealt with in Section 9.4.2.3.

To some extent, the small particles may form aggregates and transform into larger particles with improved settling rates. However, several small particles are relatively stable in urban drainage flows and may, because of the potential surface reactions, include a relatively high amount of pollutants (e.g., heavy metals, organic micropollutants, and phosphorus).

The distribution of a substance between two different phases, in this case a liquid phase and a solid phase, is under equilibrium conditions expressed in terms of a partitioning coefficient:

$$K_{s,w} = \frac{C_{\text{solid}}}{C_{\text{water}}}, \tag{3.21}$$

where

$K_{s,w}$ = partitioning coefficient (–)
C_{solid} = equilibrium concentration of a substance in the solid phase (g m^{-3})
C_{water} = equilibrium concentration of a substance in the water phase (g m^{-3})

Partitioning coefficients are therefore, in principle, equilibrium constants (cf. Section 3.2.3).

Referring to Section 3.2.4.1, the Langmuir isotherm for adsorption can be explained by a phenomenon where the sites correspond to active groups to which the adsorbed substances are bound by electrostatic forces or electron-pair bonds.

In addition to the solid–liquid equilibrium conditions as described by Equation 3.21, the dynamics in terms of adsorption rate kinetics is important (e.g., when filter media are used for treatment of stormwater and in case of its infiltration in soils, cf. Sections 9.3.2.2 and 9.4.2.4, respectively). Adsorption kinetics relevant for a process like Equation 3.20 is typically formulated by a second-order rate expression (cf. Section 3.6.2 and Table 3.5). The application of the kinetics in the case of adsorption is dealt with in Section 9.4.2.4.

3.2.4.3.3 *Hydrophilic and Hydrophobic Characteristics*

In principle, substances are either hydrophilic or hydrophobic, water friendly or water repellent, respectively. Water-soluble species are thereby hydrophilic and fat and oil that can exist in suspension—and to some extent also in the dissolved phase—are hydrophobic substances, also named lipophilic compounds. Such characteristics are important for the partitioning between particles and between particles and the substances that occur in the surrounding water phase.

Hydrophilic substances, to which a group of substances water belongs, form permanent dipoles whereas hydrophobic substances (e.g., an organic solvent like gasoline) consists of molecules with only dipoles induced by the movement of electrons (cf. Section 3.2.4.3.2). The intermolecular forces between the water molecules are therefore much stronger than corresponding forces between organic molecules of a similar molecular size. As an example, water (H_2O) with a molecular weight of 18 g $mole^{-1}$ has (at 1 atm) a boiling point equal to 100°C whereas methane (CH_4) with a molecular weight of the same magnitude (16 g $mole^{-1}$) under corresponding conditions has a boiling point as low as –161°C. The reason for a difference in the boiling point of 260°C is caused by these intermolecular forces. The dipoles of water are also responsible for the fact that ionic species can be dissolved. It is not possible for the organic solvents.

This example describes what is the case in solution. A similar interaction may take place across a water–particle interface whereby the phenomena concerning interactions between soluble species and particles can be described (cf. Sections 3.2.4.1 and 3.2.4.3.2).

A measure for the distribution of a given substance between a hydrophobic phase and a hydrophilic phase is a partioning coefficient under equilibrium conditions (cf. Equation 3.21). For that purpose, water is selected as the hydrophilic solvent and *n*-octanol (a linear structured alcohol, $CH_3(CH_2)_7OH$) as a component with hydrophobic (less hydrophilic) characteristics. The equilibrium partitioning of a given substance between these two solvents are defined as follows:

TABLE 3.2
Octanol-Water Partitioning Coefficients for Selected Micropollutants

Component	Log K_{OW}
DMP (di-methyl phthalate)	1.53
DBP (di-*n*-butyl phthalate)	4.61
DEHP (di(2-ethylhexyl)) phthalate	7.48

$$K_{OW} = \frac{C_O}{C_W} \quad (3.22)$$

where

K_{OW} = octanol-water partitioning coefficient (–)
C_O = equilibrium concentration of a substance in *n*-octanol (g m^{-3})
C_W = equilibrium concentration of the same substance in water (g m^{-3}).

The octanol-water partitioning coefficient can be used for quantification of the distribution of several types of substances between water and octanol (i.e., as a measure of the distribution between a hydrophilic and a hydrophobic phase, respectively). Therefore, K_{OW} is closely related to sorption (cf. Section 3.2.4.1). A low K_{OW} value for a substance corresponds to high solubility and mobility in water media whereas a high K_{OW} value shows a tendency of sorption to a solid phase (suspended solids, sediments, and soils) and thereby a general low mobility.

Within the area of urban drainage, K_{OW} is particularly relevant for assessment of the potential distribution of micropollutants between water and particles (i.e., the micropollutants' affinity to occur in the water phase or the opposite). The octanol-water partitioning coefficient is thereby an indicator for the occurrence and transport characteristics of micropollutants in urban drainage flows. As an example, Table 3.2 depicts values of K_{OW} for selected micropollutants (phthalates) with different solubility in water and corresponding association to particles.

Compared with DMP, DEPH is therefore a micropollutant that is associated with the particulate phase (i.e., partitioning to suspended solids, soils, and sediments). It is also seen that log K_{OW} increases with molecular weight.

The accumulation of micropollutants in living organisms often takes place in the hydrophobic part (i.e., the fat tissue). The magnitude of the K_{OW} value is therefore an indicator showing to what extent bioaccumulation may take place.

3.2.5 SPECIATION

Speciation of a substance in an aquatic system accounts for the fact that the substance depending on external conditions will appear in different chemical forms (species) with different characteristics. Speciation is relevant for substances in both soluble and solid (particulate) forms and basically concerns equilibrium conditions.

The bioavailability and the biological effect of a substance depend to a great extent on the form in which the substance appears. In case of a heavy metal, bioavailability and biological effects are to a great extent associated with species that exist as free ions (i.e., as noncomplex bound metal ions that are surrounded by water molecules). Also substances occurring in a molecular form (e.g., water-soluble xenobiotic organic compounds) may exert an effect. Compared with these forms, pollutants (metals) associated with solids (particles) or occurring as complex ions show generally reduced bioavailability and biological effects.

A simple illustration of speciation is the appearance of ammonia in the water phase depending on pH:

$$NH_3 + H_2O \leftrightarrows NH_4^+ + OH^-. \tag{3.23}$$

As an example of different characteristics for the different species of ammonia, it is well known that the nondissociated molecular form (NH_3) can be emitted from a water phase into the overlaying atmosphere, which is not the case for the ionic form. Furthermore, NH_3 has a harmful effect on fish, which is not the case for NH_4^+.

Within urban drainage speciation of heavy metals are of particular interest. Before going into details with these aspects, the concept of speciation will be illustrated and exemplified. The important thing is that among a number of external conditions that affect speciation of heavy metals, the following three are considered main factors:

1. Acidity and alkalinity, $pH = -\log(C_{H+})$
2. The concentration, C, or $pC = -\log C$, (for soluble species) of a substance
3. The redox potential, E, or $p\varepsilon = -\log E$ (cf. Section 3.6.1)

In the following, the importance of these three factors will be illustrated by selecting iron (Fe) as an example. In this way, pC–pH and E–pH diagrams are shown in Figures 3.6 and 3.7, respectively. Correspondingly, E–pC (or pε–pC) diagrams can also be constructed, however, they are less used.

As an illustration, the relation between the two first factors mentioned will be shown at a relatively high (constant) redox potential in an aqueous system (i.e., under aerobic conditions). In such an aqueous system ($H_2O \leftrightarrows H^+ + OH^-$) where iron hydroxy species exist, the equilibrium reactions shown in Table 3.3 are relevant (cf. Section 3.2.4.1.1).

As seen from Table 3.3, the four equilibrium equations all depend on pH. Via the formulation of the corresponding four equations of the solubility product and applying Equation 3.16, the solubility of the iron species can be illustrated in a diagram with $pC = -\log C$ versus pH where C represents the concentration of each of the four soluble iron species, Figure 3.6. Equation 3.3 is an example of such a solubility product corresponding to equilibrium reaction I in Table 3.3. The four solid lines in Figure 3.6 corresponding to equations I–IV delimit areas (i.e., pH and pC values), where either soluble iron (i.e., Fe^{3+} or iron hydroxy-components) or the solid form ($Fe(OH)_3$) of iron is dominating. The solubility and the speciation of iron in the aqueous system is thereby determined under the conditions that are illustrated.

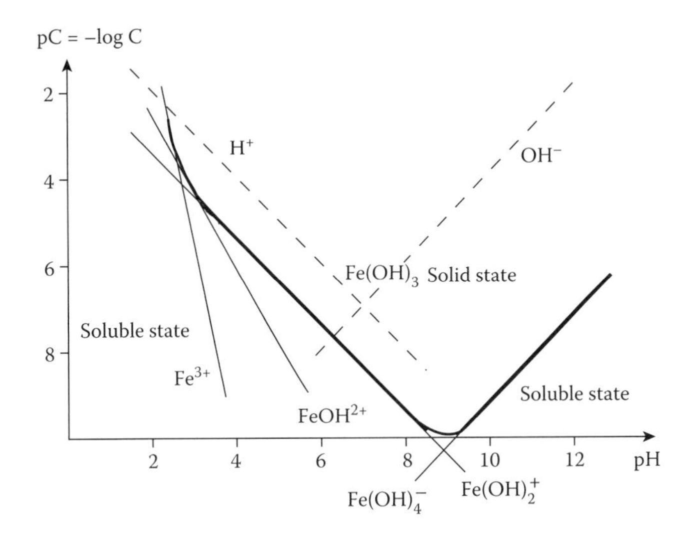

FIGURE 3.6 Solubility diagram for iron hydroxide in an aqueous system at a high (aerobic) redox potential. The equilibrium conditions for the dissociation of H_2O are indicated in terms of H^+ and OH^- species.

Figure 3.6 illustrates that the solubility can be quantified and depicts what species become stable at given conditions depending on pH and pC. The curves in Figure 3.6 show that the solid form of Fe(III), $Fe(OH)_3$, with a minimum of solubility at a pH of about 9 becomes increasingly soluble as the pH decreases or increases.

The formation of metal complexes is an important aspect of metal speciation. In general, a complex consists of a central atom or central ion that is surrounded by molecules or ions called ligands. Important examples of ligands in urban drainage flows are inorganic ions (e.g., hydroxy ion, carbonate, chloride, and sulfate). Examples of Fe complexes are shown in Table 3.3. Other examples are the following soluble Pb complexes with Pb as the central ion and chloride as the ligand that also exemplify that different charges of the complex are possible:

$$PbCl_4^{2-}, PbCl_2^0, \text{ and } PbCl^+.$$

Metal complexes affect the solubility of the metal and the toxicity. Further details in this respect are dealt with in Section 6.5.1, compare the examples in Table 6.4 and Figure 6.16.

The importance for the speciation of the third main factor (i.e., the redox conditions expressed in terms of the redox potential E or $p\varepsilon = -\log E$), can be illustrated in a diagram with E or pε versus pH at constant concentration C of the soluble species. This diagram is sometimes referred to as a Pourbaix diagram (Pourbaix 1963). The theoretical background for its quantification will be given in the following. The starting point is a relevant half-reaction (cf. Section 3.6.1):

$$\text{red} \leftrightarrows \text{ox} + n\,e^- \tag{3.24}$$

TABLE 3.3
Equilibrium Reactions I–IV and Corresponding Equations for an Aqueous System Where Iron Hydroxide Exists as a Precipitate

Equilibrium Reactions	Equilibrium Equation at 25°C
I: $Fe(OH)_3 \leftrightarrows Fe^{3+} + 3OH^-$	$\log C = 3.2 - 3\,pH$
II: $Fe(OH)_3 \leftrightarrows Fe(OH)^{2+} + 2OH^-$	$\log C = 1.0 - 2\,pH$
III: $Fe(OH)_3 \leftrightarrows Fe(OH)_2^{+} + OH^-$	$\log C = -2.5 - pH$
IV: $Fe(OH)_3 + OH^- \leftrightarrows Fe(OH)_4^-$	$\log C = -18.4 + pH$

Source: Data from The constants in the equations originate from Pankow, J. F. 1991., *Aquatic chemistry concepts*. Chelsea, MI: Lewis Publishers.

The quantification of the redox conditions of a half-reaction is expressed in terms of the Nernst's equation known from physicochemistry (cf. e.g., Atkins and de Paula 2002). Referring to Equation 3.24, the general formulation of the Nernst equation at 25°C is

$$E = E'_o + \frac{0.059}{n} \log \frac{C_{ox}}{C_{red}}, \tag{3.25}$$

where

E = redox potential of the half-reaction (V)

E'_o = standard redox potential of the half-reaction (25°C, 1 atm and standard activities: concentrations 1 mole L^{-1} and 1 for pure substances, water and solids; V)

C_{ox} = activity (concentrations) of the oxidized components (mol L^{-1} or 1 for pure substances)

C_{red} = activity (concentrations) of the reduced components (mol L^{-1} or 1 for pure substances).

The redox potential as expressed by the Nernst equation is closely related to the energy transformations of the corresponding half-reaction (cf. Equations 3.63 and 3.64).

As an example, the half-reaction for the equilibrium between Fe^{2+} and $Fe(OH)_3$ is

$$Fe^{2+} + 3H_2O \leftrightarrows Fe(OH)_3 + 3H^+ + e^- \tag{3.26}$$

With the actual values for E'_o and n, Nernst's equation for Equation 3.26 is

$$E = 1.06 + 0.059 \log \frac{(C_{H^+})^3}{C_{Fe^{2+}}}. \tag{3.27}$$

At constant concentration of the soluble iron (II), $C_{Fe^{2+}}$, Equation 3.27 determines the redox potential, E, as a function of pH. It is therefore possible to depict the equilibrium between Fe^{2+} and $Fe(OH)_3$ in a diagram with the redox potential E versus

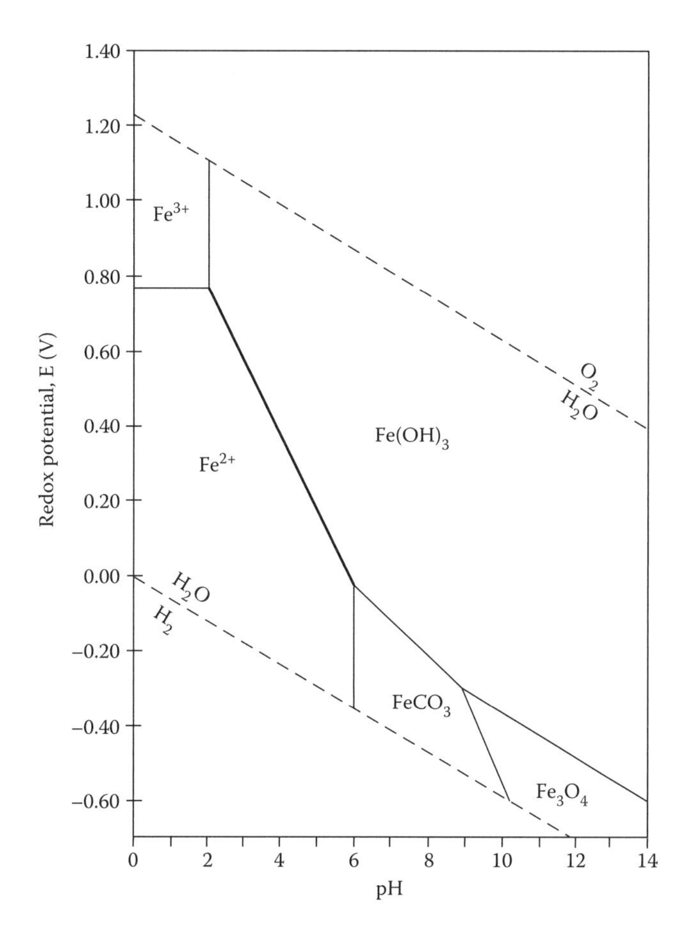

FIGURE 3.7 Redox potential, E, versus pH for iron species in an aqueous system. The total concentration of soluble iron is 2×10^{-3} mole L^{-1} and the concentration of carbonate is 10^{-3} mol L^{-1}.

pH at constant pC for iron (II), compare Figure 3.7. This equilibrium is in the figure shown by a solid straight line together with corresponding lines for other equilibrium half-reactions that are relevant for iron components in an aqueous system where inorganic carbon (carbonate) occurs.

In general, the speciation of a pollutant plays a role when dealing with effects, transport, and treatment. The speciation of heavy metals therefore requires specific attention for the following reasons:

- The different species of heavy metals show different toxicity in the environment because they possess the ability to form complexes with organic and inorganic substances as ligands.
- Removal processes in both natural and treatment systems depend on the nature of the different species (e.g., the charge of a given metal species that might affect the sorption at the surface of a particle).

In addition to the three main parameters mentioned, pH, concentration (pC), and redox potential (pε), a number of other parameters affect the formation of the different species and complexes of a substance. Most relevant for heavy metals in urban wet weather flows are the following water quality parameters:

- Inorganic ionic components as ligands (e.g., Cl^-, SO_4^{2-}, and HPO_4^{2-})
- The inorganic carbon system (in particular the alkalinity and HCO_3^-)
- Colloidal substances (e.g., humic substances (humic and fulvic acids) and dissolved organic matter (DOM) that can all act as ligands)

The MINTEQ-model (MINeral Thermal EQuilibrium) is widely used for analysis and prediction of metal speciation and formation of metal complexes (Allison, Brown, and Novo-Grodae 1991). It is a chemical equilibrium model for the calculation of metal speciation and solubility in natural waters. The model is therefore a valuable prediction tool in case of quality assessment of urban wet weather flows. Different versions of the model can be downloaded via the Internet.

3.3 MASS BALANCES

A mass balance is a basic engineering approach. Except for nuclear processes, a mass balance expresses the fact that mass will neither be created nor will it disappear, however, it can undergo transformations and be accumulated. A mass balance is a requirement for any type of model formulation in engineering.

A mass balance is a method to manage transport and transformation of substances and thereby identify where these substances occur, to which extent and in which forms they exist. Such a “household” requires that it is a well-defined system that is relevant for the mass balance (i.e., that it has clearly defined boundaries). In addition, we must seek information on system characteristics (e.g., in terms of volume, in- and outflow, and initial conditions, select what substance should be considered, and establish information on the rates of transformations and accumulations). It is the degree of detailed knowledge on the system that determines to what extent the mass balance can be established. It is crucial that this requirement is seriously observed.

Under these constraints, the general expression of a mass balance for a substance within the boundary of a system is

$$\text{Accumulation rate} = \text{inflow rate} - \text{outflow rate} +/- \text{transformation rate}. \tag{3.28}$$

Equation 3.28 can mathematically be formulated in general terms as follows:

$$\frac{d(\mathrm{CV})}{dt} = \mathrm{Q_{in}C_{in}} - \mathrm{Q_{out}C_{out}} +/- \text{transformation rate}, \tag{3.29}$$

where

C = concentration of a substance (g m^{-3})
V = volume of the system (m^3)
t = time (s)
Q = volumetric flow rate (m^3 s^{-1}).

The transformation rate must be expressed in kinetic terms relevant for the actual substance (cf. Section 3.6.2).

In Equation 3.29, the inflow and outflow rate terms are expressed as volumetric flows. In case of transport from one phase to another, corresponding terms can be expressed as a transport of substances across the boundaries (e.g., a water–solid or a water–air interface):

$$\text{Transport rate} = J_S\, A \tag{3.30}$$

where

J_S = flux rate of a substance S (g m^{-2} s^{-1})
A = the interfacial area (m^2).

The general formulation of the mass balance in Equation 3.28 is exemplified in the following example. The models for transport and transformations dealt with in Section 10.2 are also examples of mass balances.

Example 3.2: Mass Balance of Flow Reactors

The two extremes of flow reactors or vessels that, in contrast to batch reactors, have both inflow and outflow during their operation are the continuous stirred tank reactor (CSTR)—also referred to as a completely mixed flow reactor (CMFR)—and the plug flow reactor (PFR), see Figure 3.8.

The mass balance of the two flow reactor types will be illustrated for a 1′ order reaction under steady-state flow conditions.

CSTR Mass Balance

Under steady-state conditions and for a 1′ order reaction with the rate constant *k*, Equation 3.29 results in:

$$0 = Q\, C_{in} - Q\, C - k\, C\, V, \tag{3.31}$$

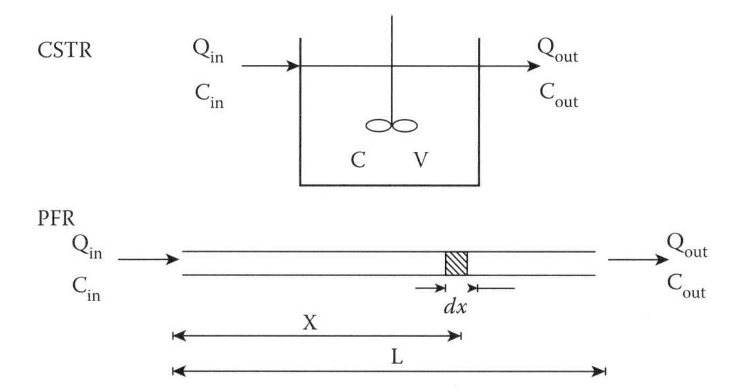

FIGURE 3.8 Principle of the two Reactor types, the CSTR and the PFR, respectively.

or

$$C = \frac{QC_{in}}{Q + kV}. \tag{3.32}$$

If a mean residence time $T_{CSTR} = V/Q$ is introduced, Equation 3.32 is expressed as follows:

$$C = \frac{C_{in}}{1 + kT_{CSTR}}. \tag{3.33}$$

It should be noted that due to complete mixing in the reactor, $C = C_{out}$.

PFR Mass Balance

In contrast to the CSTR where the constituent concentration, C, is uniformly distributed within the reactor, there is a gradual change of C in the PFR from the inlet to the outlet. Under steady-state conditions, C is a function of the distance, x, from the inlet. The residence time, $T_{PFR,x}$, of a water volume at this point depends on the flow velocity, u:

$$T_{PFR,x} = \frac{x}{u} \tag{3.34}$$

$$u = \frac{Q}{A_{PFR}} \tag{3.35}$$

where

A_{PFR} = cross-sectional area of the pipe (m^2)

The mass balance over a slice with a thickness dx at the distance x is therefore (cf. Figure 3.8):

$$\frac{dC}{dT_{PFR,x}} = -kC. \tag{3.36}$$

Integrated over the pipe length, Equation 3.36 results in the following entire mass balance:

$$\ln C_{out} = \ln C_{in} - kT_{PFR,L} = \ln C_{in} - k\frac{L}{u} \tag{3.37}$$

Comparison Between a CSTR and a PFR

Under the conditions given (i.e., for a 1′ order reaction and steady-state conditions), the reaction in a PFR compared with a CSTR is more complete for a fixed mean residence time. If a number of CSTRs are connected in series, the entire performance approaches a PFR. Even 3 to 4 CSTRs (compartments) in series will result in an outlet concentration rather close to that of a PFR.

If the conditions are different than defined in this example, CSTRs and PFRs may perform differently. As an example, there will be no difference in outlet concentration for these two types of reactors in case of a 0′ order reaction.

A mass balance plays a central role for the assessment of treatment processes and systems. With regard to this aspect, measures of the degree of treatment are typically needed. In the following, often used terms are briefly outlined and defined:

- Removal efficiency
  Removal efficiency is defined as follows:

$$E_r = \frac{C_O - C_U}{C_O} \tag{3.38}$$

where
E_r = removal efficiency (–)
C_O = concentration of a constituent at the inlet to a treatment facility ($g\ m^{-3}$)
C_U = concentration of a constituent at the outlet from treatment ($g\ m^{-3}$)

The removal efficiency is probably the most frequently used measure of treatment. In practice it is often transformed to a percentage. The removal efficiency is mostly used in case of stormwater runoff but can also be used for CSOs. The removal efficiency can be calculated for specific events or stated as an average or median value for a treatment system.

- Load reduction
  The removal efficiency is defined based on concentration values. A corresponding measure for assessment of load reduction (e.g., into an adjacent water course) can be defined

$$R_r = \frac{L_{WO} - L_W}{L_{WO}}, \tag{3.39}$$

where
R_r = load reduction (–)
L_{WO} = annual load of a constituent without treatment ($kg\ yr^{-1}$)
L_W = annual load of a constituent with treatment ($kg\ yr^{-1}$)

A number of models for calculation of L_{WO} from urban surfaces and roads are shown in Section 4.5.

3.4 PHYSICAL PROCESSES: WATER AND MASS TRANSPORT

In general, problems associated with urban drainage require that environmental process engineering principles based on fundamentals from chemistry and biology be integrated with subjects within hydrology and hydraulics. Transport and transformations of the pollutants can thereby be dealt with as an entity. Although this text focuses on the quality aspects of urban wet weather pollution, it is fully appreciated that knowledge on the quantification of the transport phenomena for both water and those soluble and particulate constituents that occur in the water phase have a central role to play.

The transport processes are given limited space in the text, sufficient to understand the basic phenomena when dealing with the quality aspects of urban drainage.

For those readers who are trained in hydrology and hydraulics, this section probably will just be a refresher. Further and more detailed description of the physical processes in particular related to urban drainage can be found in several texts Wanielista (1990), Mays (2001), Debo and Reese (2003), and Ashley and others (2004).

When dealing with urban wet weather pollution, transport of pollutants plays a role in the transport systems, in treatment devises, and when the flows reach natural environment. It is the case for the discharges into receiving waters as well as into soils. It should be noticed that transport becomes important at both small, molecular scale and at large, macroscopic scale.

3.4.1 Advection, Diffusion, and Dispersion

The physical processes in terms of both transport phenomena of the water itself (the hydraulics) and transport of the constituents, soluble as well as particulate forms, are important. These transport processes are the basis to determine where we can find the pollutants in both technical and natural systems. Transport-related aspects in terms of the particle characteristics dealt with in other sections of this text (e.g., in Section 3.2.4) are important as well.

In addition to the general transport-related phenomena, hydraulics play a specific role in a number of cases when dealing with the impact of pollution from urban and road runoff. Such direct hydraulic impacts are flooding and a number of phenomena that exert an effect on water quality. Such examples are erosion of sediments or sediment deposition in receiving waters that may cause adverse effects by changing the habitat for both plants and animals. Furthermore, scouring of sediments in receiving waters caused by extreme rainfall runoff may exert release of pollutants into the water phase and, as an example, reduced components might cause an increased DO depletion. These flow related impacts are particularly important during extreme runoff conditions.

The time scale effect of pollutant effects is discussed in Section 1.3.2. When dealing with the hydraulic effects, phenomena like flooding, habitat changes, and sediment scouring are by definition acute effects because they are associated with the impact from single events. Particularly, rainfall events that are extreme in terms of both intensity and duration are critical in this respect.

A fundamental characteristic of flowing waters is related to the mode of movement defined as either being laminar or turbulent, and a transition state between these two types of flow regimes. Laminar flow generally exists at low-flow velocities whereas at increased flow, the movement of the water changes from a calm to a whirling motion. When laminar flow exists, it is possible to determine the velocity field in time and place as a transport of the water and its associated constituents in just one direction and with a constant flow velocity over time. In turbulent flow, however, the velocity profile of the fluid varies caused by a mutual exchange of the water elements. Contrary to laminar flow, the turbulent flow velocity is determined by two terms: a mean value and a component that varies stochastically. At turbulent flow, the velocity vector changes both magnitude and direction. In Section 3.4.2.1, the flow regime related to the settling of particles is further dealt with in terms of the dimensionless Reynolds number, R.

In the following sections, a number of fundamental hydraulic phenomena that are important for the movement of soluble, as well as particulate pollutants, are briefly defined and described:

- Advection
- Molecular diffusion (Fick's first law)
- Dispersion (eddy diffusion)

3.4.1.1 Advection

Advection, also called convective transport, describes a mode of transport where a constituent is transported by the net flow of the water phase. Advection thereby describes a situation where no spreading of a constituent takes place. Advection is quantified in terms of a flux. The flux, J, of a constituent is defined as a transport (i.e., as an amount of a constituent transported per unit of time and per unit of a cross-sectional area). It is basically a vector (i.e., it includes both the magnitude of the phenomenon and a direction). The magnitude of the flux, the flux rate, is

$$J = C\,u \tag{3.40}$$

where
J = flux (rate) of a constituent ($g\ m^{-2}\ s^{-1}$)
C = constituent concentration ($g\ m^{-3}$)
u = net flow velocity of water ($m\ s^{-1}$)

Advection describes a mode of flow where all soluble or particulate units are exposed to a uniform velocity that is, the flux vector is equal for all constituents in the water phase, soluble species as well as suspended particles.

3.4.1.2 Molecular Diffusion

Diffusion describes a disordered movement of the constituents that takes place at the molecular scale. The molecular scale movement of a constituent is caused by a temperature induced mutual impact of the molecules. This temperature-induced movement is also known as the Brownian movement. The net movement of the constituents always (in average) takes place from a high concentration to an area of lower concentration.

At the macroscopic level, molecular diffusion is expressed by Fick's first law of diffusion. The driving force for the flux of a constituent is the concentration difference per unit of distance (i.e., the concentration gradient):

$$J = -D\frac{dC}{dx}, \tag{3.41}$$

where
D = molecular diffusion coefficient ($m^2\ s^{-1}$)
x = distance (m).

The magnitude of the diffusion coefficient depends on both the properties of the constituent that is transported and the fluid (water). The temperature has an influence on the magnitude of D but it is by definition independent of the mode of transport, because it fundamentally describes a phenomenon that takes place at the molecular scale.

Molecular diffusion is a phenomenon that causes slow movement of a constituent and the molecular diffusion coefficient for small molecules is typically in the order of 10^{-5} $cm^2\ s^{-1}$. The characteristic distance of travel versus time is

$$x = (2\ D\ t)^{0.5} \tag{3.42}$$

where
t = time (s).

Molecular diffusion may become important when dealing with transport in biofilms, transport in porous media, and transport across interfaces (e.g., the air–water interface). However, within the water phase it is generally exceeded by several orders of magnitude by dispersion that will be described in the following section.

3.4.1.3 Dispersion

Dispersion of a constituent describes a phenomenon that is related to its spreading in a fluid (water). In contrast to molecular diffusion, dispersion is a movement at the macroscopic scale and is caused by flow velocity variations in time and place. Dispersion is therefore related to turbulent conditions and typically exceeds molecular diffusion by several orders of magnitude. In general, it always occurs in urban wet weather flows. Dispersion is also known as eddy diffusion.

Dispersion is a random process that by nature is quite different compared with molecular diffusion. However, dispersion also concerns transport of constituents from areas of high concentration to areas of low concentration. The empirical description of dispersion follows a similar form as shown by Fick's first law, Equation 3.41:

$$J = -\varepsilon \frac{dC}{dx} \tag{3.43}$$

where
ε = dispersion coefficient ($m^2\ s^{-1}$)

Contrary to values for D, it is basically not possible to produce tables that give values for ε. The reason is that the actual flow pattern is crucial for the magnitude of the dispersion coefficient. Determination of ε therefore requires flow measurements followed by a calibration procedure.

3.4.1.4 Flow and Mass Transport in Channels

The one-dimensional flow and mass transport that take place in channels and rivers are important for several aspects of urban drainage. A great number of both simple formulas and complex computational methods exist for prediction of channel flow and mass transport. A number of such methods will be briefly dealt with in this text. Further details can be found in several texts and handbooks in hydraulics (e.g., Mays 2001).

Formulas exist for the calculation of flow in both full and partly filled pipes and in open channels. Several semiempirical models that were developed some 100 to 150 years ago are still useful and widely applied. These formulas are typically based on few and central parameters describing the physical system and often expressed in terms of a power function. One of these formulas, the so-called Manning formula, is widely applied and will be briefly described in the case of flow in an open channel.

$$V = \frac{Q}{A} = \frac{1}{n} R^{2/3} I^{1/2}, \tag{3.44}$$

where
- V = average flow velocity (m s^{-1})
- Q = flow rate (m^3 s^{-1})
- A = cross-sectional area of the channel (m^2)
- n = Manning coefficient of roughness (s $m^{-1/3}$)
- R = A/P = hydraulic radius (m)
- P = wetted perimeter of the channel (m)
- I = slope of the water surface or the bottom (m m^{-1})

The Manning coefficient of roughness, n, is a parameter that expresses the resistance to the channel flow. A channel with smooth concrete surfaces has a Manning coefficient of roughness equal to 0.010–0.015 s $m^{-1/3}$ in contrast to a value of about 0.03 s $m^{-1/3}$ for a channel constructed with soil-covered surfaces. In case of high roughness in the channel (e.g., caused by extensive vegetation), n is typically increased to 0.04 s $m^{-1/3}$ or even higher values.

As already mentioned, the Manning formula is only valid for steady state and uniform flow conditions. In case of unsteady and one-dimensional flow in open channels, the two so-called Saint Venant equations, Equations 3.45 and 3.46 are widely used as the basis for computation (cf. Figure 3.9):

The continuity equation:

$$\frac{dQ}{dx} + b\frac{dy}{dt} = q. \tag{3.45}$$

The momentum equation:

$$\frac{dQ}{dt} + \frac{d}{dt}\left(\frac{Q^2}{A}\right) + gA\frac{dy}{dx} = -gAS_f, \tag{3.46}$$

where
- Q = volumetric flow rate (m^3 s^{-1})
- x = distance downstream channel (m)

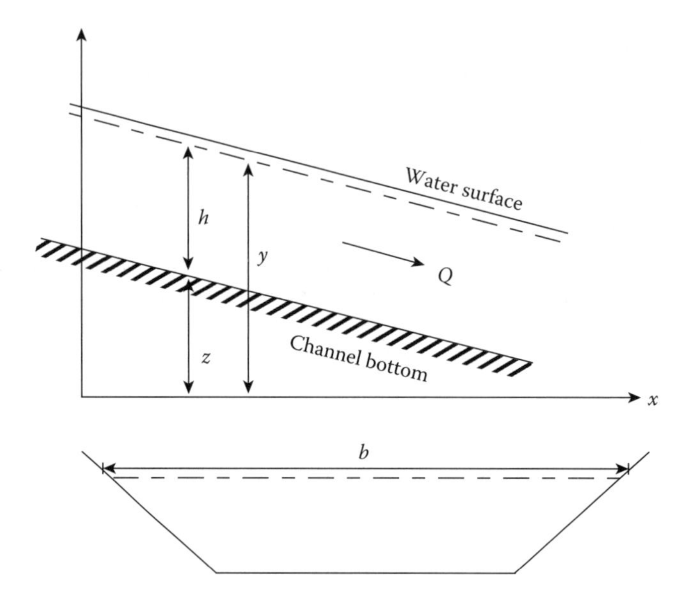

FIGURE 3.9 Longitudinal and cross-sectional characteristics of an open channel (cf. Equations 3.45 and 3.46).

b = channel width at the water surface (m)
$y = z + h$ water level coordinate (m)
z = vertical bottom coordinate (m)
h = water depth (m)
t = time (s)
q = lateral (constant) inflow of water ($m^3\ m^{-1}\ s^{-1}$ i.e., $m^2\ s^{-1}$)
A = cross-sectional area (m^2)
g = gravitational acceleration ($m\ s^2$)
S_f = friction slope, approximately equal to slope of water surface (–)

As already dealt with in Sections 3.4.1.1 and 3.4.1.3, advection (convective transport) caused by the mean flow and dispersion due to turbulent diffusion are the two major processes that determine transport of a constituent at the macroscopic scale. The mass balance of a constituent that is uniformly mixed over the cross-section of the channel is, according to Equations 3.40 and 3.43, in time (t) and place (x) formulated as follows:

$$\frac{d(\mathrm{A}\mathrm{C}_x)}{dt} + \frac{d(\mathrm{Q}\mathrm{C}_x)}{dx} = \frac{d}{dx}\left(\varepsilon \mathrm{A}\frac{d\mathrm{C}_x}{dx}\right) + q\mathrm{C}_q - s \tag{3.47}$$

where
C_x = constituent concentration at distance x ($g\ m^{-3}$)
ε = dispersion coefficient ($m^2\ s^{-1}$)
C_q = constituent concentration in the lateral inflow ($g\ m^{-3}$)
s = sink (process rate) term for the constituent per unit length ($g\ m^{-1}\ s^{-1}$).

As an example and if a 1′ order reaction in the water phase proceeds, the sink term can be formulated as follows:

$$s = A\,k\,C_x, \tag{3.48}$$

where
k = 1′ order reaction rate constant in the water phase (s^{-1}).

3.4.2 Sedimentation, Deposition, and Erosion

The variability in flow conditions that exists in wet weather flows has a crucial impact on the transport of solids. Solids can, under high flow conditions, be eroded, resuspended in the water phase, and transported to a new site where it can settle and be deposited under less turbulent and relatively quiescent conditions. The processes involved in the transport of particulate materials play a major role in both technical systems and in the receiving water systems. Important examples are related to the removal of particulate pollutants in collection, transport, and treatment systems and accumulation of such solid components in the receiving waters.

3.4.2.1 Sedimentation and Deposition

In the context of urban drainage, sedimentation (settling) and deposition of particles account for those processes and phenomena that are related to removal of particles from a water phase and accumulation as sediments and deposits. Sedimentation of discrete particles is a physically well-defined process that in terms of particle size characterization is briefly dealt with in Section 3.2.4.3. Deposition is more vaguely defined but can be understood as a transfer of particulate matter from a suspended water phase into a sediment phase. Deposition therefore also concerns accumulation and fixation of the particles in the sediment phase (where they become "deposits") irrespective of the fact that deposition occurs in flowing or stagnant waters. Sedimentation and deposition of solids and associated pollutants originating from urban wet weather flows are important processes in both technical systems and the receiving environment.

In addition to the movement of particulate matter from suspension into the sediments, it is important in flowing waters and pipes to take into account a near-bed, highly concentrated layer of materials moving along with the flow, however, with a reduced velocity. This fact makes the whole picture of sedimentation and deposition rather complex.

Within the area of urban drainage, an important case of sedimentation takes place as gravitational settling. As depicted in Figure 3.10, a particle in suspension is subject to the impact from two forces: the gravitational force (F_g) and the frictional resistance in terms of the frictional drag force (F_r).

The theoretical expression for F_g is given by

$$F_g = (\rho_p - \rho_w)\,g\,V \tag{3.49}$$

where
F_g = gravitational force ($kg\ m\ s^{-2}$) or (N)
ρ_p = specific density of the particle ($kg\ m^{-3}$)

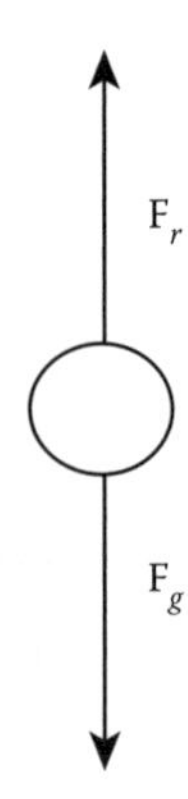

FIGURE 3.10 A spherical particle subject to a gravitational force, F_g, and a frictional drag force, F_r.

ρ_w = specific density of water (kg m^{-3}); 10^3 kg m^{-3} at 15°C
g = gravitational acceleration = 9.81 m s^{-2}
V = volume of the particle (m^3).

The frictional drag force is

$$F_r = 0.5 \text{ C A } \rho_w \, v^2, \tag{3.50}$$

where
F_r = frictional drag force (kg m s^{-2}) or (N)
C = drag coefficient (–)
A = cross-sectional area of the particle perpendicular to the direction of the movement (m^2)
v = settling velocity of the particle (m s^{-1}).

For spherical particles, the drag coefficient, C, depends on the flow regime. The drag coefficient is described in terms of the dimensionless Reynolds number, R, which for spherical particles is formulated as follows:

$$R = v \, d \, \rho_w \, \mu^{-1} = v \, d \, \upsilon^{-1}, \tag{3.51}$$

where
R = Reynolds number (–)
d = diameter (spherical) of the particle (m)
μ = dynamic viscosity of water (kg m^{-1} s^{-1}) or (N s m^{-2}) or (Pa s); 10^{-3} kg m^{-1} s^{-1} at 15°C
υ = kinematic viscosity of water (m^2 s^{-1}).

An empirical relation of the drag coefficient, C, for spherical particles versus Reynolds number, R, is shown in Figure 3.11. For $R < 1$ the flow regime is laminar and for $R >$ about 2 10^3 is turbulent. For R within this interval, transitional flow regimes exist.

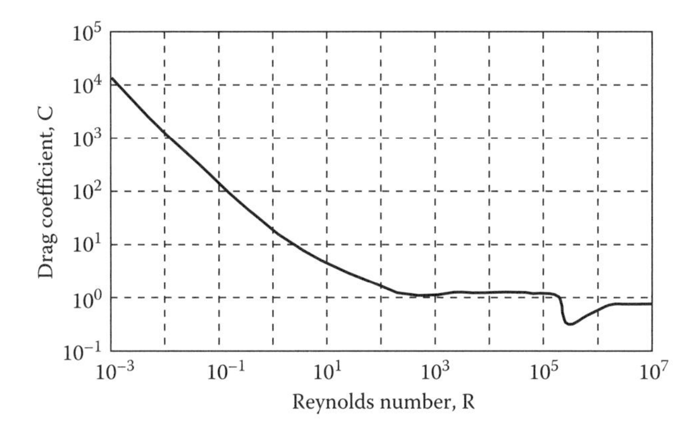

FIGURE 3.11 The drag coefficient, C, for a spherical particle versus Reynolds number, R. For R < 1, under laminar flow conditions, $C = 24\ R^{-1}$, which is a straight line in the double logarithmic depiction.

Settling of particles under steady-state conditions result in $F_g = F_r$. According to this equation and for spherical particles and laminar flow (cf. Figure 3.11), the settling velocity of a particle, v, is based on Equations 3.49 through 3.51 described as follows:

$$v = gd^2 \frac{\rho_p - \rho_w}{18\mu}. \tag{3.52}$$

Equation 3.52 is named Stokes's law. In theory, Stokes's law in water is valid for spherical particles with diameters between 1 and 100 μm. In real systems, however, diffusion in the suspended water phase will reduce the settling velocity of particles for both stormwater and CSOs. For real systems, Stokes's law is only approximately valid for particles with a diameter between 40 and 100 μm (cf. Figure 3.12). Settling of particles < 10 μm will hardly take place (cf. Figure 3.5).

As already discussed, there are several limitations in the use of Stokes's law for real systems within the area of urban drainage. The complex mixtures of organic and inorganic particles of different shapes and flow regimes with R > 1 are just examples of constraints for applying a simple concept to predict sedimentation and deposition. Although theories like Newton's law exist for settling under turbulent flow conditions, empirical relationships are generally needed for both technical and natural systems. Empirical descriptions of pollutant deposition will therefore be included and applied in the following chapters when considered appropriate.

Further considerations on settling relevance for constituents dealt with in urban drainage are found in Eckenfelder (1989), Nazaroff and Alvarez-Cohen (2001), Tchobanoglous, Burton, and Stensel (2003), and Gregory (2006).

3.4.2.2 Erosion and Resuspension

Erosion is a process that is related to the release of solid and typically loosely bound materials deposited at a solid–liquid interface, for example, dust at urban surfaces,

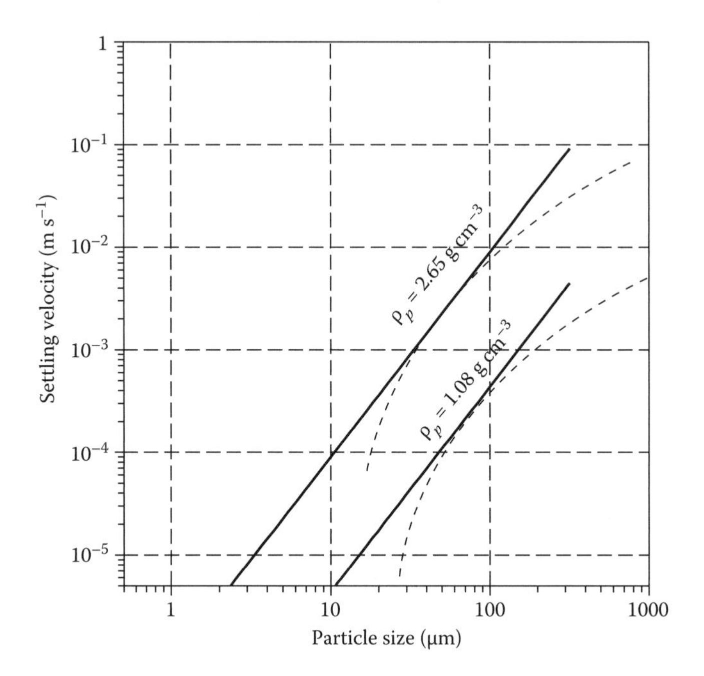

FIGURE 3.12 Theoretical settling velocities shown as solid lines for discrete spherical particles. The velocities are given for both high (inorganic matter, $\rho_p = 2.65$ g cm^{-3}) and low (organic rich matter, $\rho_p = 1.08$ g cm^{-3}) particle density at 15°C. The tendency for settling in real systems is shown by the dashed lines.

sewer solids (sediments and biofilms), and sediments in receiving waters. Erosion is caused by the transfer of energy from a flowing water phase to the surface of a solid, stationary phase. Resuspension describes the process that typically follows erosion when the released particulate materials are transferred into the flowing bulk water phase.

The transfer of energy from a laminar flowing water phase to a plane solid–water interface can be described by Newton's formula for the shear stress, τ, at a cross-section parallel with a plane surface (cf. Figure 3.13)

$$\tau = \mu \frac{\partial u}{\partial x} \tag{3.53}$$

where

τ = shear stress (kg m^{-1} s^{-2}) or (N m^{-2})
μ = dynamic viscosity of water (kg m^{-1} s^{-1}) or (N s m^{-2}); 10^{-3} kg m^{-1} s^{-1} at 15°C
u = flow velocity (m s^{-1})
x = distance from the solid surface (m)

Although Equation 3.53 is derived for laminar flow, it is also pragmatically used for turbulent flow conditions.

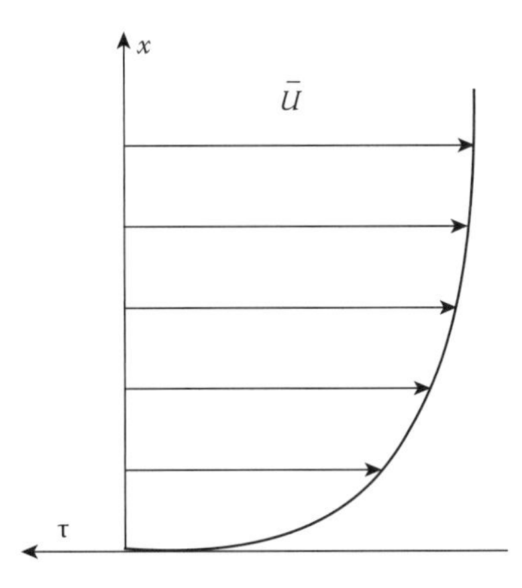

FIGURE 3.13 The velocity profile in a laminar flowing water phase at a plane, stationary solid surface.

The shear stress, τ, is according to Newton's formula, Equation 3.53, proportional with the velocity gradient, $\partial u/\partial x$. The τ is a vector in a direction determined by the tangent to the velocity profile, Figure 3.13. At the water-solid interface where $u = 0$, τ is parallel with and pointing in a direction opposite to the flow velocity.

In terms of erosion, the numerical value of τ is a measure of the force per unit area of the interface exerted by the flowing water. It is a measure of the impact onto those materials that are deposited at the surface. The shear stress is therefore a central parameter when assessing the erosion process.

The flow regime of extreme runoff events generated in sewer pipes, on urban surfaces, and on roads may cause erosion followed by resuspension of the eroded materials. The critical value of τ (τ_{crit}) that causes erosion depends on the characteristics of the materials and the system in question. In case of particulate materials deposited, it is the particle size, shape, and specific gravity that are important. Furthermore, potential interactions between the particles will play a significant role for the magnitude of τ_{crit}. Materials where such internal interactions exist are considered cohesive materials and are typically found in sediments with contents of organic matter that may form a kind of interstitial glue. Theoretical approaches for determination of τ_{crit} are not available and empirical determination of τ_{crit} is therefore needed.

The following two empirical relationships developed by Shields (1936) and Mehta (1988), respectively, can be applied for estimation of τ_{crit}:

$$\text{Shields: } \tau_{\text{crit}} = \gamma(\rho_b - \rho_w)\, g\, d \tag{3.54}$$

$$\text{Metha: } \tau_{\text{crit}} = \alpha(\rho_b - \rho_w)^{\beta} \tag{3.55}$$

where

ρ_b = specific bulk density of the deposited materials (kg m^{-3})
ρ_w = specific density of water (kg m^{-3}); 10^3 kg m^{-3} at 15°C
d = particle size, diameter (m)
g = gravitational acceleration = 9.81 m s^{-2}
α, β, and γ = constants

3.4.3 Transport in Porous Media: Soils and Filters

In the context of urban drainage, transport in porous media concerns the movement of water and associated constituents in both soils and filter media. Transport in soils occurs during natural infiltration of runoff water from the soil surface as well as in infiltration systems designed for discharge and treatment of urban and highway runoff. Transport of water and associated pollutant removal in filter media are particularly relevant in the case of treatment of the runoff.

In the following, basic characteristics related to the transport of water and associated constituents in porous media will be dealt with. Further details that are particularly relevant in cases of infiltration of stormwater with focus on treatment of the runoff are dealt with in Section 9.4.

3.4.3.1 Basic Characteristics of Porous Media

Porous media are heterogeneous materials that, in principle, constitute particles surrounded by air and water. In case of soils, particularly in the top layer, mineral components, organic matter (humic substances), biota, gases, and water will typically dominate. The gases are those originating from the overlying atmosphere and those produced in the soil itself (e.g., CO_2 produced from the microbial breakdown of organic materials).

The particle size is an important characteristic of soil or filter media, compare the classification of particles including different soil types in Figure 3.5. Furthermore, the porosity, ε, determines the relative magnitude of the internal void space (volume) between the soil particles that is available for water and gas (cf. Table 9.6).

3.4.3.2 Water Transport in Porous Media

Transport of water and associated constituents in porous media depends on the extent of saturation. Saturated and unsaturated systems are defined as systems where the pore volume is totally and only partially filled with water, respectively. If saturated conditions in soils exist, Equation 3.56, referred to as Darcy's law, states the relation between central parameters determining the transport of water in porous media:

$$u = \frac{Q}{A} = -K\frac{\partial h}{\partial l}, \tag{3.56}$$

where

u = one-dimensional flow velocity (m s^{-1})
Q = volumetric flow rate (m^3 s^{-1})
A = cross-sectional area (m^2)
K = hydraulic conductivity (m s^{-1})
h = pressure head or hydraulic head (m)
l = length of travel (m).

The term $\partial h/\partial l$ in Equation 3.56 is named the hydraulic gradient. The hydraulic conductivity, K, is a parameter that quantifies the permeability of a porous media (i.e., the ability to transport water through the material). The hydraulic conductivity

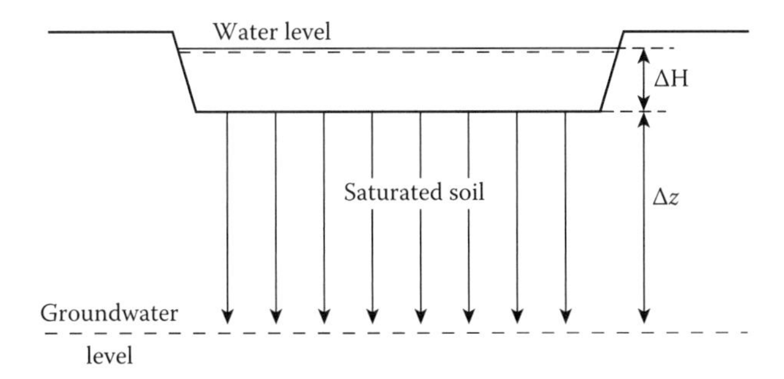

FIGURE 3.14 Schematic illustration of the flow of water from an infiltration pond through a saturated soil column to the groundwater.

is a central parameter in case of filtration and infiltration. It is high for gravel and low for silt and clay.

Figure 3.14 exemplifies Equation 3.56 in the case of vertical flow of water through a soil from an infiltration pond provided that the flow from the pond to the groundwater takes place under saturated conditions.

In relation to the case depicted in Figure 3.14, Darcy's law, Equation 3.56 is formulated in the following equation with $h = \Delta H + \Delta z$ and $l = \Delta z$:

$$u = \frac{Q}{A} = -K\frac{\Delta H + \Delta z}{\Delta z}. \tag{3.57}$$

Further details concerning transport of both water and associated pollutants in porous media—also concerning transport through a clogged layer of deposits at the bottom of an infiltration pond—are dealt with in Section 9.4.

3.5 PHYSICOCHEMICAL PROCESSES

Related to the urban wet weather phenomena, it is important to mention that several aspects dealt with in Sections 3.2.3 through 3.2.5 in principle concern physicochemical processes and phenomena. In the following, three physicochemical processes that are central for both treatment of pollutants and for their environmental effects will be dealt with:

- Mass transfer across interfaces, reaeration (air–water mass transfer), adsorption, and desorption
- Coagulation
- Flocculation

The character of the first point, mass transfer across interfaces, depends on what phases and what constituents it concerns. The following section deals with the air-water interface whereas sorption (liquid–solid interactions) is dealt with in Section 3.2.4.1.2

3.5.1 Mass Transfer Across an Air–Water Interface

Transport of substances from one phase to another (i.e., transport across an interface), is often an integral part of the movement of pollutants from one location to another. Often it is the case that the resistance of transport across the interface is limiting the rate of transport or transformation and therefore important to quantify. Air–water and water–solid interfacial transport are important phenomena within the area of urban drainage. Particularly reaeration, transport of oxygen across the air–water interface is in several cases (e.g., in natural surface water systems) a very central process. In the following, mass transfer across the air–water interface will be exemplified by reaeration.

Interfacial mass transport is a subject of any comprehensive textbook on physicochemistry. Furthermore, a detailed description of these processes is found in Thibodeaux (1996). Air–water mass transport processes related to sewer networks are dealt with in Hvitved-Jacobsen (2002).

Different theoretical approaches exist for the understanding of interfacial mass transport. In case of air-water transport, the simple two-film theory developed by Lewis and Whitman (1924) is still a sound basis for its understanding and quantification. This theory is based on molecular diffusion of a substance through two stagnant films, a liquid and a gas film, at the air–water interface formulated under equilibrium conditions at the interface, see Figure 3.15.

Description of the equilibrium for a substance at an air–water interface is found in Section 3.2.4.1.3 and expressed in terms of Henry's law. Under nonsteady-state conditions (i.e., in case of a net transport of mass across the interface), the volumetric rate of oxygen transfer is

$$F_{O_2} = K_L aO_2(S_{OS} - S_O) \tag{3.58}$$

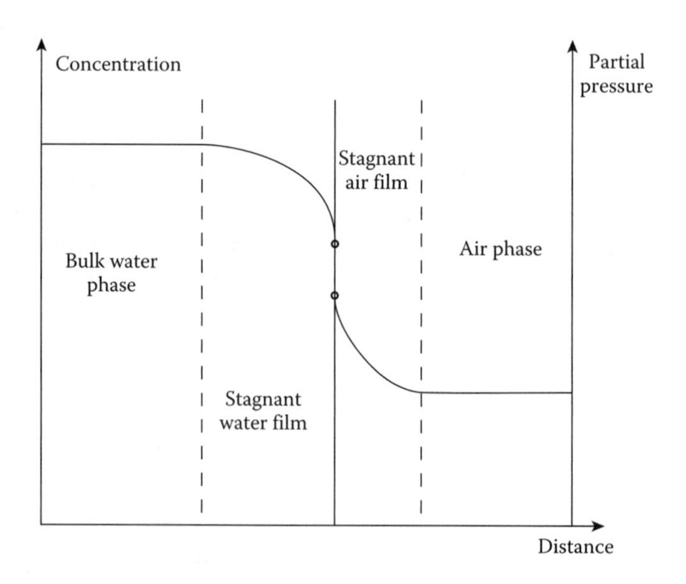

FIGURE 3.15 Equilibrium depicted for a volatile substance at the air–water interface according to the two-film theory.

where

F_{O_2} = volumetric rate of oxygen transfer (g m^{-3} s^{-1}, g m^{-3} h^{-1} or g m^{-3} d^{-1})
$K_La O_2$ = overall volumetric oxygen transfer coefficient (reaeration rate coefficient) (s^{-1}, h^{-1} or d^{-1})
S_{OS} = DO saturation concentration in bulk water phase (in equilibrium with the overlaying atmosphere; g m^{-3})
S_O = DO concentration in bulk water phase (g m^{-3}).

Based on Equation 3.58, the DO flux, J_{O_2}, can be determined

$$J_{O_2} = A\frac{F_{O_2}}{V} = \frac{F_{O_2}}{d_m}, \tag{3.59}$$

where

J_{O_2} = flux of DO (g m^{-2} s^{-1}, g m^{-2} h^{-1} or g m^{-2} d^{-1})
A = surface area of water (m^2)
V = water volume (m^3)
d_m = hydraulic mean depth (m).

A number of semiempirical expressions for estimation of the oxygen transfer coefficient, $K_La O_2$, have been proposed for water surfaces in lakes, streams, and sewer pipes (e.g., Krenkel and Orlob 1962; Tsivoglou and Neal 1976; Thibodeaux 1996; Hvitved-Jacobsen 2002). A summary of equations for prediction of the reaeration coefficient for lowland rivers is found in Cox (2003).

Although Equation 3.58 is semiempirical and formulated for transfer of oxygen, it is the basis for determination of transfer coefficients for other volatile substances. Equation 3.60 is in this case central because it relates a mass transfer coefficient for a substance to its diffusion coefficient and corresponding characteristics for oxygen. Knowledge on the oxygen transfer, in general based on experimental techniques, is crucial for determination of air–water transfer of other volatile substances. Constraints for the use of Equation 3.60 will be discussed in the following:

$$\frac{K_La}{K_La O_2} = \left(\frac{D}{D_{O_2}}\right)^n \tag{3.60}$$

where

D = molecular diffusion coefficient of a volatile substance in water (m^2 s^{-1}, m^2 h^{-1}, or m^2 d^{-1})
D_{O_2} = molecular diffusion coefficient of DO in water (m^2 s^{-1}, m^2 h^{-1}, or m^2 d^{-1})
n = a number between 0.5 and 1 (–).

The value of n depends on what theory is applied for the mass transfer across the air–water interface (Hvitved-Jacobsen 2002). From a pragmatic point of view, it appears that n is about 1 in a slow-flowing water body whereas it approaches 0.5 under turbulent conditions.

As stated in Section 3.4.1.2, molecular diffusion coefficients for small molecules are in the order of 10^{-5} cm^2 s^{-1}. According to the two-film theory it is also expected that volatile substances have mass transfer coefficients (K_La values) in the same order of magnitude. Due to different Henry's law constants, soluble molecular substances may still show different levels of volatility (cf. Equation 3.58).

Equation 3.60 refers to the water phase. The reason is that the resistance to transport of oxygen across the air–water interface according to Liss and Slater (1974), at least at a degree of about 98%, occurs in the water film and not in the air film. Therefore, Equation 3.60 is only valid for volatile substances with a Henry's constant, H, larger than 250 atm (mol fraction)$^{-1}$ corresponding to about 4.5 atm (mole)$^{-1}$.

As already mentioned, interfacial reactions may also occur between a solute and a solid substance. The phenomenon is, in this case, described as sorption (cf. Section 3.2.4.1.2). Interactions that in particular relates to particles in suspension are included in Sections 3.2.4.3.

3.5.2 Coagulation

In relation to urban wet weather pollution, coagulation is a process whereby suspended small particles, often colloidal particles, in water or wastewater aggregate. The cause of aggregation is the occurrence of both attractive and repulsive forces between the particles, occurring between charged particles in the suspended water phase but also between other types of surface solids (Figure 3.16).

Around 1940, the stability of coagulated particles was independently explained by two research groups, Deryagin–Landau and Verwey–Overbeek, and now referred to as the DLVO theory. In brief, the theory involves the occurrence of an equilibrium minimum point of energy caused by the attraction and repulsion forces, thereby creating a more or less stable distance between the colloidal particles. The so-called van der Waals attraction forces play an important role in this respect. These forces are the result of the formation of nonpermanent, induced dipoles within a molecule or within a colloidal particle. Dipole–dipole interactions result in intermolecular attraction forces (forces between molecules or particles). The van der Waals forces are important although they are weak compared with covalent bonds that typically form stable and strong bonds

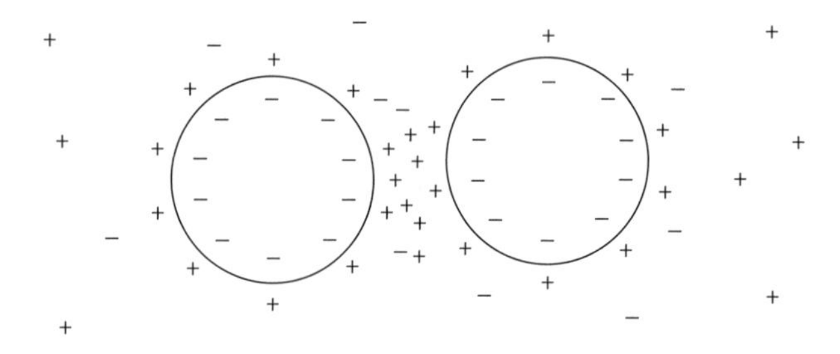

FIGURE 3.16 Schematic of coagulation showing two negatively charged particles surrounded by positive counter ions.

between atoms in molecules. According to the DLVO theory, the repulsive forces are caused by the configuration of ions around the particles forming a so-called diffuse double layer (i.e., a cloud of ions outside the surface of the particle, cf. Figure 3.16). The relatively high concentration of ions between two particles means that water molecules from the bulk water phase tend to penetrate into this space, thereby trying to push the particles apart. A number of external conditions affect the stability of coagulation (e.g., pH, ionic strength, temperature, and the degree of stirring).

In general, small particles (colloids) in wastewater and urban runoff have a negative charge and are repellent. Coagulation is therefore a process that is required as an initial step in the formation of larger particles from small ones (e.g., as a preprocess required for flocculation). A substance that enhances coagulation is called a coagulant. Often inorganic chemicals like iron or aluminum salts (e.g., ferric chloride and alum) are selected for that purpose but also organic polymers can act as coagulants. Generally, an increase in the ionic strength caused by increased concentrations of ions in solution may enhance coagulation (e.g., caused by the use of deicing agents on roads).

3.5.3 Flocculation

Flocculation is a process whereby electrically neutralized particles collide and agglomerate to form larger particles (flocs) that can be separated from a suspension by settling, flotation, or filtering. Within the area of urban drainage, flocculation is particularly important when dealing with treatment of the wet weather flow. However, it is also a process that is relevant when pollutants are discharged into receiving waters. Typically, flocculation requires coagulation. The use of polymers (polyelectrolyte compounds) may enhance flocculation by causing interparticle bridging.

The rate at which flocculation takes place is controlled by the mixing process, the input of energy in terms of the intensity of energy dissipation in the system. This input of energy affects the size and the manner in which the flocs are formed. In this respect, mixing in flocculation reactors is generally expressed in terms of a velocity gradient, G, also termed average shear rate:

$$G = \left(\frac{P}{\mu \, V} \right)^{0.5} \tag{3.61}$$

where

G = mean velocity gradient or average shear rate (s^{-1})
P = power dissipated (W) or (N m s^{-1})
μ = dynamic viscosity of water (kg m^{-1} s^{-1}) or (N s m^{-2}) or (Pa s), for water 10^{-3} kg m^{-1} s^{-1} at 15°C
V = water volume in the reactor (m^3)

The product of the velocity gradient, G, and the hydraulic retention time (HRT), *t*, in the system (reactor) is often used as a measure for the extent of flocculation and therefore a central design criterion. For successful flocculation, Fair and Geyer (1954) proposes values of G to range from 10 to 100 s^{-1} and $t\,G$ to be within the limits of 10^4 and 10^5.

The shear stress on the particles during mixing (flocculation) is proportional to the average shear rate (cf. Equation 3.53):

$$\tau = \mu G, \tag{3.62}$$

where

τ = shear stress (kg m^{-1} s^{-2}) or (N m^{-2}).

In order to ensure sufficient particle collisions with an appropriate power input, flocculation requires that the particles be subject to a shear stress. However, not to an extent that causes destruction of flocs that are already formed.

3.6 CHEMICAL PROCESSES

Although the heading of this section is Chemical Processes, it is to some extent relevant to include biological processes as well. The reason is that a specific chemical process is not necessarily limited to occurring in an abiotic environment but can also proceed in a biological system. A chemical process that is initiated by a biological system is sometimes referred to as a biochemical process. The point is that the basic characteristics of a chemical process in terms of stoichiometry (mass balance) and energy content of the constituents are fundamental qualities that are not determined by the environment—being biotic or abiotic. What can be different, however, are the pathways of the process and thereby the occurrence of intermediates.

3.6.1 Stoichiometry and Electron Transfer of Redox Processes

Chemical as well as biochemical processes that proceed within the area of urban drainage are reduction–oxidation (redox) processes. This section has therefore a wider perspective than just being related to the classification of system characteristics. The terms developed in the following are generally applicable.

Redox processes include two process steps: oxidation of a component (A) and reduction of another substance (B), see Figure 3.17. A redox process basically proceeds caused by transfer of electrons from the component that is oxidized (the electron donor) to the component that is reduced (the electron acceptor).

Balancing of redox processes in specific cases according to the concept shown in Figure 3.17 follows a number of steps that will not be described in this text. Details

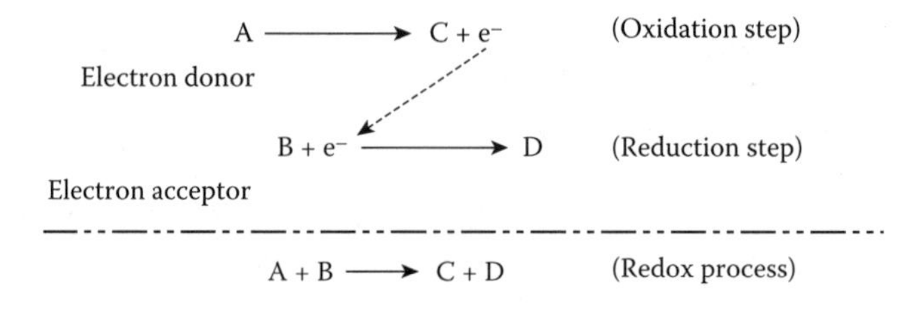

FIGURE 3.17 Oxidation and reduction steps and electron transfer involved in a redox process. The oxidation process and the reduction process are often named half-reactions.

can be found in Hvitved-Jacobsen (2002). The description is based on the definition of an oxidation level (OX) of the elements and the application hereof in terms of calculation of the electron equivalent (e-eq) number for a reduction or an oxidation step. The fundamental concept for the stoichiometry of any redox process is an electron mass balance (i.e., that the number of electrons produced by the oxidation step equals the number of electrons consumed by the reduction step).

The energy transformations that are related to the redox processes follow the general rules of thermodynamics (Atkins and de Paula 2002). The major thermodynamic energy function in this respect is the Gibbs's free energy, G, that defines the state and the potential for a change in state of a redox processes. The G is a measure of the driving force (the work-producing potential) of a redox process. For naturally occurring processes, the change in Gibbs's free energy (ΔG) is therefore negative, the redox process looses work-potential when it proceeds. At constant temperature and pressure, ΔG equals the maximum work that can be produced by the redox process. Therefore, ΔG is a measure of the tendency for the redox process to proceed:

$$\Delta G = \Delta H - T\Delta S. \tag{3.63}$$

where

G = Gibbs's free energy (kJ $mole^{-1}$)
H = enthalpy (kJ $mole^{-1}$)
T = temperature (K)
S = entropy (kJ $mole^{-1}$)

The Gibbs's free energy equals the work potential that is lost by transfer of the electrons from the oxidation to the reduction step. The difference in electron potential between these two half-reactions is therefore related to ΔG for the redox process:

$$\Delta G^{\circ\prime} = -nF\Delta E'_o = -nF(E'_{o,red} - E'_{o,ox}) \tag{3.64}$$

where

$G^{\circ\prime}$ = Gibbs's free energy at standard conditions (25°C, pH 7 and 1 atm)
n = number of electrons transferred according to the reaction scheme (–)
F = Faraday's constant equal to 96.48 (kJ $mole^{-1}$ V^{-1})
$\Delta E'_o$ = redox potential of electron acceptor, $E'_{o,red}$, minus redox potential of electron donor, $E'_{o,ox}$ (V)

The redox potentials, E'_o, for a large number of half-reactions relevant for both chemical and biochemical reactions can be found in CRC (2004). The redox potentials in order of magnitude for selected half-reactions are shown in Figure 3.18. For simplicity reasons, only the redox pairs and not the half-reactions are shown in this figure. As an example, the equilibrium for oxygen between the oxidized form (O_2) and the reduced form (H_2O) is

$$\frac{1}{4}O_2 + H^+ + e^- \leftrightarrows \frac{1}{2}H_2O \tag{3.65}$$

Redox potentials are of general importance for evaluation in the course of redox processes. The selection of redox pairs in Figure 3.18 is in the context of urban

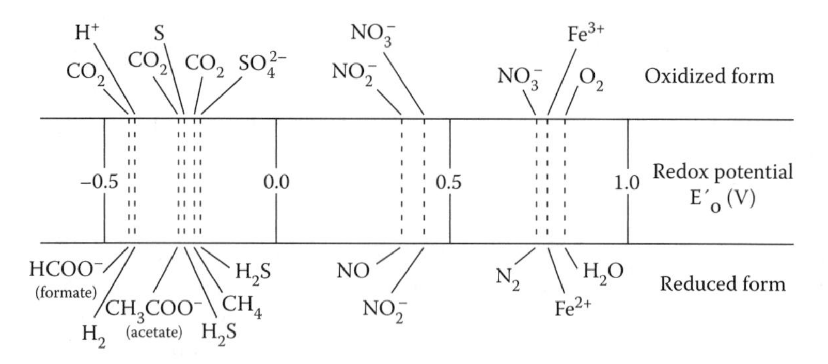

FIGURE 3.18 Order of magnitude of potentials for redox pairs (electron tower) of particular importance for microbial processes. The redox potentials are given at standard conditions, 25°C, pH 7, 1 atm and 1 molar concentration of relevant components (except for H^+ and OH^- for which components pH = 0 and pH = 14, respectively, are standard values). The Nernst equation can be used to account for different external conditions (cf. Equation 3.25).

TABLE 3.4
Electron Acceptors and Corresponding Conditions for Microbial Redox Processes

Process Conditions	Electron Acceptor	Examples
Aerobic	Dissolved oxygen is available	Conditions required by higher plants and animals
Anoxic	Nitrate (or nitrite) but not dissolved oxygen is available	Microorganisms that can respire using nitrate can exist (i.e., denitrifying organisms)
Anaerobic	Neither dissolved oxygen nor nitrate is available	Sulfate reducing microorganisms (sulfate is available) and fermenting microorganisms (biodegradable organic matter is available) can exist

drainage and of particular relevance when dealing with biochemical processes. The reason is that process conditions in terms of aerobic, anoxic, and anaerobic conditions are closely related to the availability of a specific electron acceptor and thereby to the redox potential. Table 3.4 shows this relation.

The process conditions referred to in Table 3.4 are relevant in both the technical part of the drainage system and in the receiving environment.

3.6.2 Process Kinetics

Process kinetics concern aspects related to the rate of reactions including the formulation of rate expressions. Process kinetics is therefore basic for any quantitative description of transformations that take place in a system. Such systems are both chemical and biological by nature. As previously stressed, there is basically no difference between chemical and biological processes. Briefly described, processes of living organisms just combine transport with transformations of chemical components, often in an interaction with a number of other processes and affected by catalysts (enzymes). The conditions for a process in terms of the kinetics are typically different in a biological system compared with the conditions that exist in a clean chemical environment.

Processes are either homogeneous or heterogeneous, they proceed within one phase or between phases, respectively. Since biological processes generally include transport of constituents across a membrane (cell wall), they are by nature basically heterogeneous. For practical reasons, however, microbial processes that proceed in suspension are typically considered homogeneous. Even a process that proceeds in a biofilm may, in terms of the impact onto the chemical composition in the water phase outside the film, be interpreted as homogeneous.

It is the tradition in chemistry and chemical engineering to describe process rate equations in terms of their reaction order that is defined as the sum of the exponents (empirical coefficients) of the concentrations in the rate expression. The following is a general rate expression of a water phase (homogeneous) reaction between two reacting components, A and B:

$$r = k\,(C_A)^a\,(C_B)^b \tag{3.66}$$

where

r = reaction rate; the rate of change of concentration, dC/dt

k = rate constant (unit depends on the expression of the reaction rate, r, in terms of a and b)

C_A and C_B = concentrations of the components A and B, respectively, typically in units of g m^{-3} or g $mole^{-1}$

$a + b$ = reaction order (–)

A large number of processes are relevant within the area of urban drainage. Such processes are chemical or biological and homogeneous or heterogeneous by nature. The simple distinction between rate expressions in terms of reaction order is still useful, however, not always the most relevant characteristic. Table 3.5 shows examples of some central process rate expressions.

The kinetics of chemical processes depend on a number of external conditions with temperature and pH as characteristic parameters. Furthermore, catalysts can increase the reaction rate.

The temperature dependence of the reaction rate can be expressed via its rate constant, k, (or μ_{max}) as follows:

$$k_2 = k_1\,\alpha^{(T2-T1)} \tag{3.67}$$

where

α = temperature coefficient (–)

T = temperature (°C)

Equation 3.67 is valid in a temperature interval where α can be considered constant.

TABLE 3.5
Characteristics of Selected Rate Expressions for Processes That Are Relevant within the Area of Urban Drainage

Rate Expression, r	Characteristics
$r = k\,C^0 = k$	Zero-order reaction (e.g., the maximum reaction rate at high and nonlimiting concentrations)
$r = dC/dt = k\,C^{0.5}$	Half-order reaction* (e.g., the reaction rate, surface flux, referring to the concentration, C, in the water phase outside the surface of a partly penetrated biofilm)
$r = dC/dt = k\,C$	First-order reaction; an often applied approach for transformations in a water phase
$r = \partial C_A/\partial t = k\,C_A\,C_B$	Second-order reaction; a reaction that can be used for description of sorption (cf. Equation 9.9) (Liu, Sansalone and Cartledge 2005)
$r = \partial X/\partial t = \mu_{max}\dfrac{S}{K+S}X$	Monod kinetics;* microbial biomass (X) growth rate (first-order description) limited by a substrate (S)

Note: K and k are constants.
*Typically relevant for biological processes.

Further details concerning kinetics of chemical and biological processes in terms of both basic theory and applications relevant for this text can be found in a number of books, Snoeyink and Jenkins (1980), Atkins and de Paula (2002), Hvitved-Jacobsen (2002), and Tchobanoglous, Burton, and Stensel (2003).

3.7 MICROBIOLOGICAL PROCESSES

When dealing with urban drainage, microbial processes are particularly important in relation to the following two phenomena:

- Transformation of pollutants (e.g., related to their biodegradability)
- DO mass balances and DO depletion (aerobic microbial activity)

DO depletion is closely related to the activity of microorganisms in terms of degradation of organic matter, see Section 6.3, and DO mass balances of environmental systems, see Section 9.3.1.8. Several subjects dealt with in Sections 3.2.2 and 3.6 are also directly related to these phenomena.

The activity of microorganisms is therefore relevant in two ways:

- Generally as influencing both transformation and accumulation of a number of substances and thereby the mass balances
- Particularly as being responsible for the transformation of biodegradable organic matter (under aerobic conditions)

The last mentioned aspect, the biodegradability of organic matter, leads to a discussion of how to characterize organic matter (cf. Section 3.2.1.1). The traditional way is to use the two parameters BOD (biological oxygen demand) and COD (chemical oxygen demand). In both cases, the organic matter is defined in terms of its ability to consume oxygen. A number of other parameters are also applied, particularly TOC (total organic carbon) and DOC. Each of the parameters has different characteristics and advantages: they are all well-known worldwide, they are relatively easily determined analytically, BOD refers to what is biodegradable (in e.g., five days as BOD_5), and COD is well suited when dealing with mass balances for organic matter (Schaarup-Jensen and Hvitved-Jacobsen 1991).

All the parameters mentioned are, however, bulk parameters that give no answer on details concerning the active biomass concentration or the information on the biodegradability. Further details on the composition of organic matter may therefore be required in terms of the following classification:

- Microbial active (heterotrophic) biomass concentration
- Substrate concentration (soluble and particulate)
- Biologically nonactive part of organic matter (inert organic materials)

A classification of wastewater according to this concept and applied in the activated sludge model (ASM) is described in Henze et al. (2000). A corresponding concept for wastewater applied under sewer conditions is found in Hvitved-Jacobsen (2002). Determination of the OUR is the analytical central tool for measurement of the biodegradability and thereby the analytical basis for more detailed classification of organic matter (Vollertsen and Hvitved-Jacobsen 2001).

The classification of organic matter is the basis for the description of central microbial heterotrophic transformations (cf. Hvitved-Jacobsen 2002):

- Growth of biomass (aerobic, anoxic, and anaerobic), utilization of substrate for growth of biomass coupled with utilization of the substrate for energy purposes
- Maintenance energy requirement, utilization of a substrate for energy purposes without biomass growth
- Hydrolysis (aerobic, anoxic, and anaerobic), a breakdown process for relatively large molecules to smaller molecules
- Fermentation, oxidation of one substrate (organic matter) coupled with the reduction of another

In addition to these processes that are central for the carbon cycle, the microbial initiated transformations of nitrogen and phosphorus play a significant role when dealing with urban drainage pollution.

Briefly, the microbial processes for nitrogen are

- N-assimilation: microbial uptake of nitrogen, particularly as ammonia
- Ammonification: different processes for transformation of organic nitrogen intermediate products ending with the formation of ammonia

- Nitrification: the formation of nitrate from ammonia via nitrite
- Denitrification: activity of anoxic organisms whereby nitrate (or nitrite) is reduced to N_2 (gaseous nitrogen)
- N-fixation: a process whereby N_2 is the nitrogen source for production of organic nitrogen products

Concerning phosphorus, P-uptake in organisms as orthophosphate and polyphosphate may take place. Contrary, the mineralization of organic bound phosphorus may result in the release of inorganic P-species.

Several textbooks deal with basic aspects of microbiological processes (e.g., Stanier et al. 1986).

3.8 PLANTS AND EUTROPHICATION

It is not a central objective of this text on urban wet weather pollution to deal with growth of aquatic plants. However, the subject becomes relevant in this context in particularly two related cases:

- Nutrient discharges from both combined sewer overflows (CSOs) and SWR into receiving waters can result in excessive growth of both algae and macrophytes, the higher (multi cell) plants. The complex phenomenon related to this type of impact is eutrophication (cf. Section 6.4.2).
- Eutrophication can cause unpleasant conditions and malfunctioning of detention ponds and constructed wetlands for management of stormwater runoff. This fact can also become a delimiting factor for the use of such facilities and that at the same time may serve as a recreational element in urban areas.

The negative impact of nutrients originating from wet weather sources can therefore be a major reason why such discharges should be controlled, see Chapter 9.

Aquatic plants and algae include a great number of species. A grouping of such species that is relevant to surface waters receiving wet weather discharges from urban areas and roads is

- Algae suspended in the water phase
- Rooted plants
- Free floating plants including both algae and macrophytes

It is often seen that these species adapt to specific environmental conditions. Some species are mainly found in streams, other types in ponds. Furthermore, the salt content (salinity) in the receiving water is of great importance.

The rooted macrophytes consist of species that are both submerged and emergent. The emergent aquatic plants perform a major part of their photosynthesis, their production or organic matter from CO_2 in the air whereas the submerged plants use the dissolved and ionic inorganic carbon, particularly HCO_3^-. The submerged plants can therefore affect the alkalinity and the pH of the water phase. They also directly interact with the water phase in terms of a potential for uptake of constituents like heavy metals and organic micropollutants.

Further details on the effect of nutrient discharges and eutrophication are dealt with in Section 6.4 and particularly in Section 6.4.2.

3.9 FINAL REMARKS

Several basic characteristics of urban wet weather pollution in terms of its event-based nature and the stochastic and dynamic phenomena are quite unique compared with the different types of dry weather pollution. The methods for analysis of the wet weather phenomena and the corresponding techniques and mitigation methods for solving adverse environmental impacts are developed under these constraints. Urban wet weather pollution is therefore not just a topic under something more superior, it is an independent discipline in environmental engineering studies.

The physical, chemical, and biological phenomena dealt with in this chapter are an integral part and play a central role for analysis of and solutions to problems related to urban drainage. It is important to stress that under external conditions, these physical, chemical, and biological phenomena are valid irrespective of the system they refer to. These phenomena are in general valid within the area of urban wet weather pollution. Observing the objectives of the text, the subjects addressed in this chapter, and the way in which they are presented, are, however, formulated in the context of and in accordance with the nature and the needs of the wet weather phenomena.

REFERENCES

Allison, J. D., D. S. Brown, and K. J. Novo-Grodae. 1991. *MINTEQA2 and PRODEFEA2: A geochemical assessment model for environmental systems*. Version 3.0 users manual. U.S. EPA/600/3-91/021.

APHA-AWWA-WEF. 1995. *Standard methods for the examination of water and wastewater*, 19th ed. Washington, DC: APHA (American Public Health Association), AWWA (American Water Works Association), WEF (Water Environment Federation).

Arnbjerg-Nielsen, K., T. Hvitved-Jacobsen, A. Ledin, K. Auffarth, P. S. Mikkelsen, A. Baun, and J. Kjølholt. 2002. Bearbejdning af målinger af regnbetingede udledninger af NPo og miljøfremmede stoffer fra fællessystemer i forbindelse med NOVA 2003 [Combined sewer overflow loads of conventional pollutants, heavy metals and organic micropollutants]. Environmental Project Series No. 701, Danish National Agency of Environmental Protection.

Ashley, R. M., and W. Dabrowski. 1995. Dry and storm weather transport of coliforms and faecal streptococci in combined sewage. *Water Science and Technology* 31 (7): 311–20.

Ashley, R. M., M. Verbanck, T. Hvitved-Jacobsen, and J.-L. Bertrand-Krajewski, eds. 2004. *Solids in sewers*. IWA (International Water Association) Scientific & Technical Report No. 14.

Atkins, P., and J. de Paula. 2002. *Physical chemistry*. Oxford, UK: Oxford University Press.

Burton, G. A., and R. E. Pitt. 2002. *Stormwater effects handbook: A toolbox for watershed managers, scientists, and engineers*. Boca Raton, FL: Lewis Publishers/CRC Press.

Connell, D. W. 2006. *Basic concepts of environmental chemistry*, 2nd ed. Boca Raton, FL: CRC Press.

Cox, B. A. 2003. A review of dissolved oxygen modelling techniques for lowland rivers. *The Science of the Total Environment* 314–16 (October): 303–34.

CRC. 2004. *Handbook of chemistry and physics*, 85th ed. Boca Raton, FL: CRC Press.

Debo, T. N., and A. J. Reese. 2003. *Municipal stormwater management.* Boca Raton, FL: Lewis Publishers/CRC Press.

Duke, C. V. A., and C. D. Williams. 2008. *Chemistry for environmental and earth sciences.* Boca Raton, FL: CRC Press.

Eckenfelder, W. W. 1989. *Industrial water pollution control.* New York: McGraw-Hill.

Ellis, J. B., and W. Yu. 1995. Bacteriology of urban runoff: The combined sewer as a bacterial reactor and generator. *Water Science and Technology* 31 (7): 303–10.

Eriksson, E., A. Baun, P. S. Mikkelsen, and A. Ledin. 2005. Chemical hazard identification and assessment tool for evaluation of stormwater priority pollutants. *Water Science and Technology* 51 (2): 47–55.

Fair, G. M., and J. C. Geyer. 1954. *Water supply and waste-water disposal.* New York: John Wiley & Sons, Inc.

Field, R. 1993. Use of coliform as an indicator of pathogens in storm-generated flows. Proceedings of the 6th International Conference on Urban Storm Drainage, Niagara Falls, Ontario, Canada, September 12–17, 78–84.

Gregory, J. 2006. *Particles in water: Properties and processes.* Boca Raton, FL: CRC Press/Taylor & Francis.

Hahn, H., E. Hoffmann, and M. Schäfer, 1999. Managing residuals from wastewater treatment for priority pollutants. *European Water Management* 2 (1): 49–56.

Henze, M., W. Gujer, T. Mino, and M. V. Loosdrecht. 2000. Activated sludge models ASM1, ASM2, ASM2d and ASM3. Scientific and Technical Report No. 9, IWA (International Water Association).

Hvitved-Jacobsen, T. 2002. *Sewer processes: Microbial and chemical process engineering of sewer networks*. Boca Raton, FL: CRC Press.

Krenkel, P. A., and G. T. Orlob. 1962. Turbulent diffusion and the reaeration coefficient. *J. Sanitary Engineering Division, ASCE (American Society of Civil Engineers)* 88 (SA2): 53–84.

Lewis, W. K., and W. G. Whitman. 1924. Principles of gas absorption. *Journal of Industrial and Engineering Chemistry* 16 (12): 1215.

Liss, P. S., and P. G. Slater. 1974. Flux of gases across the air–sea interface. *Nature* 247:181–84.

Liu, D., J. J. Sansalone, and F. K. Cartledge. 2005. Adsorption kinetics for urban rainfall-runoff metals by composite oxide-coated polymeric media. *Journal of Environmental Engineering* 131 (8): 1168–77.

Mays, L. W. ed. 2001. *Stormwater collection systems design handbook.* New York: McGraw-Hill.

Mehta, A. J. 1988. Laboratory studies on cohesive sediment deposition and erosion. In *Physical processes in estuaries*, eds. J. Dronkers and W. Van Leussen, 427–445. New York: Springer-Verlag.

Nazaroff, W. W., and L. Alvarez-Cohen. 2001. *Environmental engineering science*. New York: John Wiley & Sons, Inc.

Ono, Y., I. Somiya, T. Kawaguchi, and S. Mohri. 1996. Evaluation of toxic substances in effluents from a wastewater treatment plant. *Desalination* 106 (1–3): 255–61.

Pankow, J. F. 1991. *Aquatic chemistry concepts*. Chelsea, MI: Lewis Publishers.

Pitt, R., and M. Bozeman. 1982. *Sources of urban runoff pollution and its effects on an urban creek,* USEPA-600/S2-82-090. Cincinnati, OH: U.S. Environmental Protection Agency.

Pourbaix, M. 1963. *Collection of electrochemical equilibrium.* Paris: Gauthier-Villars.

Schaarup-Jensen, K., and T. Hvitved-Jacobsen. 1991. Simulation of dissolved oxygen depletion in streams receiving combined sewer overflows. In *New technologies in urban drainage*, ed. C. Maksimovic, 273–82. New York: Elsevier Applied Science

Shields, A. 1936. Anwendung der Ähnlichkeitsmechanik und der Turbulenzforschung auf die Geschiebebewegung [Application of hydraulic model concepts and turbulence research on drag force induced movement]. Berlin: *Mitteilungen der Preußischen Versuchsanstalt für Wasserbau und Schiffbau*, H.26 (in German).
Snoeyink, V. L., and D. Jenkins. 1980. *Water chemistry.* New York: John Wiley & Sons, Inc.
Stanier, R. Y., J. L. Ingraham, M. L. Wheels, and P. R. Painter. 1986. *The microbial world.* Englewood Cliffs, NJ: Prentice Hall.
Stumm, W., and J. J. Morgan. 1996. *Aquatic chemistry: Chemical equilibria and rates in natural waters*, 3rd ed. New York: John Wiley & Sons
Tchobanoglous, G., F. L. Burton, and H. D. Stensel. 2003. *Wastewater engineering: Treatment and reuse*. New York: McGraw-Hill.
Thevenot, D. R. ed. 2008. *Daywater: An adaptive decision support system for urban stormwater management.* IWA (International Water Association).
Thibodeaux, L. J. 1996. *Environmental chemodynamics.* New York: John Wiley & Sons, Inc.
Trapp, S., and M. Matthies. 1998. *Chemodynamics and environmental modeling: An introduction.* New York: Springer-Verlag.
Tsivoglou, E. C., and L. A. Neal. 1976. Tracer measurement of reaeration: III. Predicting the reaeration capacity of inland streams. *Journal of Water Pollution Control Federation* 48 (12): 2669–89.
Vollertsen, J., and T. Hvitved-Jacobsen. 2001. Biodegradability of wastewater: A method for COD-fractionation. *Water Science and Technology* 45 (3): 25–34.
Wanielista, M. 1990. *Hydrology and water quantity control.* New York: John Wiley & Sons, Inc.

4 Stormwater Runoff: Sources, Transport, and Loads of Pollutants

Chapter 4 aims at quantification of generation, transport, and loads of constituents in stormwater runoff originating from urban impervious surfaces and roads. This chapter will deal with pollutants in terms of their origin and transport and their characteristics at the point where they are discharged and enter into the environment or treatment facilities. It is considered important to focus on methods for estimation of the loads in this respect. This chapter will also give examples of what are considered characteristic levels of the pollutants well knowing that such levels are subject to a considerable variability (cf. Section 2.4). Last, it is important to include expressions and formulations that can be applied for modeling the pollutant loads. The subjects dealt with here are a continuation of the more general descriptions and concepts dealt with in Chapters 2 and 3.

4.1 ATMOSPHERIC CONSTITUENTS AND DEPOSITION

The atmosphere serves as a transport—and to some extent also transformation—medium for pollutants prior to their occurrence on urban or road surfaces and in the runoff water. Transport of the pollutants takes place in the gas phase as well as in and associated with the different phases of H_2O that might occur in the atmosphere. The pollutants themselves might occur as gases but are typically associated with different types of particles. Transport of the pollutants is relevant over long distances but also takes place over a few meters and can even be relevant in the magnitude of a few centimeters.

4.1.1 Basic Characteristics of the Atmosphere

Excluding H_2O as vapor, liquid, or snow, the main components that constitute the (dry) atmosphere are nitrogen (N_2, 78.1%), oxygen (O_2, 20.9%), and argon (Ar, 0.9%). Furthermore, a number of components are found—often in relatively low concentrations—that affect the environment in different ways, however. In terms of their concentration, carbon dioxide (CO_2, 380–385 ppm), neon (Ne, 18 ppm), helium (He, 5 ppm), and methane (CH_4, 1–2 ppm) are following the most important gases that, in general, occur in the atmosphere. Both CO_2 and CH_4 are greenhouse gases with an impact on the heat balance of the earth.

In addition to these gases, a number of substances, of which several can be characterized as pollutants, occur in the atmosphere in solid, liquid, or gaseous forms. These components can originate from natural processes as well as from man-influenced sources (i.e., anthropogenic activities). Combustion products from fuel and emissions from industry are examples that contribute to the level of atmospheric pollutants. Under both dry and wet weather conditions these substances can be transferred to urban and road surfaces for being a part of the pollutant load of runoff.

The potential sources for the pollutants in the atmosphere are numerous. A pragmatic way to distinguish between the origins is

- *Stationary point sources*
  Emissions from both local and regional sources (e.g., industries, power plants, and buildings)
  Corrosion products (e.g., from buildings and infrastructure)
- *Surface related sources*
  Particles (e.g., salt) from the sea
  Soil particles
  Materials from corrosion and wear of surfaces (e.g., roofs and roads)
  Chemicals (e.g., deicing agents and chemicals for weed control)
- *Mobile sources*
  Vehicles (e.g., corrosion, wear, combustion, and spills)
- *Specific events*
  Accidents (e.g., vehicle collision and fire)
  Inappropriate handling of materials (e.g., spill of wastes from industries)

The list indirectly shows that both long- and short-distance transport can occur.

A different overview of sources might show the potential origin of the different pollutants in the atmosphere (cf. Table 4.1).

Table 4.1 shows what main pollutants might occur in the atmosphere, not necessarily those associated with the most adverse effects (e.g., other heavy metals and volatile organic compounds, VOCs, belonging to the group of micropollutants).

The pollutants in the atmosphere can be associated with both particles and be absorbed in droplets like clouds and rain or associated with snow particles. Atmospheric particles include both solid and liquid (aerosol) particles that are suspended in the air. Such particles typically range in diameter between 2 nm and 10–100 μm and are more or less subject to deposition and association with urban and road dust (cf. Section 4.1.3). Major particulate constituents in the lower atmosphere are mineral dust and inorganic salts (e.g., sulfates, nitrates, carbonates, and chlorides). In near-sea areas, chloride can temporarily be a dominating airborne substance.

4.1.2 Acid Rain and Constituents Associated with the Rain

Although acid rain is not directly a subject of this text, the acidification of the rain is important in terms of the quality of urban runoff. First of all, it may directly exert an effect onto the alkalinity and pH value of the runoff. Secondly, a number of constituents (e.g., sulfide, nitrate, and chloride) are transferred from the atmosphere to

TABLE 4.1
Main Types of Pollutants in the Atmosphere and Their Origin

Pollutants	Sources
Particles	Traffic, particularly diesel engines Dust from streets and buildings Industry Marine particles (salts)
PAH, polyaromatic hydrocarbons	Wood-burning stoves Diesel engines
Nitrogen dioxide, NO_2	Gasoline and diesel engines Power plants Secondary pollution
Ozone, O_3	Secondary pollution
Sulfur dioxide, SO_2	Combustion of coal and oil
Lead, Pb	Leaded gasoline (in general considerably reduced compared with previous use)
Gasoline	Cars (evaporation and spills)
Carbon monoxide, CO	Cars

Note: The table also shows major secondary sources for pollutants (i.e., pollutants that are produced via chemical transformations in the atmosphere).

the runoff in case of acid rain. Indirectly the acidification can affect the solubility and speciation of heavy metals and thereby their effect onto the environment (cf. Sections 3.2.5 and 6.5.1).

As depicted in Example 3.1, the pH value of clean rainwater is in the case of equilibrium between $CO_2(g)$ and $CO_2(aq)$ about 5.6. Volatile products emitted into the atmosphere from the combustion of fossil fuel for electric power and heat generation and gasoline or diesel combustion in automotive engines can further lower this value. A number of gases in the atmosphere that have a natural origin from volcanic activities can also contribute to acidification of the rain. However, in this context, the anthropogenic components are focused on. The following three strong acids are in this respect the dominating components that can affect the pH value of rainwater:

- *Sulfuric Acid, H_2SO_4*
  Sulfuric acid in the atmosphere has its origin in sulfur (S) from fossil fuels. Sulfur dioxide (SO_2) produced as a result of combustion is further oxidized to SO_3 that reacts with water to form H_2SO_4.
- *Nitric Acid, HNO_3*
  To some extent, combustion of fossil fuels results in the formation of NO and NO_2 (NO_x). These substances can produce HNO_3.
- *Hydrochloric Acid, HCl*
  Chlorine containing substances like polyvinyl chloride (PVC) can, by incineration, result in the formation of HCl.

TABLE 4.2
Typical Atmospheric Concentration Levels for Gases Associated with Acid Rain

Constituent	Typical Concentrations in Nonpolluted Air (ppb)	Typical Concentrations in Polluted Air (ppb)
SO_2	0.2–10	10–200
NO_x	0.5–5	5–20
NH_3	5–20	5–20

A net production of H^+ originating from these acids may lead to the phenomenon known as acid rain in terms of both dry and wet deposition (cf. Section 4.1.4). In case of acid rain (also including snow) the alkalinity of the runoff is considerably reduced resulting in pH values that can easily fall to between 4.0 and 4.5. The SO_2, NO_x, and HNO_3 are examples of dry deposited components whereas H^+, SO_4^{2-}, NO_3^-, and NH_4^+ are characteristic for wet deposition. A number of cations dominated by sodium, calcium, magnesium, and potassium also occur in the precipitation.

The occurrence of components affecting acidification of the rain can be quantified in different ways. A common way is to produce emission estimates for selected pollutants (e.g., SO_2, NO_x, and NH_3). Such estimates can be expressed for the different sources, often at a national level and may serve as the objective of a background for implementation of control. Table 4.2 shows levels of concentration for selected gases associated with acid rain. As seen from this table, the variability is considerable.

The relative occurrence of the strong acids in the atmosphere varies considerably with sulfuric acid being typically dominating followed by nitric acid. These acids in the atmosphere can be transferred to the rainwater or be neutralized by bases like carbonate containing particles and ammonia in the atmosphere or at the ground. The acid–base reactions and their impact on both the terrestrial and the aquatic ecosystems are site dependent and rather complex (cf. Stumm and Morgan 1996).

In addition to the effects that are previously mentioned in this section and that are directly related to the quality of the runoff water, the following impacts of acid rain are expressed in general terms:

- Reduced alkalinity and corresponding fall of pH in surface waters. A pH < about 5 may result in waters without fish species and increased concentrations of dissolved metals (e.g., aluminum).
- Negative effects on growth of trees.
- Corrosion of materials, (alkaline) building materials.
- Effect on the quality of groundwater in terms of increased levels of bioavailable (dissolved) heavy metals.

4.1.3 Characteristics of Atmospheric Particles

In addition to being a source for pollutants in the stormwater runoff, particles in the urban atmosphere constitute a major health risk for humans. The direct impact and the sources of the particles are closely related to their size characteristics. Particularly the small particles are important in this respect subdivided into the following three groups:

- *Particles < 0.1 μm*
  These particles are mainly generated at high temperatures (e.g., in combustion engines and particularly in diesel engines). The particles mainly exist in the urban atmosphere close to the site where they are generated. They will typically, by very fast coagulation, produce larger particles and they are not, as small particles, transported over long distances. They may move to the lungs and from there into the blood and probably affect the white blood cells.
- *Particles 0.1–2.5 μm*
  Such particles have different origin and contrary to the very small particles, they are transported over long distances. They can be adsorbed in the lungs.
- *Particles 2.5–10 μm*
  Particles belonging to this group are typically whirled up from the soil and originate from wear of road materials and automobile tires. These particles slowly settle, however, they may when inhaled affect the upper part of the lungs.

All types of particles have a potential to end up in the stormwater runoff.

4.1.4 Deposition of Pollutants

Particle bound pollutants from the atmosphere are deposited on urban surfaces and roads and they contribute to the content of pollutants in the surface runoff. The bulk of the deposition consists of the following three processes:

- *Dry deposition*
  Dry deposition takes place during dry weather conditions. Sedimentation as well as adsorption processes are responsible for the accumulation of substances at the urban surfaces.
- *Wet deposition*
  Wet deposition takes place by absorption of particles in raindrops during their transport through the atmosphere. This type of deposition also includes pollutant removal from the atmosphere by snow particles.
- *Occult deposition*
  This type of pollutant removal takes place by absorption of pollutants in the small droplets in clouds and fog.

In addition to the deposition of particles, gaseous components like carbon dioxide, ammonia, and acids can be absorbed in raindrops and cloud droplets (cf. Section 4.1.2).

As mentioned in Section 4.1.3, pollutants in the atmosphere originate from both locally and regionally distributed sources and the atmospheric transport takes place over both long and short distances. In general, it is expected that the deposition of pollutants in cities is larger than what is observed for the rural areas as exemplified in Table 4.3.

Table 4.3 demonstrates that there is a general tendency of increased atmospheric pollution and corresponding deposition rates for the measured heavy metals with

TABLE 4.3
Characteristic Heavy Metal Bulk Deposition Rates (mg m^{-2} yr^{-1}) and Atmospheric Concentrations (ng m^{-3}) for Rural and Urban Sites in Denmark

Substance	Rural Sites*	Suburban Areas	Cities	Center of Big Cities
Copper, Cu				
Deposition	1.4	3.3	6.2	12.7
Atmospheric concentration	3.5	10.0	15.0	73.0
Lead, Pb				
Deposition	8	22	48	80
Atmospheric concentration	45	93	221	625
Zinc, Zn				
Deposition	15	41	88	118
Atmospheric concentration	35	64	70	176

Source: Data from Hovmand, M. F., Atmosfaerisk metalnedfald i Danmark [Atmospheric deposition of metals in Denmark]. PhD dissertation, Technical University of Denmark, 1980; Ellermann, T., Andersen, H. V., Bossi, R., Christensen, J., Frohn, L. M., Geels, C., Kemp, K., Lofstrom, P., Mogensen, B. B., and Monies, C., *Atmosfaerisk deposition, 2006* (Atmospheric deposition 2006), Danmarks Miljøundersøgelser (NERI, National Environmental Research Institute), Technical Report No. 645, 2007.

*The measurements at the rural sites (6–7 stations) have been continued. During the period 2000–2006, the bulk deposition rates for Cu, Pb, and Zn have been reduced to about 0.8, 1.0, and 8 mg m^{-2} yr^{-1}, respectively, for the rural sites. It is assumed that the substantial reduction in the deposition rate for Pb gradually seen during the past 30 years is mainly caused by leaded gasoline being replaced by unleaded gasoline.

increased population density of the site. It is also interesting to notice that stepwise there is approximately a doubling in the deposition rates when moving from a rural area to the center of a big city.

4.2 SNOW EVENTS: POLLUTANT ACCUMULATION AND RELEASE

Precipitation in regions of cold climate, including the temperate climate zones, may during winter periods occur as snowfall. The resulting snowpack is typically accumulated close to the urban surfaces and roads. The details of the pollutant accumulating processes in snow are basically not known. However, it is well known that the snow becomes contaminated with pollutants during both snowfall and during the snow accumulation period. Due to heating, less efficient motor vehicle operation, and use of deicing agents, the contribution of pollutants from the different diffuse sources is often higher in winter than during snow-free periods. Furthermore, it is evident that both soluble and particulate pollutants are adsorbed during the snow accumulation period.

It is likely that soluble and hydrophilic pollutants in snow are first released during snowmelt leaving hydrophobic substances until the end of the melting period. Furthermore, the more coarse particles may to some extent remain on the surfaces after melting is complete. The different chemical and physicochemical processes and the associated transport processes during snow accumulation and melting affect the release of pollutants to the runoff water from the snowmelt. Speciation of heavy metals influenced by chloride used as a deicing agent may affect their solubility and toxicity whereas polycyclic aromatic hydrocarbons (PAHs) in general remain in the snow until the very end of melting (Marsalek et al. 2003). Therefore, it is expected that the first flush of soluble and toxic hydrophilic pollutants occur during snowmelt whereas a last flush of hydrophobic and solid species occur toward the end of the snowmelt event. The quality of the runoff from snowmelt therefore depends on the stage of melting and is not identical to what will be determined when analyzing samples of the snowpack.

Increased impervious surfaces caused by frozen soil areas adjacent to a paved catchment or road may also contribute to the runoff from snowmelt. The runoff coefficient may therefore, depending on the structure and slope of surrounding areas, turn out to be quite different from what is the case from runoff of rainwater in the very same area. Such complex aspects must be judged in each specific case.

The typically high concentrations of pollutants in snowmelt compared with the runoff from rain events is illustrated in Section 4.6.2 (compare the results in Tables 4.8 and 4.9). Such quality characteristics are subject to a considerable variability, which is also the case for the dissolved and particle bound fractions in the snowmelt (Viklander 1999; Marsalek et al. 2003). A comparison of pollutant concentrations in snowmelt and stormwater runoff from an interstate highway site in the United States showed concentration levels that were 4–6 times larger in the snowmelt (FHWA 1987). Other investigations show less increased concentrations in snowmelt, however, still about two times higher values than those for stormwater runoff.

4.3 STATIONARY AND MOBILE SOURCES FOR POLLUTANTS

The sources for pollutants in stormwater runoff are legion (cf. Sections 2.3 and 4.1.1). In general, each site has its specific and dominating contributions. In a broad context, the following phenomena affect the occurrence of pollutants in the runoff from urban areas and roads:

- Climate conditions in terms of rainfall pattern, occurrence of snow during winter, wind, and temperature
- Land use (e.g., corresponding to residential, commercial, and industrial catchments)
- Maintenance and management characteristics (e.g., street sweeping, use of deicing agents, use of herbicides, and procedures for collection and management of urban solid waste)
- Material use in construction (buildings) and vehicles (e.g., copper roofs, brake linings, and zinc-coated lamp standards and highway structures)
- Spills in urban areas and on roads (e.g., type, frequency, and procedures for maintenance)
- Traffic volume, speed, and traffic culture
- Vehicles (e.g., their age and maintenance)
- Laws and regulations

For practical purposes, these nonpoint source characteristics can be dealt with in different ways. As an example, the traffic and vehicle contributions with PAH indirectly include potential contributions from sources like diesel vehicle exhaust and abrasion of tire and asphalt pavement (Pengchai, Nakajima, and Furumai 2005).

The variability that exists for the pollutant parameters will in general not directly allow a quantitative use of the characteristics mentioned above. Qualitatively, it is possible to select realistic parameter levels for the pollutant loads with the information on such contributions by comparison with pollutant loads from similar catchments. As will be illustrated in Section 4.5.5, it depends on the actual case to some extent possible to relate pollutant loads to the traffic, in terms of average daily traffic (ADT) and the land use.

Based on a large number of investigations in different countries, a number of specific sources that are considered important for a given pollutant are outlined in Table 4.4. Since local conditions play a significant role for what is important, the table will never turn out to be a list for quantitative use. Furthermore, the use of new materials and products and the implementation of source control measures make the list subject to changes over time.

Table 4.4 is only a brief list that illustrates the source-related contributions for selected pollutants. A large number of other metals and organic micropollutants are relevant in the wet weather flows. As an example, the wear of automobile brakes also contributes with trace metals like antimony (Sb) and molybdenum (Mo). The number of organic micropollutants transported with the runoff is numerous (cf. Section 3.2.1.4).

TABLE 4.4
Outline of Sources Considered of Particular Importance for a Given Pollutant

Pollutant	Sources
Cadmium	Tire wear
Chromium	Metal plating, vehicles (engine parts and brakes)
Chloride	Deicing salt
Copper	Vehicles (engine parts and brakes), fungicides, copper roofs
Dioxins	Combustion processes
Lead	Leaded gasoline, tire wear, lubricating oil, brake wear
Nickel	Diesel fuel, lubricating oil
Nitrogen	Fertilizer, atmosphere
Oil and grease	Vehicle spills
PAHs	Incomplete combustion; particles related to (diesel) vehicle exhaust; atmospheric fallout (stationary combustion); abrasion of tire and asphalt pavement
Particles (TSS)	Several different sources: pavement wear, construction sites, atmosphere
Pesticides and herbicides	Weed and pest control
Phosphorus	Fertilizer, atmospheric fallout
Zinc	Lamp standards, road structures, tire wear

TABLE 4.5
Examples of Size Distribution, Organic Matter and Heavy Metal Contents in Street Sweeping Sediments Originating from Two Cities in France

Origin of Street Sweeping Sediments	Size Distribution, d_{10}, d_{50}, d_{90} (μm)	Size < 63 μm (%)	Organic Matter (%)	Copper (μg g^{-1})	Lead (μg g^{-1})	Zinc (μg g^{-1})
Bordeaux	74, 901, 4000	9	6.1	65	122	281
Lille	7, 231, 3610	35	5.9	97	106	356
Dutch target value			–	36	85	140

Source: Data from Pétavy, F., Ruban, V., Conil, P., and Viau, J. Y., Reduction of sediment micro-pollution by means of a pilot plant. Proceedings of the 5th International Conference on Sewer Processes and Networks, Delft, the Netherlands, August 29–31, 297–306, 2007.

Note: The pollutant contents are compared with Dutch standards for polluted soils. (From Spierenburg, A., and Demanze, C., Soil pollution: Comparison and Application of the Dutch Standards, *Environnement et Technique,* 146, 79–81, 1995.)

An essential part of the pollutants in the runoff from urban surfaces and roads has temporarily been accumulated in street dust. In general, the small and organic-rich particles typically show a relatively high concentration of pollutants. As an example, Zhao et al. (2008) found that street dust particles of a diameter < 63 μm had higher contents of PAHs than fractions of higher particle size. Table 4.5 shows particle size distribution and selected chemical characteristics of sediments from street sweeping (Pétavy et al. 2007).

As seen from Table 4.5, the variability of both size distribution and pollutant content is, as expected, rather high. Corresponding size distributions and pollutant contents for sediments from wet detention ponds receiving urban runoff are shown in Table 9.4.

4.4 DRAINAGE SYSTEMS: COLLECTION AND TRANSPORT OF RUNOFF WATER

A drainage system for urban and road runoff water operates between the point where the rain hits the surface and the point where it is either discharged into a receiving water body or into a facility for its detention or treatment. The function of the drainage system is to collect the runoff water and transport it in an efficient and safe way. The technical installations that collect and transport the runoff vary from one region to another and even from one site to another determined by both the local climate conditions and the engineering tradition. Such facilities should observe the basic hydraulic characteristics by taking into account the variability of the runoff flow and be able to manage the flows under extreme conditions.

The hydrology and the hydraulics related to the flow of runoff water from urban areas and roads are not core subjects of this text. Basic characteristics in this respect are dealt with in Section 3.4.1. The hydraulic performance in terms of sedimentation and deposition of pollutants and erosion and resuspension of materials in the drainage system are subjects of Section 3.4.2. Types and performance of drainage systems for urban and road runoff are dealt with next in terms of transport characteristics of the constituents. Specific construction details of the transport systems are not focused on.

4.4.1 Collection and Transport Systems for Urban and Road Runoff

The basic characteristics of the drainage networks are dealt with in Section 1.4.1. At the very top of the system, a characteristic feature is that the collection of the runoff water takes place at a surface (roofs, paved urban areas, and roads) from where it is diverted into an inlet structure. Such structures can be rather simple, directly leading the stormwater into a manhole or structures like gully pots and catch basins that to some extent also act as sediment traps.

The transport systems for the runoff water are either underground pipes or open channels and swales. In terms of transport of the particulate constituents, the underground pipes must be self-cleansing, whereas in some cases it is intended that open transport systems like swales remove pollutants (cf. Chapter 9). As further dealt with in Chapter 9, best management practices (BMPs) like wet detention ponds and infiltration ponds serve as both stormwater transport systems and facilities for pollution control and mitigation. Such systems can also be designed to observe recreational objectives by being an element of "water in the city."

From this very brief description of the transport system for urban and road runoff, it is readily seen that the performance of the system (e.g., in terms of handling the particulate constituents in the runoff water) plays a central role. The particles must either be kept in suspension during transport or they must be deliberately removed for pollutant control purposes.

4.4.2 Sedimentation and Erosion in Open Channels and Pipes

Sedimentation in pipes and channels, if not occurring under controlled conditions as a part of an intended pollutant removal, is a potential problem because it may result in a reduced flow capacity. Pipes and channels must therefore be self-cleansing (cf. also Section 5.2.3). Removal in sediment traps of coarse materials from the runoff flow might be appropriate. Basic characteristics of particles are dealt with in Section 3.2.4.3. Particle characteristics in terms of sedimentation and deposition are subjects of Section 3.4.2.1.

In general, particles with a diameter of > 100 μm will easily settle in runoff water under quiescent flow conditions. Theoretically according to Stokes's law and depending on the density of the particles, the settling velocity of such particles is between 4×10^{-4} m s^{-1} and 8×10^{-3} m s^{-1} (cf. Equation 3.52 and Figure 3.12). Considering the mass of suspended particles with a diameter of < 100 μm, particles with a diameter of < 50 μm is dominating (Andral et al. 1999). Based on this study it was also shown that 60% of the particle mass smaller than 100 μm settled with a velocity with less than 1.1×10^{-3} m s^{-1}. It is expected that the major part of what settles in this case is inorganic matter.

The opposite phenomenon of settling, erosion, and resuspension of deposits (scouring), is a potential problem too, particularly when it occurs in open channels.

Erosion of open channels must be avoided to protect the structure of the channel and keep it well functioning. Furthermore, a resulting siltation of downstream network systems or receiving waters caused by settling of eroded materials should also be avoided. In general, natural channels have typically reached a level of stability that is not necessarily the case for artificially constructed channels. The risk of erosion is assessed by the magnitude of the shear stress, τ, acting on the channel bottom (cf. Section 3.4.2.2). In a rectangular channel under steady state flow, the average bottom shear stress and the hydraulic radius can be expressed as follows:

$$\tau = \gamma R I = g \rho R I \tag{4.1}$$

$$R = \frac{A}{P} \tag{4.2}$$

where

τ = shear stress (kg m^{-1} s^{-2}) or (N m^{-2})
γ = specific weight of water (kg m^{-2} s^{-2})
g = gravitational acceleration (m s^{-2})
ρ = density of water (kg m^{-3})
R = hydraulic radius (m)
A = cross-sectional area of the channel (m^2)
P = wetted perimeter of the channel (m)
I = channel slope (m m^{-1}).

A basic criterion for stable functioning of a channel in case of extreme urban runoff flows is that a critical shear stress, τ_{crit}, at the bottom is not exceeded. Table 4.6 shows examples of such critical shear stresses for different materials applied for

TABLE 4.6
Values of Critical Shear Stresses for Erosion of Open Channels and Corresponding Manning Numbers for Selected Materials of Construction

Material	Manning Number, $M = 1/n$, (m$^{1/3}$ s^{-1})	Critical Shear Stress, τ_{crit}, (N m^{-2})
Fine sand	50	1.3
Loamy sand	50	1.8
Loamy silt	50	2.3
Firm loam	50	3.6
Fine gravel	50	3.6
Coarse gravel	40	14
Grass; low to high shear stress values	–	20–200
Rock riprap; 150 to 300 mm	–	100–200

Source: Data from Fortier, S., and Scobey, F. C., *Transactions of the American Society of Civil Engineers (ASCE)*, 89, No. 1588, 940–84, 1926.

construction of open channels and a corresponding typical value of the channel roughness. The channel roughness in Table 4.6 is expressed in terms of a Manning number, M, which is equal to the reciprocal of the Manning coefficient of roughness, n (cf. Equation 3.44 in Section 3.4.1.4).

Example 4.1: Scouring of an Open Channel

A channel is designed to receive inflow of stormwater runoff resulting in a design flow of totally 1.2 $m^3\ s^{-1}$. The cross-sectional area of the channel is rectangular, it is 10 m wide, the slope is $I = 0.004$ and the bottom material consists of loamy silt with characteristics as shown in Table 4.6. The question is if there is a risk of scouring under these wet weather flow conditions.

A combination of the Manning formula and the continuity equation is the starting point for answering this question (cf. Equation 3.44 in Section 3.4.1.4). Equation 4.3 is formulated with the Manning number, $M = 1/n$ (cf. Table 4.6):

$$V = \frac{Q}{A} = M R^{2/3} I^{1/2} \tag{4.3}$$

where

V = flow velocity (m s^{-1})
Q = flow ($m^3\ s^{-1}$).

With b and y as channel width and water depth, respectively, and applying Equation 4.2, Equation 4.3 is formulated as follows:

$$\frac{Q}{yb} = M\left(\frac{yb}{2y+b}\right)^{2/3} I^{1/2} \tag{4.4}$$

Except for the water depth, y, the values of the parameters in Equation 4.4 are all known. Since y cannot be directly isolated, iteration is applied for its determination. Starting with $y = 1$ m, 14 iteration steps result in a fairly constant value of $y = 0.143$ m. Based on Equation 4.1, the shear stress, τ, can be calculated:

$$\tau = g\rho RI = 9.81 \times 1 \times 10^3 \frac{0.143 \times 10}{2 \times 0.143 + 10} 0.004 = 5.45\,\mathrm{N\,m^{-2}}$$

This value of the shear stress is significantly larger than τ_{crit}, = 2.3 N m^{-2} as shown in Table 4.6. It is therefore concluded that there is a risk of scouring in the channel and it is proposed that the channel be protected (e.g., by rock riprap or concrete).

4.5 POLLUTANT BUILDUP AND WASH-OFF AT URBAN AND ROAD SURFACES

In principle, pollutants accumulate on urban and road surfaces during dry weather periods and wash-off processes following remove pollutants during runoff events, see Figure 4.1. Buildup thereby includes a number of complex processes and phenomena (e.g., deposition, wind erosion, and street cleaning) that occur during the

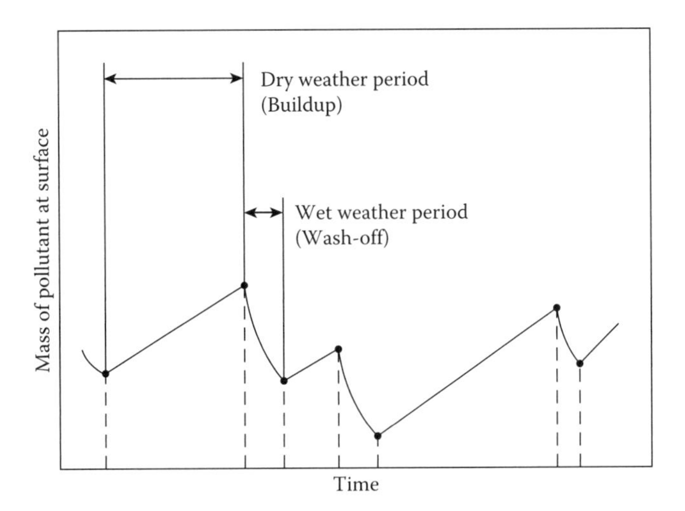

FIGURE 4.1 The principle of (linear) pollutant buildup and (exponential formulated) pollutant wash-off at urban and road surfaces during dry and wet weather periods, respectively.

interevent dry weather periods and that lead to accumulation of dust and pollutants on the urban or road surfaces. Wash-off is the erosion, resuspension, and transport processes of the dust that occur during runoff.

Knowledge on the buildup and wash-off processes are fundamentally important when predicting the load and the variability in load from a catchment. A number of empirical approaches have been developed for characterization and simulation of pollutant buildup and wash-off at urban and road surfaces. Sections 4.5.1 through 4.5.3 give examples of such empirical equations. As will be discussed in this text, only the wash-off, and not the buildup of pollutants on impervious surfaces, has a potential to be formulated in conceptual terms. Empirical approaches are therefore often the only relevant approach in practice. Further details in this respect are found in Huber (1986).

4.5.1 Pollutant Mass Load Models

The following empirical model depicts an overall load-runoff relationship during an event in relative simple terms:

$$M = A\,\alpha\left(\frac{y}{A}\right)^{\beta} \tag{4.5}$$

where

M = pollutant mass load from a catchment during a runoff event (g)
A = impervious catchment area (ha)
y = average runoff flow rate during the event (m hr^{-1})
α and β = empirical coefficients

Due to the empirical nature of Equation 4.5, α and β are not given a unit.

The equation refers to a specific catchment and it is expected that both α and β will vary from one site to another, which was also confirmed by Lee and Bang (2000) who used an equation similar to Equation 4.5. The regression coefficients in such empirical equations should therefore be determined based on investigations performed at the site in question and not transferred from one location to another.

4.5.2 Pollutant Buildup Models

In contrast to Equation 4.5, a buildup equation aims at predicting the pollutant accumulation during a period of time at an impervious surface. Pollutant buildup is a complex phenomenon that includes contributions from several sources like atmospheric fallout, inputs from surface erosion, and surrounding vegetation, spills, automobile emissions and decay, snow accumulation, and constructions (cf. Section 4.3). Furthermore, a number of dry weather removals reduce the pollutant buildup (e.g., wind erosion, degradation at the surface, and street sweeping). It is evident that such complex elements of pollutant buildup in reality only make it relevant to formulate empirical equations that should be calibrated on relevant local data.

In a review of pollutant buildup at impervious surfaces, Huber (1986) refers that equations for the amount of pollutant build-up, M, versus time, t, fall into one of four functional forms:

- Linear $M = \alpha t$
- Power function $M = \alpha t^{\beta}$
- Exponential $M = M_{\text{limit}}(1 - e^{-\beta t})$
- Michaelis-Menton $M = M_{\text{limit}}(t/\alpha + t)$,

where

α and β = empirical parameters to be calibrated.

The fact that it is relevant to consider such different approaches indicates that the buildup process is complex and subject to considerable variability.

A simple linear buildup model is

$$M(t) = M_0 + r_{b-u} A \, \Delta t \tag{4.6}$$

where

$M(t)$ = accumulated pollutant mass at a surface at time t (g)
M_0 = initial mass of a pollutant accumulated (g)
r_{b-u} = pollutant buildup rate (g ha^{-1} d^{-1})
A = impervious catchment area (ha)
Δt = time of accumulation, antecedent dry weather period (d).

Various methods based on monitored data exist for the determination of M_0 and r_{b-u} (Kobriger et al. 1981). They described such methods for highway systems in the United States and found for the buildup of total solids (TS), a relationship between r_{b-u} and the ADT, see Figure 4.2.

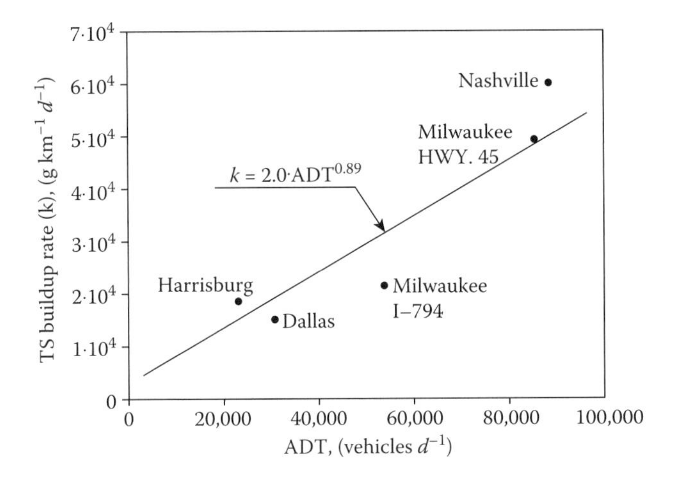

FIGURE 4.2 Total solids (TS) buildup rates versus average daily traffic (ADT).

In Figure 4.2, the pollutant buildup rate, k, is expressed in terms of the length of the highway. The relation between the buildup rates k and r_{b-u} is expressed in Equation 4.7:

$$k = b_1\, r_{b-u} = 10\, b_2\, r_{b-u} \tag{4.7}$$

where
k = pollutant buildup rate (g km^{-1} d^{-1})
b_1 = specific road surface (ha km^{-1})
b_2 = width of road surface (m)

Equation 4.8 shows a slightly extended version of Equation 4.7 for pollutant buildup by taking into account removal processes for pollutants at an urban or road surface (Sartor, Boyd, and Agardy 1974).

$$M(t) = A\frac{r_{b-u}}{r_d}\left(1 - e^{-r_d t}\right) + M_0\, e^{-r_d t} \tag{4.8}$$

where
r_d = decay (removal) rate coefficient for a pollutant (d^{-1})
t = antecedent dry weather period, ADWP (d)

Mourad, Bertrand-Krajewski, and Chebbo (2005) used a pollutant buildup concept similar to the model presented in Equation 4.8. In this case it is also crucial that buildup and decay rates can be determined either directly, based on data originating from experiments, or indirectly, based on model calibrations. As mentioned by Bertrand-Krajewski (2006), the two parameters r_{b-u} and r_d are correlated and cannot be calibrated independently when applied in a model.

4.5.3 Pollutant Wash-Off Models

In this context, a pollutant wash-off model is defined as a model that, based on results from a pollutant buildup model, predicts the load that is associated with a runoff event. Although Equation 4.5 predicts the mass load from a runoff event, with this definition it is not considered a wash-off model.

Until now, only a few attempts have been made to describe the wash-off processes in a conceptual way. An example is a model complex developed by Svensson (1987) based on theories for sediment transport in channels. Svensson related the detachment of particles to the energy of the raindrops hitting the urban surface and the associated transport of the particles in the runoff being proportional to the surface flow and the bottom shear stress. Svensson also found that the transport rate could be limited by both the detachment rate and the particle size distribution. Although this description of the wash-off seems correct and sound, further development is needed, and not yet performed, to make the model description suitable for practical applications.

An empirical formulation of the pollutant wash-off from an urban or road surface can be expressed in terms of a first-order removal process (Sartor and Boyd 1972; Sartor, Boyd, and Agardy 1974). When applying the information from Equation 4.6, the following empirical expression describes the wash-off:

$$M(t) - M(t + \Delta t) = M(t)\,(1 - e^{-k\,\Delta t}) \tag{4.9}$$

where

k = first-order removal coefficient (hr^{-1})
Δt = runoff period (hr)

Equation 4.9 is formulated for a runoff event but can also be expressed on a time-step basis within an event. The equation can be reformulated by taking into account the cumulative runoff depth during a runoff event:

$$\Delta M = M_0\,(1 - e^{-\alpha y}) \tag{4.10}$$

where

ΔM = cumulative amount of a pollutant transported within an event (g)
α = wash-off coefficient (m^{-1})
y = cumulative runoff depth (m)

A number of formulations of the wash-off process that follows the principles of Equations 4.9 and 4.10 have been developed by Jewell and Adrian (1978). In these equations, Egodawatta and Goonetilleke (2008) propose to introduce a dimensionless capacity factor, C_F, determining the ability of a rainfall to mobilize pollutants from the impervious surfaces. They found that C_F typically varied between 0.3 and 0.9, particularly depending on the rainfall intensity and the density of the particles rather than their size.

4.5.4 Regression Models for Prediction of Pollutant Loads

The models described in Sections 4.5.1 through 4.5.3 are examples that aim to predict pollutant mass loads, pollutant buildup on impermeable surfaces, and wash-off during runoff events. Being empirical of nature, it is important to remember that these models are only relevant in the case that adequate data sets exist and can be applied for the determination of the coefficients in the equations. In general, it is therefore not possible to transfer such coefficients from one site to another. The models are, however, widely applied. It is the case as part of more complex stormwater management models (e.g., the SWMM model). Careful model calibration is therefore a crucial requirement for a meaningful model prediction.

The models described in the preceding sections are formulated in terms of the contributing area (e.g., per hectare). Correspondingly, the pollutant loads from roads and highways are often described per unit length of the road.

An alternative to the model formulations dealt with in Sections 4.5.1 through 4.5.3 is to estimate the pollutant load based on a site mean concentration (cf. Section 2.6.2):

$$M_t = \text{SMC}\ V_t \tag{4.11}$$

where

M_t = pollutant mass load during a period of time, t, from a runoff event (g)
SMC = site mean concentration for the runoff water (g m^{-3})
V_t = volume of runoff water during a period of time, t, from a runoff event (m^3)

By applying this approach, it is possible to include information on the stochastic nature of SMC as well as historical data for rainfall and runoff.

As an alternative to Equation 4.11 with a single parameter, SMC, the variability of the event mean concentration (EMC) can be taken into account. The following estimation of an EMC originates from the French CANOÉ model. The use of this model requires that an appropriate data set of EMCs exists for calibration (Bertrand-Krajewski, 2006):

$$\text{EMC} = a\ H^b\ (I_{\text{max}5})^c\ t^d \tag{4.12}$$

where

EMC = event mean concentration (g m^{-3})
H = rainfall depth (mm)
$I_{\text{max}5}$ = maximum rainfall intensity in five minutes (mm hr^{-1})
t = antecedent dry weather period, ADWP (hr)
a, b, c, and d = parameters to be determined by calibration

Equation 4.13 is a pragmatic way of estimating an annual pollutant load from an urban area:

$$L = P\ \varphi\ \text{SMC}\ A \tag{4.13}$$

where

L = annual pollutant load from an urban area (g yr^{-1})
P = average annual rainfall depth (m yr^{-1})
φ = runoff coefficient, see Equation 2.7 (–)
A = contributing catchment (m^2)

4.5.5 Models for Pollutant Wash-Off from Roads

The traffic density of a road or highway is intuitively considered a central parameter that will determine the corresponding pollutant level in the runoff. The following regression model is, from this standpoint, expected as a first linear approximation:

$$\text{SMC} = a\ \text{ADT} + b \tag{4.14}$$

where
 ADT = average daily traffic (vehicles d^{-1})
 a and b = empirical parameters to be calibrated

Values of the parameters a and b for various pollutants that showed a statistically significant correlation in the analysis are found in FHWA (1987). Figure 4.2 is an example that also relates to Equation 4.14.

Due to the variability that is expected for pollutant concentrations in road runoff, it is clear that the correlation stated in Equation 4.14 should be weak. Equation 4.14 is therefore most useful to weight the prediction of the SMC value toward the higher or lower end of a probability distribution. It has also been seen that traffic-related loads are only sorted into two groups: roads with ADT > 30,000 (urban highways) and those with ADT < 30,000 vehicles per day (rural highways), resulting in 2–4 times higher runoff concentrations in the first group (FHWA 1987).

4.6 POLLUTANT CONCENTRATIONS AND LOADS

4.6.1 General Information

Pollutant concentrations and loads related to urban and road runoff are subject to a considerable variability in time and place (cf. Section 2.4). Typically, the coefficients of variation (COVs) for EMC or SMC values are in the order of 0.8–1.2. A standard deviation can therefore easily reach the same level of magnitude as a mean value. The pollutant concentrations and loads given in the following tables must therefore only be considered as "order of magnitude" and examples. Although such literature values are widely used in practice, there is no reliable substitute for sampling and monitoring to determine a correct pollutant level in a specific case.

The information available in terms of quantification of pollution from urban and road runoff is often given in the following units:

- g m^{-3}: For EMC and SMC values in stormwater runoff from urban catchments and roads
- g ha^{-1} yr^{-1} or g m^{-2} yr^{-1}: Yearly unit area loads (export coefficients) from a catchment
- g km^{-1} yr^{-1} or g m^{-1} yr^{-1}: Yearly load from a unit length of a road or a highway

The characterization of a catchment in terms of pollution is given by the sources for the pollutants (cf. Section 4.3). For practical purposes, the loadings of pollutants can pragmatically be defined by a land use. The following terms referring to a catchment are in this respect most often used to distinguish between or characterize different levels of pollution:

- Residential areas (different densely populated urban areas)
- Industrial areas
- Commercial areas
- Parking lots
- ADT for roads and highways

4.6.2 Specific Pollutant Concentration and Load Characteristics

Characteristic pollutant concentrations (SMC values) and corresponding COV values for the runoff from residential areas are shown in Table 2.2. These results originate from the Nationwide Urban Runoff Program (NURP) performed in the United States during the period 1979–1982 (U.S. EPA 1983). The NURP was an extensive monitoring program for wet weather pollution and reflects, in general, the situation in the United States around 1980 and to some extent the time after that. During the period after 1980, however, the use of unleaded gasoline has considerably reduced the diffuse pollution with lead to a concentration level that 20 years later was about 10 times lower than stated in Table 2.2. Results from more recent investigations in the United States indicate that, other than lead, most pollutants did not vary significantly from the NURP information (Debo and Reese 2003). Although data from the National Stormwater Quality Database (NSQD) also show that other pollutants, in particular the heavy metals, have been reduced, the differences between the median NURP and NSQD observations are likely due to the random nature of stormwater data and not from significant trends (Pitt and Maestre 2005). This database includes stormwater runoff measurements from 66 agencies and municipalities in the United States over almost 10 years until 2003. A comparison of selected median pollutant concentrations in stormwater runoff originating from the NURP program and the NSQD database, respectively, is outlined in Table 4.7.

Similar levels of stormwater runoff concentrations, as shown in Table 4.7, have been observed in other countries too. As an example, Table 5.2 shows SMC values considered characteristic for stormwater runoff from urban surfaces and roads in Northern Europe. Although there are differences between the pollutant concentrations shown in Tables 2.2, 4.7, and 5.2, they cannot be considered conflicting when taking into account the variability of stormwater quality data.

A number of other interesting results can be drawn from the NSQD database (Pitt and Maestre 2005). In general it was observed that land use was important with freeway locations showing the highest median stormwater runoff concentrations although industrial and institutional sites for some pollutants also showed high concentrations. As might be expected, geographical variations were observed to be more important than seasonal variations.

Rather than applying values for pollutant concentration and loads originating from the general available literature, it is recommended finding information from the local (regional) area where the conditions for pollutant load might be considered

TABLE 4.7
Comparison of NURP and NSQD Median Stormwater Runoff Concentrations

	Overall		Residential	
Pollutant (Unit)	**NURP**	**NSQD**	**NURP**	**NSQD**
BOD_5 (g m^{-3})	9	8.6	10	9
COD (g m^{-3})	65	53	73	55
TSS (g m^{-3})	100	58	101	48
Total Kjeldahl Nitrogen, TKN (g m^{-3})	1.5	1.4	1.9	1.4
Total P (g m^{-3})	0.33	0.27	0.38	0.3
Filtered P (g m^{-3})	0.12	0.12	0.14	0.17
Total Pb (mg m^{-3})	144	16	144	12
Total Cu (mg m^{-3})	34	16	33	12
Total Zn (mg m^{-3})	160	116	135	73

Source: Data from Pitt, R. E., and Maestre, A., Stormwater quality as described in the national stormwater quality database (NSQD). Proceedings of the 10th International Conference on Urban Drainage, Copenhagen, Denmark, August 21–26, 2005.

more comparable to the site in question. Several municipalities and highway authorities around the world have—although to a less extent than in the United States, collected and disseminated such information.

In the following tables a number of results from different sampling and monitoring programs will exemplify the importance of land use for the load of pollutants associated with urban and road runoff. Furthermore, these tables will exemplify the different units for pollutant load as given in Section 4.6.1.

As an example, Table 4.8 summarizes results for micropollutants found in urban runoff in Norway with a rather cold temperate climate.

The results shown in Table 4.8 can be compared with the results shown in Table 4.9 on snowmelt runoff, also originating from Norway. As indicated by comparison, the pollutant concentrations in runoff from snowmelt are considerably higher than those obtained from the runoff of rain (cf. Section 4.2).

The stormwater runoff loadings shown in Table 4.10 are, by urban land use in the United States, given per unit area of catchment and time (e.g., Horner et al. 1994; Burton and Pitt 2002). Such data are export coefficients. The main part of the results originates from a period of time when unleaded gasoline was commonly used. The table only serves as an example and it is crucial to be aware that such values are subject to considerable variability in time and place although similar results have been published by other authors (e.g., Gilbert and Clausen 2006).

Table 4.10 indicates that stormwater pollutant loads from freeways are relatively high compared with the contribution from other types of land use. This fact is in agreement with a statement by Ellis, Revitt, and Llewellyn (1997) who concluded that available data indicate that highway runoff pollution is of concern (toxicity) in case of urban highways with a traffic load of more than 30,000 vehicles d^{-1} (cf. particularly Section 4.5.5).

TABLE 4.8
Typical Minimum and Maximum SMC Values for Heavy Metals and PAH Based on Stormwater Runoff Measurements from Seven Catchments in Norway

Pollutant (mg m^{-3})	Central Urban Areas	Residential Areas	Commercial Areas
Cd	0.1–0.5	0.1–0.4	0.1–0.5
Hg	0.2–1.2	< detection limit	< detection limit
Pb	1–33	1–8	1–19
Ni	3–190	1–10	1–11
Cr	1–170	1–12	1–7
Zn	10–300	5–140	8–92
Cu	6–120	3–15	4–31
PAH	0.1–2.7	0.1–0.8	0.01–0.3

Source: Data from Storhaug, R., Miljøgifter i overvann [Micropollutants in stormwater runoff]. *Statens Forurensningstilsyn* [the Norwegian EPA], TA 1373, 96,18, 1996.

TABLE 4.9
Concentrations of Heavy Metals and PAH from Snowmelt Runoff in Norway

Pollutant (mg m^{-3})	Snowmelt Runoff
Cd	3.9–26
Hg	0.19–13.2
Pb	max. 690
Ni	42–106
Cr	30–150
Zn	200–740
Cu	13–430
PAH	1500–11,600

Source: Data from Baekken, T. and Joergensen, T. 1994. *Vannforurensning fra veg: Langtidseffekter* [Pollution from road runoff: Long term effects]. Oslo, Norway: Statens vegvesen, Vegdirektoratet (The Norwegian State Highway Administration).

In general, the wet weather loads of pollutants onto receiving waters from urban areas and roads must be added to the loads from other sources. Such pollutant sources do not only include the dry weather inputs via wastewater treatment plants but also those associated with rural runoff. This is illustrated in Example 4.2 showing a comparison between wet weather urban and rural loads of phosphorus and nitrogen, both loads originating from runoff. In case of a lake, the runoff from the urban impervious surfaces normally takes place directly via inflow structures from storm sewers. In contrast, the

TABLE 4.10
Typical Unit Area and Unit Time Loadings (Export Coefficients) of Pollutants from Stormwater Runoff by Land Use

Pollutant ($kg\ ha^{-1}\ yr^{-1}$)	Commercial	Residential High-Density	Residential Medium-Density	Residential Low-Density	Industrial	Freeway	Parking
TSS	1,100	450	270	10	550	1,000	450
Total P	1.7	1.1	0.4	0.05	1.5	1.0	0.8
TKN	7.5	4.7	2.8	0.3	3.7	8.9	5.7
BOD	70	30	15	1	NA	NA	53
COD	470	190	60	10	230	NA	300
Pb	3.0	0.9	0.06	0.01	0.2	5.0	0.9
Zn	2.3	0.8	0.1	0.05	0.4	2.3	0.9
Cu	0.4	0.03	0.03	0.01	0.1	0.4	0.07

Source: Data from Horner, R. R., Skupien, J. J., Livingston, E. H., and Shaver, E. H., *Fundamentals of urban runoff management: Technical and institutional issues.* Washington, DC: Terrene Institute and U.S. Environmental Protection Agency, 1994; Burton, G. A., and Pitt, R. E., *Stormwater effects handbook: A toolbox for watershed managers, scientists, and engineers.* Boca Raton, FL: Lewis Publishers/CRC Press, 2002.

Note: NA = Not available.

inputs from rural runoff typically occur more or less continuously from the catchment via creeks, rivers, and direct diffuse inputs. In case of nutrient loads into more or less stagnant waters like a lake, the mode of input is because of the accumulative effect of the nutrients of minor importance compared with the annual load (cf. Section 6.4).

Example 4.2: Discharges of Phosphorus and Nitrogen from Urban and Rural Runoff Into a Lake

Following what is mentioned above, Figure 4.3 exemplifies the relative importance of the inflow of phosphorus and nitrogen into a lake from urban and rural

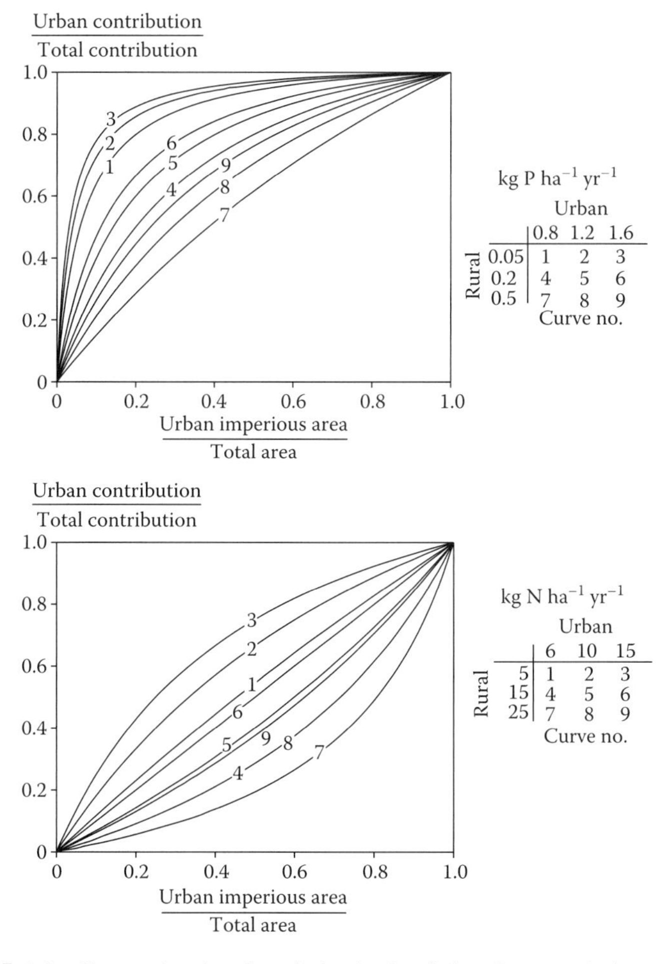

FIGURE 4.3 Curves showing the relative loads of phosphorus and nitrogen into a lake from urban sources relative to the total load from both the urban and the rural land. These loads are depicted versus the relative magnitude of impermeable area for the entire contributing land. The values of low, mean, and high specific nutrient load in the two matrices are based on data for Northern European countries.

sources, respectively. The export coefficients used in this example originate from Northern Europe and are, in the example, given as low, median, and high values.

The result shown in Figure 4.3 reflects the fact that nitrogen inflow to surface waters from rural areas in contrast to phosphorus, because of inputs from farmland, is relatively high when compared with the consumption for aquatic plant growth. The curves show that even a relatively small urban impervious area, 20% of the total catchment area, on average (curve No. 5) will contribute with about 60% of a phosphorus load. Due to the fact that lakes are often phosphorus limited for algal growth, a reduced input of phosphorus from urban areas can be relatively important for control of eutrophication (cf. Section 6.4).

Examples of pollutant concentrations for SWR from urban catchments and roads are published in a large number of papers and books and also available on Web sites, Burton and Pitt (2002) and www.bmpdatabase.org. Concerning data for local uses, reports from authorities and governmental agencies are often valuable.

REFERENCES

Andral, M. C., S. Roger, M. Montréjaud-Vignoles, and L. Herremans. 1999. Particle size distribution and hydrodynamic characteristics of solid matter carried by runoff from motorways. *Water Environment Research* 71 (4): 398–407.

Baekken, T., and T. Joergensen. 1994. *Vannforurensning fra veg: Langtidseffekter* [Pollution from road runoff: Long term effects]. Oslo, Norway: Statens vegvesen, Vegdirektoratet (The Norwegian State Highway Administration).

Bertrand-Krajewski, J.-L. 2006. Influence of field data sets on calibration and verification of stormwater pollutant models. Proceedings from the 7th International Conference on Urban Drainage Modeling (UDM), April 2–7, Melbourne, Australia.

Burton, G. A., and R. E. Pitt. 2002. *Stormwater effects handbook: A toolbox for watershed managers, scientists, and engineers*. Boca Raton, FL: Lewis Publishers/CRC Press.

Debo, T. N., and A. J. Reese. 2003. *Municipal stormwater management,* 2nd ed. Boca Raton, FL: Lewis Publishers/CRC Press.

Egodawatta, P., and A. Goonetilleke. 2008. Understanding road surface pollutant wash-off and underlying physical processes using simulated rainfall. *Water Science and Technology* 57 (8): 1241–46.

Ellermann, T., H. V. Andersen, R. Bossi, J. Christensen, L. M. Frohn, C. Geels, K. Kemp, P. Lofstrom, B. B. Mogensen, and C. Monies. 2007. *Atmosfaerisk deposition, 2006* (Atmospheric deposition 2006), Danmarks Miljøundersøgelser (NERI, National Environmental Research Institute), Technical Report No. 645.

Ellis, J. B., D. M. Revitt, and N. Llewellyn. 1997. Transport and the environment: Effects of organic pollutants on water quality. *Journal of Water Environmental Management* 11: 170–77.

FHWA. 1987. *Methodology for analysis of pollutant loadings from highway stormwater runoff.* Washington, DC: U.S. Department of Transportation, Federal Highway Administration (FHWA), FHWA/RD-87/086.

Fortier, S., and F. C. Scobey. 1926. Permissible canal velocities. *Transactions of the American Society of Civil Engineers (ASCE)* 89, No. 1588: 940–84.

Gilbert, J. K., and J. C. Clausen. 2006. Stormwater runoff quality and quantity from asphalt, paver, and crushed stone driveways in Connecticut. *Water Research* 40 (4): 826–32.

Horner, R. R., J. J. Skupien, E. H. Livingston, and E. H. Shaver. 1994. *Fundamentals of urban runoff management: Technical and institutional issues*. Washington, DC: Terrene Institute and U.S. Environmental Protection Agency.

Hovmand, M. F. 1980. Atmosfaerisk metalnedfald i Danmark [Atmospheric deposition of metals in Denmark]. PhD dissertation, Technical University of Denmark.

Huber, W. C. 1986. Deterministic modeling of urban runoff quality. In *Urban runoff pollution*, eds. H. C. Torno, J. Marsalek, and M. Desbordes, 167–242. Germany: Springer-Verlag.

Jewell, T. K., and D. D. Adrian. 1978. SWMM stormwater pollutant washoff functions. *Journal of Environment Engineering Division* 104 (5): 489–99.

Kobriger, N.P, T.L. Meinholz, M.K. Gupta and R.W. Agnew. 1981. *Constituents of highway runoff: Volume III, Predictive procedurefor determining pollutant characteristics in highway runoff*, Federal Highway Administration (FHWA), Office of Research and Development, FHWA/RD-81/044.

Lee, J. H., and K. W. Bang. 2000. Characterization of urban stormwater runoff. *Water Research* 34 (6): 1773–80.

Marsalek, J., G. Oberts, K. Exall, and M. Viklander. 2003. Review of operation of urban drainage systems in cold weather: Water quality considerations. *Water Science and Technology* 48 (9): 11–20.

Mourad, M., J.-L. Bertrand-Krajewski, and G. Chebbo. 2005. Stormwater quality models: Sensitivity to calibration data. *Water Science and Technology* 52 (5): 61–68.

Pengchai, P., F. Nakajima, and H. Furumai. 2005. Estimation of origins of polycyclic aromatic hydrocarbons in size-fractionated road dust in Tokyo with multivariate analysis. *Water Science and Technology* 51 (3–4): 169–75.

Pétavy, F., V. Ruban, P. Conil, and J. Y. Viau. 2007. Reduction of sediment micro-pollution by means of a pilot plant. Proceedings of the 5th International Conference on Sewer Processes and Networks, Delft, the Netherlands, August 29–31, 297–306.

Pitt, R. E., and A. Maestre. 2005. Stormwater quality as described in the national stormwater quality database (NSQD). Proceedings of the 10th International Conference on Urban Drainage, Copenhagen, Denmark, August 21–26.

Sartor, J. D., and G. B. Boyd. 1972. *Water pollution aspects of street surface contaminants.* Washington, DC: U.S. Environmental Protection Agency, EPA-R2-72-081.

Sartor, J. D., G. B. Boyd, and F. J. Agardy. 1974. Water pollution aspects of street surface contaminants. *Journal of Water Pollution Control Federation* 46 (3): 458–67.

Spierenburg, A., and C. Demanze. 1995. Pollution des sols: Comparaison et application de la norme hollandaise [Soil pollution: Comparison and application of the Dutch standards]. *Environnement et Technique* 146:79–81.

Storhaug, R. 1996. Miljøgifter i overvann [Micropollutants in stormwater runoff]. *Statens Forurensningstilsyn* [the Norwegian EPA], TA 1373, 96: 18.

Stumm, W., and J. J. Morgan. 1996. *Aquatic chemistry: Chemical equilibria and rates in natural waters,* 3rd ed. New York: John Wiley & Sons, Inc.

Svensson, G. 1987. Modelling of solids and metal transport from small urban watersheds. PhD thesis, Chalmers University of Technology, Department of Sanitary Engineering, Gothenburg, Sweden.

U. S. EPA (U.S. Environmental Protection Agency). 1983. Results of the nationwide urban runoff program, Volume I: Final report. Washington, DC: Water Planning Division, NTIS No. PB 84-185552.

Viklander, M. 1999. Dissolved and particle-bound substances in urban snow. *Water Science and Technology* 39 (12): 27–32.

Zhao, H., C. Yin, M. Chen, and W. Wang. 2008. Runoff pollution impacts of polycyclic aromatic hydrocarbons in street dusts from a stream network town. *Water Science and Technology* 58 (11): 2069–76.

WEB SITES

www.bmpdatabase.org: A database with information to be applied for stormwater management projects, in terms of calculation and evaluation of pollutant loads and design of control measures (BMPs).

5 Combined Sewer Overflows: Characteristics, Pollutant Loads, and Controls

A combined sewer network serves the dual purpose in the same network to collect and transport both the daily flow of wastewater and the wet weather runoff from an urban catchment. Corresponding high flow rates must therefore be managed during rain events. When the hydraulic capacity of an interceptor or a trunk sewer is exceeded during a runoff event, overflow of excess water from the network into adjacent receiving water must take place upstream of this point. Combined sewer networks are therefore equipped with structures where overflow can take place. Such overflows, combined sewer overflows (CSOs), mean that a mixture of untreated wastewater, runoff water from the catchment, and materials eroded from deposits in the network itself is discharged. The consequence of CSOs is a potential pollution and deterioration of downstream receiving water systems.

This chapter aims at giving a description and quantification of the performance, management, and design of CSOs in terms of pollutant load characteristics and control of such discharges. This Chapter 5 will be restricted to the sewer network and the interface between the sewer system and the receiving water. The effects onto receiving water systems from CSOs are subjects dealt with in Chapter 6.

An overview of basic characteristics of combined sewers and particularly the relation to the wet weather pollution is given in Sections 1.4 and 2.3.

5.1 OVERFLOW STRUCTURES AND CSO CHARACTERISTICS

In principle it is not possible to avoid wet weather discharges of untreated wastewater from a combined sewer network. If such discharges are considered nonacceptable in a given case, a combined system is not the correct choice of sewer network. In terms of reduction of pollutant loads from CSOs it is, however, important that the following possibilities exist:

- Wet weather pollutant loads into receiving waters from combined sewer networks can be reduced by both structural and nonstructural measures.
- It is to some extent possible to select the location where the CSO discharge takes place. As an example, a CSO structure might be relocated from an

upstream position with discharges into a sensitive receiving water system to a less sensitive location downstream.

5.1.1 Overflow Structures

An overflow structure in a combined sewer network serves the purpose of reducing the flow rate into a downstream network during an extreme runoff event. The capacity of the downstream system in terms of surcharge, flooding, or overloading of a wastewater treatment plant thereby determines what part of the flow can be routed into the interceptor and thereby defines what portion of the flow must be discharged as CSO.

Figure 5.1 shows the principle of a simple overflow structure. The capacity of the interceptor can simply be determined by a reduced pipe diameter compared with the size of the inflow pipe or a device that regulates the outflow from the overflow structure. When the capacity of the interceptor is exceeded, the mixed wastewater and runoff water will pass the overflow weir and be discharged as CSO.

The details for construction of an overflow structure are legion and depend on the specific objective, the local tradition, and the possibility and requirements for implementation in the entire sewer network. Figure 5.2 represents three examples that illustrate the possibility of detention as a part of an overflow structure, a temporary accumulation of the inflow in case the flow rate exceeds the capacity of the wastewater collection or treatment system located downstream. When the incoming flow rate is reduced to a value lower than this capacity, the wastewater in the detention basin (detention tank) can be diverted to the interceptor. Figure 5.2 illustrates the concept of both in-line (on-line) and off-line detention. The latter type often requires that the wastewater be pumped to the interceptor when the inflow rate is reduced to a level below its flow capacity. Structural details for such detention tanks are legion with the CSO outfall being located either upstream or downstream of the detention tank. Further details of the construction and performance of overflow structures are

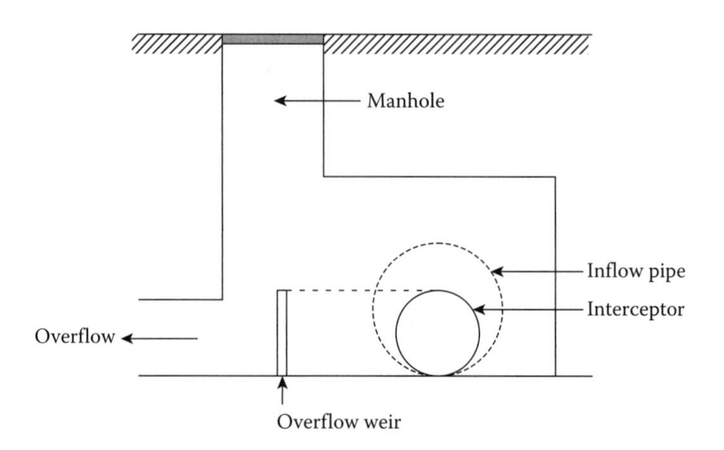

FIGURE 5.1 Schematic of a simple overflow structure.

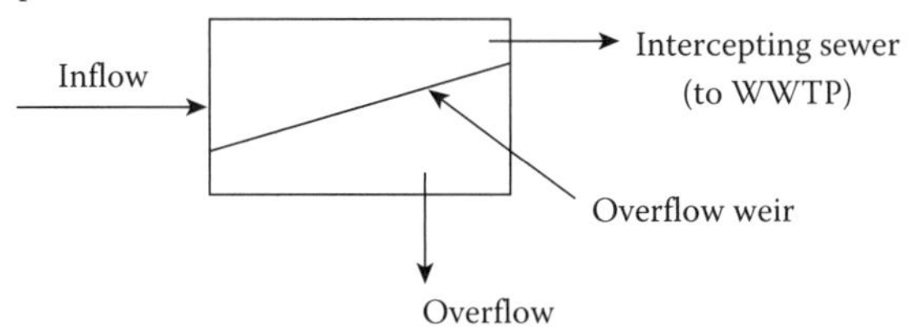

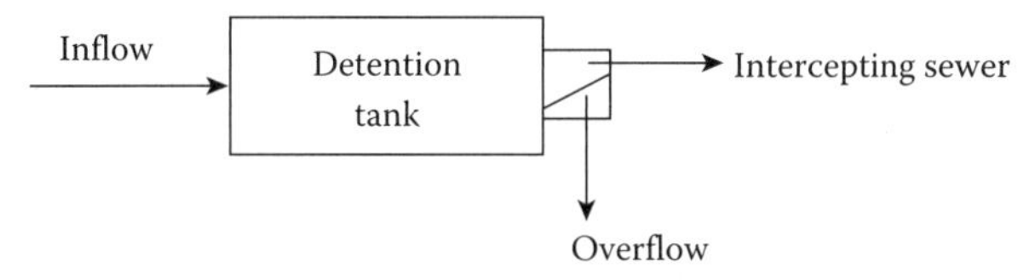

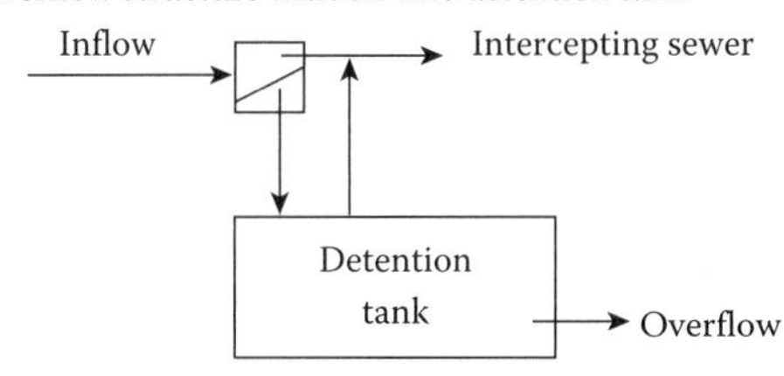

FIGURE 5.2 Selected configurations of overflow structures with and without detention tanks.

dealt with in Section 5.5. In this section it will also be illustrated how treatment of the CSOs can be managed.

A specific type of overflow, sanitary sewer overflow (SSO) may occur in a sanitary sewer network (i.e., in separate sewer catchments). Such unintended overflows can occur if the network is overloaded or pipe blockages have reduced the capacity of the network. A sanitary sewer can be hydraulically overloaded possibly caused by wrong connections that allow inflow of wet weather discharges to the sanitary sewer. Another common cause of SSOs is the occurrence of pipe blockages (e.g., grit and rocks that create bottlenecks or fats, oils, and grease, commonly known as FOG) that accumulate on pipe walls. Root intrusion is also a possible pipe blockage.

5.1.2 Flow Characteristics

As described in Section 5.1.1 and illustrated in Figures 5.1 and 5.2, an overflow structure starts to operate when the water table reaches the top of the overflow weir, corresponding to a start of overflow when the flow capacity of the downstream network is exceeded. The performance of an overflow structure during a runoff or an overflow event can therefore be illustrated by comparison of the inflow and outflow hydrographs. Figure 5.3 shows this principle in the case of an overflow structure with and without a detention tank.

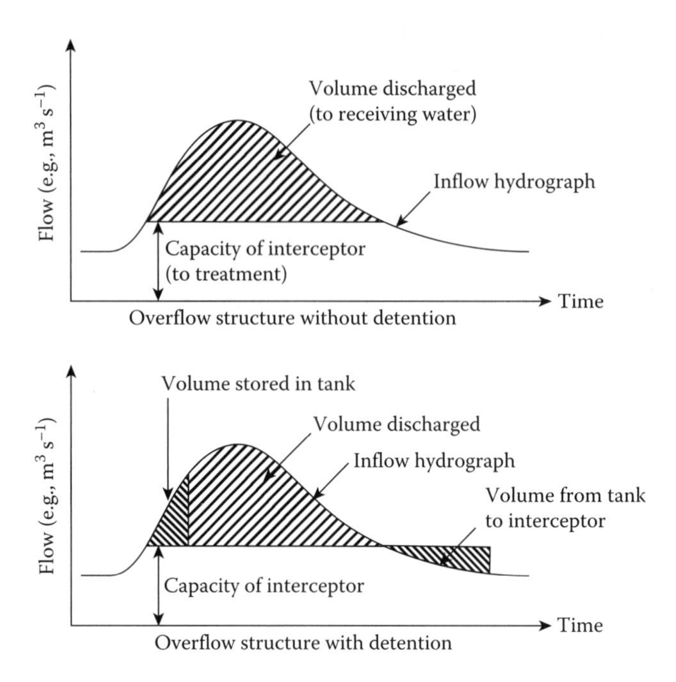

FIGURE 5.3 The principle of inflow and outflow hydrographs during a runoff event for an overflow structure without and with a detention tank.

Figure 5.3 shows that the flow capacity of the interceptor located downstream and the detention tank volume will influence the magnitude of the CSO volume that is discharged. It is evident that a detention tank extends the time a wastewater treatment plant is exposed to high inflow rates from a runoff event and also increases the volume of wastewater that must be treated. The capacity of a wastewater treatment plant (WWTP) receiving wastewater from a combined sewer catchment must therefore be designed differently compared with a treatment plant serving a separate sewer area. Particularly the capacity of the clarifiers must be designed taking into account the effect of high flow rates during extended time periods following a rain event. It is crucial to consider these complex interactions between reduced CSO impacts onto local receiving waters and increased WWTP impacts when implementing detention capacity in a sewer network.

It should be noted that the inflow/outflow relations shown in Figure 5.3 are simply described and do not take into account those detailed hydraulic phenomena that have an impact on the baseflow (dry weather flow) and the flow capacity of the interceptor during a runoff event.

5.1.2.1 Mixing Ratio

Due to the importance of dilution of the wastewater flow in terms of the mixing between runoff water and wastewater during an overflow event, the following definition of a mixing ratio, m, (also named dilution) is a central parameter used for design and as a performance indicator of a combined sewer (see Figure 5.4). The parameter

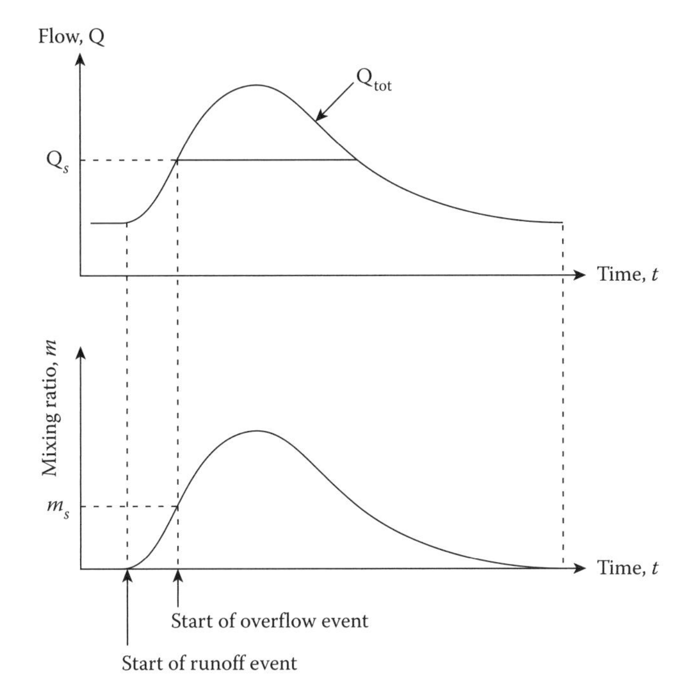

FIGURE 5.4 Principle of flow variation during a runoff event and the corresponding variation of the mixing ratio, m.

m as defined by Equation 5.1 is particularly of relevance at the point where an overflow structure is located:

$$m = \frac{Q_{\text{tot}} - Q_{\text{ww}}}{Q_{\text{ww}}} \tag{5.1}$$

where
m = mixing ratio or dilution (–)
Q_{tot} = wet weather flow rate in the inlet pipe to an overflow structure (m^3 s^{-1})
Q_{ww} = dry weather flow rate of wastewater in the inlet pipe (m^3 s^{-1})

Equation 5.1 is expressed in general terms because Q_{tot} and Q_{ww}, for a specific overflow structure, are subject to variability in time. Therefore, further specifications of the conditions under which m should be determined are needed. Equations 5.2 and 5.3 are versions that are applied in practice:

$$m_s = \frac{Q_s - Q_{\text{ww, max}}}{Q_{\text{ww, max}}}, \tag{5.2}$$

where
m_s = mixing ratio at the start of an overflow event (–)
Q_s = flow rate capacity of the interceptor (m^3 s^{-1})
$Q_{\text{ww,max}}$ = daily maximum dry weather flow rate in the inlet pipe (m^3 s^{-1})

Equation 5.2 can be applied as a first estimate to design and assess the performance of an overflow structure in terms of what level of dilution of the incoming wastewater is required before overflow to a receiving water body can be accepted. The value of $Q_{ww,max}$ selected for Equation 5.2 is typically what can be defined as a maximum daily flow rate value but other definitions can be used as well.

The annual average mixing ratio, m_y, can be expressed as follows:

$$m_y = \sum_{i=1}^{n} \frac{\sum_{j=1}^{p} \frac{(Q_{tot} - Q_{y,ww})/Q_{y,ww}}{p}}{n} \tag{5.3}$$

where

m_y = annual average mixing ratio for CSOs (–)
Q_{tot} = wet weather flow rate in the inlet pipe for $Q_{tot} > Q_s$ (m^3 s^{-1})
$Q_{y,ww}$ = annual average dry weather flow rate in the inlet pipe (m^3 s^{-1})
p = number of computation steps per event (–)
n = number of overflow events per year, for $Q_{tot} > Q_s$ (–)
$i = 1,\ldots, n$; integer (–)
$j = 1,\ldots, p$; integer (–)

A m_y-value is, in terms of a relation to receiving water, only of interest during an overflow event. Therefore only when $Q_{tot} > Q_s$, a contribution to m_y is relevant. The double Σ-sign refers to the determination of an average m_y-value for a specific overflow event and for all overflow events per year, respectively. Determination of m_y requires that a relevant historical rainfall series exists and modeling based on the corresponding runoff series. Equation 5.3 is, in this respect, simply formulated by not taking into account that the volumetric discharges vary from event to event.

As previously mentioned, Equations 5.2 and 5.3 are just examples that illustrate and specify the circumstances under which Equation 5.1 is applied. Equation 5.2 is typically valuable for design considerations because it expresses a dilution risk value with a low dilution ratio for the wastewater when the overflow starts and ends. On the contrary, Equation 5.3 includes information in terms of an average dilution ratio on a yearly basis. This formula is therefore particularly valuable for estimation of yearly average loads of pollutants into a receiving water body from an overflow structure. From the definitions it appears that $m_y > m_s$.

The value of m_s selected for design purposes vary considerably. In some countries it can be as low as two corresponding to half-full sewer pipes during high dry weather flows, whereas in other countries it is found to be as high as 10–20. The value selected will considerably affect the number of overflow events per year from an overflow structure.

5.1.2.2 Runoff Number

In addition to the mixing ratio, a flow-related parameter important for occurrence of overflow is the runoff number, also named runoff value:

$$a = \frac{Q_s - Q_{WW}}{F} \tag{5.4}$$

where

a = runoff number or runoff value (m s^{-1} or μm s^{-1})
F = contributing area of the connected catchment at point of overflow structure (m^2)

It is assumed that the runoff number can be regarded as constant throughout the runoff event and therefore in principle is valid for a box rainfall. As can be seen from the units, the runoff number is expressed in terms of a rainfall intensity.

The difference in the nominator of Equation 5.4 between the flow rate capacity of the interceptor and a characteristic dry weather flow rate is a measure of the interceptor's free capacity for the runoff water (i.e., what exceeds the daily flow rate). The runoff number, a, in principle determines what flow of surface runoff per unit area of the contributing catchment the interceptor is capable of transporting before overflow to adjacent receiving water will take place. The runoff number thereby becomes a very important design and performance parameter for an overflow structure combining basic characteristics of the sewer network and the contributing catchment. The a-value is directly a measure of the amount of runoff (often expressed as rain depth in units of μm at the catchment surface area) that per unit of time characterizes the performance of the overflow structure right before the overflow starts. The runoff number is therefore a central and simple performance and design parameter for an overflow structure.

In terms of units, μm s^{-1} is, in general, selected for practical purposes because of the magnitude of the runoff number. The magnitude of the parameter varies according to the design but is typically found in the interval 0.05–2 μm s^{-1}.

5.2 CSO POLLUTANT SOURCES

The pollutants discharged from CSOs originate from the following three major types of sources (cf. Sections 1.4.2.3 and 2.3):

- The urban surfaces
- The daily flow of wastewater
- Eroded sewer solids (originating from deposits and biofilm in the sewer network)

The relative importance of these pollutant sources for the quality of CSOs depends on a large number of catchment characteristics and how the sewer system is designed and operated. Some of these aspects have already been dealt with (e.g., the pollutant variability, first flush; cf. Section 2.4). The degree of dilution of the daily flow of wastewater before the hydraulic capacity of the interceptor is exceeded is another important factor for the amount of pollutants discharged (cf. Section 5.1.2). The actual value of the dry weather flow during a runoff event relative to the flow rate capacity of the interceptor is therefore central. This value is not solely determined by constructive design parameters of the network but will also change with time (e.g., because of different dry weather flows during day and night).

Further and more specific details concerning the three major pollutant sources for the CSOs in addition to what was included in Sections 1.4.2.3 and 2.3 will be dealt with in the following subsections.

5.2.1 Surface Runoff

The runoff from the urban and road surfaces is basically dealt with in Section 2.3.1. The quality of the surface runoff is a major subject in Chapter 4. In principle, there is no difference in the inflow of surface runoff into a combined sewer system compared with a storm sewer in a separate sewer network. However, often a combined sewer catchment is older and in several countries also typically found in a city center (i.e., densely populated and with a relatively high traffic load).

5.2.2 Wastewater

In terms of runoff from urban areas, the general characteristics of wastewater are dealt with in Section 2.3.2 and are therefore not further focused on in this chapter. A large number of textbooks on wastewater is available (e.g., Tchobanoglous, Burton, and Stensel 2003).

5.2.3 Sewer Sediments and Biofilms

In general, it is the magnitude and the dynamics of deposition of solids, biofilm growth, and sediment erosion that, in a sewer network, determines to what extent sewer solids will contribute to the pollutant levels in the CSOs. These in-sewer processes refer to both dry weather periods where deposition and biofilm growth take place and the wet weather periods where erosion and resuspension of the deposits dominates. Simply expressed, it is sedimentation of particulate materials and growth of the biofilm that, during dry weather, creates the potential for an erosion of these materials during wet weather flow. Basic details of sedimentation and erosion are found in Section 3.4.2. Figure 5.5 outlines processes and occurrence of sewer solids.

In combined sewer networks, particularly those where sewer solids accumulate, the pollutant contribution from eroded materials is often more important for the pollutant contents of the CSOs than the contributions from both the urban surfaces and the wastewater. This fact makes it very important to construct self-cleansing sewers as a means, in addition to the capacity of the network, to reduce the pollutant impacts from CSOs.

Since sewer sediments have been subject to deposition, there is a potential risk for resettlement in the receiving water after discharge of CSOs. The same potential

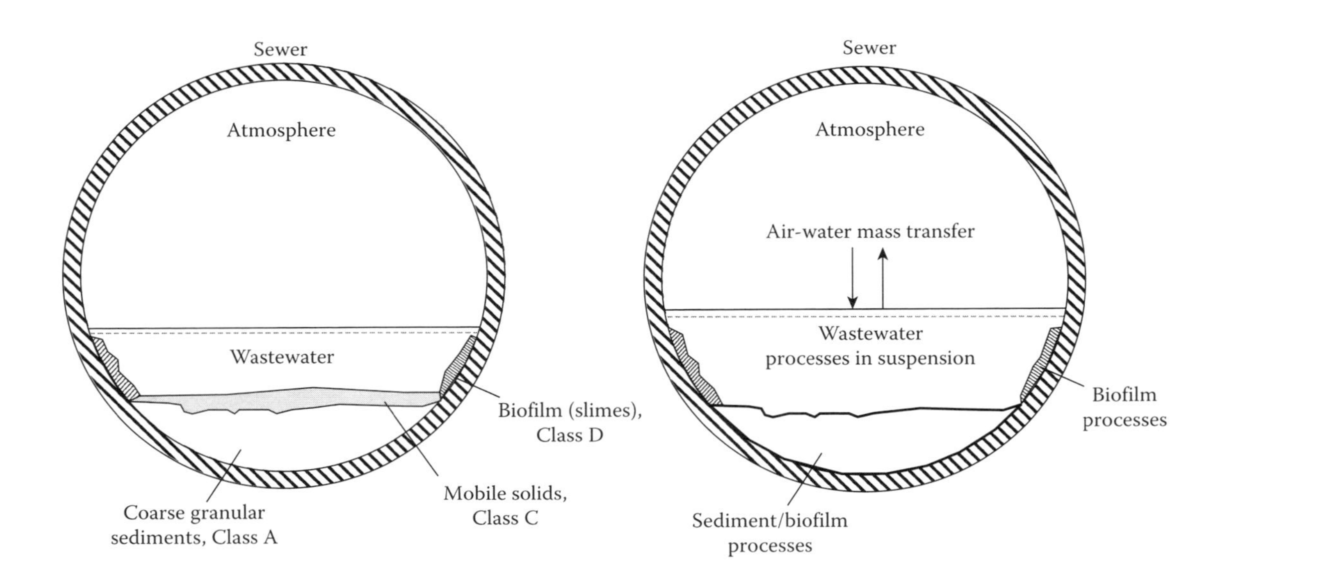

FIGURE 5.5 Outline illustrating processes and occurrence of sewer solids. (Modified from R. M. Ashley and T. Hvitved-Jacobsen, *Wet-Weather Flow in the Urban Watershed: Technology and Management*, 187–223, Lewis Publishers, Boca Raton, FL, 2003.) Further details of the sewer sediment types are dealt with in Section 2.3.3 and Table 2.4.

for settlement also exists for detached biofilms. The performance of the overflow structure is therefore crucial in terms of its ability to divert as much as possible of the sewer solids to the wastewater treatment plant and not into the overflow.

The solids in sewers play a specific and central role in the performance of a combined sewer network and therefore also for the pollution from CSOs. These solids are defined as follows:

- Solids in the water phase (i.e., suspended particles and colloids)
- Sewer sediments (i.e., deposits and near-bed materials)
- Biofilms (sewer slimes)

The transport characteristics and the occurrence of sewer solids under changing dry and wet weather conditions in the sewer network, particularly in terms of transport, deposition, and erosion, will be dealt with next.

5.3 DEPOSITION, EROSION, AND TRANSPORT OF SEWER SOLIDS

5.3.1 Sediment Transport and Characteristics

The suspended solids that are transported in a combined sewer during both dry and wet weather flows consist of inorganic particles like sand and different fractions of organic materials. The different size and density of the particles result in corresponding different transport characteristics. Depending on the flow regime and the characteristics of the particles, the flow of solids takes place in the bulk water phase as "wash load" and in the near-bottom water phase as "bed load" with a continuous exchange of solids between the bulk water phase, the bottom, and the near bed region (Ashley et al. 2004). Furthermore, the term "suspended load" is used to characterize the flow of particles that originate from deposits in the sewer. Typically, the solids concentration profile increases with increasing water depth and the dense particles are concentrated in the lower 10–25% part of the water phase where they are transported with a velocity lower than in the bulk water phase.

Basic characteristics of sewer solids—suspended solids, sediments, and biofilms—that are deposited in combined sewers are dealt with in Section 2.3.3. Sewer sediments constitute a complex mixture of different components and a simple description in terms of a mathematically formulated model of their stability and external forces when the flow of water is not possible. Sewer sediments are not like sediments in natural waters that are more or less granular and mainly inorganic and uniform in nature. Sewer sediments are to a large extent composed of cohesive materials like clay, silt, and different organic fractions. Physical, chemical, and biological processes in the sediments can consolidate and weaken it. Materials like fat and bitumen may change sewer sediments into a cemented structure whereas anaerobic biodegradation that produces gas bubbles with a content of methane and carbon dioxide might have a weakening effect (Vollertsen and Hvitved-Jacobsen 2000; Banasiak et al. 2005).

Such basic aspects characterizing sewer solids are further described in this chapter. An overview focusing on the prediction of sewer solids transport is dealt with in Banasiak and Tait (2008).

5.3.2 Deposition of Sewer Solids

In order to reduce the risk for accumulation of deposits, it is crucial to construct self-cleansing sewers. The velocity, or better the shear at a sewer pipe wall, must, therefore during dry weather flow, be sufficiently large to avoid deposition and to keep the biofilm thickness low.

Due to the importance of the sewer network and the variability of wastewater characteristics, it is clear that it is not simple to quantify deposition or erosion. Stokes's law or extended versions of it is, for practical purposes, often not applicable (cf. Section 3.4.2). A rather simple approach related to self-cleansing conditions is based on the determination of the magnitude of the shear stress (cf. Sections 3.4.2 and 4.4.2). In the case of flow in a sewer pipe, the shear stress is defined as the force exerted by the flowing water per unit area of the pipe wall and formulated by the following equation (cf. Equation 4.1 in Section 4.4.2):

$$\tau = \gamma R I \qquad (5.5)$$

where

τ = shear stress (kg m^{-1} s^{-2}) or (N m^{-2})
γ = specific weight of water, the gravitational acceleration multiplied with the density of water (kg m^{-2} s^{-2})
$R = A/P$ = hydraulic radius (m)
A = wetted cross-sectional area of the pipe (m^2)
P = wetted perimeter of the pipe (m)
I = energy gradient line; approximated with the slope of sewer bottom line (m m^{-1})

As a first approximation, the magnitude of τ determines whether deposits in a sewer network will occur or not. From a pragmatic point of view, it can be accepted that sewer solids deposit temporarily, during low flow rate conditions in the night if these deposits are eroded and brought into suspension during hours of high flow rate. According to this pragmatic approach, a sewer pipe should be maintained "clean" at least one hour each day resulting in a low level of sewer solids deposition that otherwise—and in addition to a reduced flow capacity of the pipe—might contribute to the contents of pollutants in the CSOs. The values of shear stresses shown in Table 5.1 refer to a frequency of occurrence that is determined as a value in average occurring during one hour each day.

TABLE 5.1
Empirically Based Relation between the Magnitude of Shear Stress and the Occurrence of Deposits in Sewer Pipes

Shear Stress, τ (N m^{-2})	Extent of Deposits
$\tau > 1.5$	No permanent occurrence of deposits
$0.5 < \tau < 1.5$	Moderate risk for occurrence of deposits
$\tau < 0.5$	High risk for occurrence of deposits

It must be stressed that the values shown in Table 5.1 be considered crude estimates and that such a simple approach is just an indication for occurrence of deposits. The values in Table 5.1 are, however, based on a number of investigations (e.g., by Brombach 1992) who concluded that in typically combined sewers, about 90% of normal sediments can be transported with a minimum shear stress of 1.6 N m^{-2}. It is, however, seen that higher values compared with those in Table 5.1 are recommended for design of self-cleansing sewers and referred to as critical shear stress values, τ_{crit} (Ashley et al. 2004). A corresponding, but differently expressed, approach for erosion of open channels is described in Section 4.4.2.

A substitute for shear stress as a criterion for self-cleansing of sewers can, for practical reasons, be expressed as recommended values for sewer characteristics. As an example, such characteristics can, for transport of solids that are defined by the particle diameter (e.g., sand), be expressed as a relation between pipe diameter, minimum velocity of the flow, and minimum bed slope.

Sediment deposits tend to buildup over a longer time and it is argued whether the deposits reach an equilibrium state (Schellart et al. 2007).

5.3.3 Erosion of Sewer Sediments

The basic physical parameter determining erosion is the shear stress, τ. In case of inorganic granular and uniform materials, erosion occurs when a critical value, τ_{crit}, is exceeded (cf. Section 3.4.2.2). For sewer sediments the phenomenon is more complex and erosion can start at a rather low value of τ, about 0.5 N m^{-2} in case of consolidated sediments not being completed at a value more than 10 times as high.

Formulation of the erosion phenomena in model terms is therefore difficult and can only be done empirically or semiempirically and with uncertainty. The following empirical model developed by Wotherspoon (1994) and based on analysis of 61 sediment samples, however, gave a strong correlation between the critical shear stress, τ_{crit}, and the moisture content of the sediment

$$\tau_{crit} = 9.66 \times 10^7\, m^{-3.1682} \quad (5.6)$$

where

τ_{crit} = critical shear stress (kg m^{-1} s^{-2}) or (N m^{-2})
m = moisture content of sediment in a combined sewer, mass of water per unit mass of solids (%)

According to Torfs (1995), the rate of erosion is determined by the excess shear stress, the difference between the shear stress at the sediment surface and the critical shear stress:

$$E = E_m \left(\frac{\tau - \tau_{crit}}{\tau_{crit}} \right)^{\alpha} \quad (5.7)$$

where

E = rate of erosion (g m^{-2} s^{-1})
E_m = erosion constant (g m^{-2} s^{-1})
α = constant.

Equation 5.7 is valid in case the clay particle content >3–15%. Typical values for E_m are 0.1–3.7 g m^{-2} s^{-1}. The lower values of E_m correspond to erosion of loose deposits whereas the higher values apply to erosion of consolidated sediments. Typically, α is about 1.

Based on Equation 5.7, a model for estimation of the rate of erosion was developed by Skipworth, Tait, and Saul (1999). They formulated a model for the erosion rate of fine-grained organic rich sediments in combined sewer pipes, a type C solids (cf. Table 2.4 in Section 2.3.3) based on a varying value of τ_{crit} through this weak layer of sediment.

5.3.4 Detachment of Biofilms

Sloughing of sewer biofilms (sewer slimes) occurs intermittently as a natural process of biofilm growth and aging. The specific details of this process are not known, hence it is not possible to predict. During steady-state (dry weather) conditions in sewer networks the first estimate of detachment of aerobic biofilms correspond to the growth rate, typically between 10 and 70 g COD m^{-2} d^{-1} or a thickness of 50–100 μm d^{-1}.

Particularly in the case of wet weather flows, the shear stress on the sewer walls will increase abrasion of the biofilm. In general, the detached biofilm will easily settle.

Further aspects related to solids in combined sewers are found in Ashley and Hvitved-Jacobsen (2003) and Ashley et al. (2004).

5.4 EXTREME EVENT AND ANNUAL LOAD CALCULATIONS

The variability of the different phenomena associated with urban runoff is of fundamental importance when dealing with quantification of pollutant loads and environmental impacts. This variability must therefore be taken into account when engineering the CSO-related aspects. If data are available, modeling based on long historical rainfall series and stochastic approaches are in general the best basis for design and management. Sufficient environmental and system-related data and appropriate models are, however, not always available and other more simple means must be applied. This section dealing with methods for calculation of pollution loads from overflow structures will highlight the basic understanding needed for application of models, but also aims at giving the reader methods for approximations for simple calculations. It is crucial to stress that results from the simple calculations that are performed with basic understanding and care are not necessarily less valuable than results obtained from complex models.

5.4.1 Mass Balance at an Overflow Structure

Pollutants in a combined sewer network originating from the three major sources dealt with in Section 5.2 occur as a mixture of runoff water and wastewater. From measurements under dry weather conditions, the pollutant contribution from the wastewater during a runoff event can be fairly well estimated. However, what pollutants have their origin from the urban surface runoff or the eroded sewer solids is not easy to determine. When dealing with pollutant loads from the CSOs, it is from a pragmatic point of view appropriate to operate with only two contributing sources for the pollutants:

- Pollutants originating from the daily wastewater flow
- Pollutants originating from the urban surfaces and the eroded sewer solids

This approach follows the description made in Section 2.6.1. The water transported in a combined sewer network during a runoff event is therefore—although mixed together—from a mass balance point of view composed of two parts:

$$V_{tot} = V_{ww} + V_{run} \qquad (5.8)$$

where

V_{tot} = total volume of water that during a runoff event passes a cross section of the inflow pipe into an overflow structure (m^3 per event)

V_{ww} = dry weather volume of wastewater that passes a cross section of the inflow pipe into an overflow structure during a runoff event (m^3 per event)

V_{run} = volume of runoff water into an overflow structure originating from the catchment (m^3 per event)

In terms of the mass balance shown in Equation 5.8, it is important to understand that it is the *runoff event* and not the *overflow event* that is relevant. In Figure 2.10, the difference between these two types of events is outlined. A runoff event includes all pollutants that belong to one specific event whereas the overflow event just takes a portion of them. Furthermore, the extent of an overflow event is determined by the construction of the overflow structure (e.g., the position of the overflow weir). Only a mass balance, and corresponding sampling for analysis, based on a runoff event that capture the entire amount of pollutants related to a runoff event will therefore provide results that can be applied in a meaningful way for other network systems.

The value of V_{tot} in Equation 5.8 is determined from flow measurements in the sewer network. The V_{ww} is typically estimated from corresponding daily dry weather flow measurements taking into account the daily, weekly, or even yearly variability in the flow. The V_{run} can thereby be determined from Equation 5.8. However, this value may also, and often in a better way, be estimated based on actual rainfall and catchment data.

In continuation of Equation 5.8, the following mass balance for a pollutant transported through a cross section of the inlet pipe to an overflow structure is expressed for the entire runoff event (cf. Equation 2.24 in Section 2.6.1):

$$V_{tot}\, C_{tot} = V_{ww}\, C_{ww} + (V_{tot} - V_{ww})\, C_{runoff} \qquad (5.9)$$

where

C_{tot} = concentration (EMC-value) of a constituent experimentally determined from regular sampling during the entire runoff event ($g\ m^{-3}$)

C_{ww} = estimated concentration of a constituent in the dry weather wastewater flow (g m^{-3})

C_{runoff} = concentration (EMC-value) of a constituent in the runoff water flow of the combined sewer (g m^{-3}).

All parameters in Equation 5.9 except for C_{runoff} can be determined from either measurements or, to some extent, also based on model simulations (cf. Section 2.6.1). The important conclusion is that Equation 5.9 is the basis for determination of C_{runoff}.

Determined in this way, C_{runoff} is an event-based mean concentration of a constituent originating from both the urban surfaces and the eroded sewer solids. This value refers to the (fictitious) runoff water volume ($V_{run} = V_{tot} - V_{ww}$) in the sewer network. The C_{runoff} therefore becomes *the* central pollutant concentration related to wet weather pollution from CSOs. Due to its nature, C_{runoff} is an EMC characteristic parameter (cf. Section 2.6.1). The COD values from combined sewers previously referred to in Figures 2.14 and 2.15 are all EMC values determined in the same way as described for C_{runoff}.

There is, in the international literature, a confusion concerning the details for definition and measurement of C_{runoff}. In several cases, measurements have been performed in the overflow (i.e., related to an overflow event and not a runoff event). Determined in this way, only a portion of the wet weather pollutants transported in the sewer is taken into account. Only because of specific circumstances (e.g., eroded sediments often play a dominating role compared with the contributions from the wastewater or because of a low value of the mixing ratio), this conflict may result in minor deviations. However, such deviations remain unknown and cannot be clearly identified because of other types of variability. A comparison between C_{runoff} values—and what might be expressed as C_{CSO} values—across the world is therefore difficult. Measurements performed in the overflow itself will, in principle, never be comparable from one system to another because of different system characteristics of the network, particularly due to construction details of the overflow structure (e.g., the height of the overflow weir and the capacity of the interceptor). Furthermore, occurrence of a first flush phenomenon can be observed when monitoring the runoff event, not necessarily seen in the overflow event. A source related mass balance must include pollutants from the entire runoff event.

Referring to Figure 5.6, it is crucial to understand that point #1 is relevant for sampling during the entire runoff event. Sampling at this point is therefore the best choice for comparison of data between water quality characteristics of combined

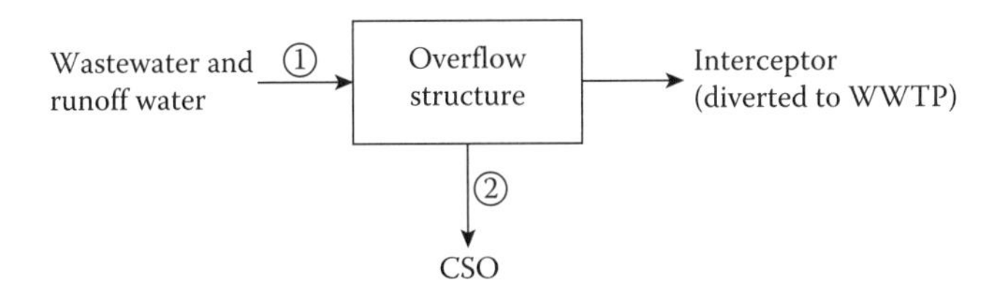

FIGURE 5.6 Points for sampling and monitoring at an overflow structure.

sewage from different network systems. However, an uncertainty is that potential solids deposition in the overflow structure is not taken into account. Sampling at point #2 refers to an overflow event and is thereby appropriate for determination of CSO loads from a specific overflow structure.

In contrast to what is the case in Scandinavian countries, there is no clearly specified phrase in British English or American English for the fictitious flow of runoff water, $V_{run} = V_{tot} - V_{ww}$, when it occurs in a combined sewer network. When occurring in this network it includes, although mixed with wastewater, the "wet weather pollutants" originating from both the urban surfaces and the eroded sewer solids. The mass balance, Equation 5.9, is therefore the basis for determination of the corresponding concentration, C_{runoff}.

5.4.2 Annual Pollutant Loads from an Overflow Structure

Determination of annual loads is of particular relevance when dealing with pollutants with accumulative effects (cf. Section 1.3.2). Furthermore, pollutant loads, during a specific period of time, may serve as an indicator for the overall performance of an overflow structure.

Equation 5.9 expresses the mass balance for a runoff event at an overflow structure and includes a concentration, C_{runoff}—an EMC value—that refers to the volume of runoff water in the combined sewer network and a concentration, C_{ww}, which refers to the contribution from the dry weather wastewater flow. Based on these values, the mass balance for a pollutant discharged via the overflow can be expressed by applying the annual average mixing ratio, m_y (cf. Equation 5.3):

$$C_y = \frac{C_{ww} + m_y C_{runoff}}{1 + m_y} \tag{5.10}$$

where

C_y = annual mean concentration of a pollutant in the CSO (g m^{-3})

To comply with this definition of C_y, the values selected for both C_{ww} and C_{runoff} must be the best available estimates of annual mean concentrations. In case of C_{runoff}, a site mean concentration (SMC) is probably the best choice.

Based on Equation 5.10, the annual amount of a pollutant that is discharged into the receiving water is

$$P_y = V_y C_y \tag{5.11}$$

where

P_y = annual amount of pollutant discharged (g)

V_y = annual volume of overflow (m^3)

In general, V_y should be determined from simulations with a runoff model based on a historical rainfall series and relevant sewer network characteristics.

In theory, V_y can be estimated from those single events that cause overflow. A rather simple estimate of the volume of water that is discharged from a single event is

$$V_0 = R\,F - Q_s\,t_r - V_b \tag{5.12}$$

where
V_0 = volume of water discharged during a single runoff event (m^3)
R = rainfall (runoff) depth of an event (m)
F = contributing area of the connected catchment at point of overflow structure (m^2)
Q_s = flow rate capacity of the interceptor (m^3 s^{-1})
t_r = duration of the runoff event; the duration of the rain event is a first estimate (s)
V_b = detention capacity in the sewer (m^3)

Equation 5.12 includes those terms that influence the magnitude of the volume of water discharged from an overflow structure. It is, however, important to stress that this approach is often too simple to give reliable results because the equation does not take into account the dynamics of the flow during a runoff event. This dynamic behavior includes both rainfall-runoff relations (cf. Section 2.2.1), and the runoff flow relative to the capacity of the overflow structure. The application of a runoff model based on a historical rainfall series is therefore the only reliable way to determine the CSO volume.

5.4.3 Pollutant Load from an Overflow Structure during a Single Event

Pollutant loads from a single event is particularly important in case of a pollutant with an acute effect (cf. Section 1.3.2). Events of importance are therefore the extreme ones.

Computation of the pollutant load from a single event can best be performed by application of a runoff model with the rainfall hyetograph for this event. The pollutant concentrations, in terms of values for both the wastewater and the runoff water, can be based on monitored results but are typically best estimates of C_{ww} and C_{runoff}, respectively. There is no statistical evidence that extreme values of overflow volumes result in corresponding extreme concentrations, high or low. From a logical point of view, low values could be expected because of dilution, however, high values might also occur because of sediment erosion. In practice, a best estimate for C_{runoff} is typically still the SMC value.

If a model computation is not performed, a simple estimate of pollutant loads from a single overflow event can take its starting point from the fact that the concentrations, C_{ww} and C_{runoff}, refer to the contributions from the wastewater and the mixed contributions from the urban surfaces and the eroded sewer solids, respectively. The volumes discharged corresponding to these two contributions in the overflow are

$$V_0 = V_{0,ww} + V_{CSO} \tag{5.13}$$

where

$V_{0,\text{ww}}$ = volume of wastewater discharged during a single overflow event (m^3)
V_{CSO} = volume of runoff water discharged during a single overflow event (m^3)

In this respect, one must remember that Equation 5.8 concerns the entire runoff event whereas Equation 5.13 only refers to the volume that passes the overflow weir. An approximation and a pragmatic approach is to express the ratio between the two partial volumes of Equation 5.13 as follows (cf. Equation 5.12):

$$\frac{V_{0,\text{ww}}}{V_{\text{CSO}}} = \frac{t_r\, Q_{\text{ww}}}{R\, F} \tag{5.14}$$

An estimated pollutant load from a single (extreme) event is therefore

$$P_0 = V_0\left(\frac{t_r\, Q_{\text{ww}}}{R\,F + t_r\, Q_{\text{ww}}} C_{\text{ww}} + \frac{R\,F}{R\,F + t_r\, Q_{\text{ww}}} C_{\text{runoff}}\right) \tag{5.15}$$

where

P_0 = amount of pollutant discharged from a single event (g)

5.5 CHARACTERISTIC POLLUTANT CONCENTRATIONS AND LOADS

Data characterizing urban drainage phenomena are subject to considerable variability in time and from site to site (cf. Section 2.4). The pollutant concentrations in runoff water from combined sewers and from stormwater runoff shown in Table 5.2 are therefore rough estimates although they are considered typical levels for Northern Europe.

It is interesting to compare the C_{runoff} concentrations that include contribution of pollutants from both urban surfaces and eroded sewer solids with the corresponding C_{SWR} concentrations that only include pollutants from the first source mentioned. The difference between the C_{runoff} values and the C_{SWR} values will, to some extent, reflect the contribution from eroded sewer solids. It is concluded that sewer solids play a dominating role for the CSO concentration level of organic matter, nitrogen, and phosphorus whereas metals mainly seem to originate from the urban surface runoff. It is again important to stress that the concentration values only refer to the runoff water and therefore does not include contributions of constituents originating from the daily wastewater flow in the combined sewer.

In several countries it was, and often still is, common to sample for analysis of CSO components in either the overflow or in the wet weather stream of the sewer and not subtracting the contribution of wastewater originating from the dry weather flow. In contrast to what is done in this text, reported pollutant concentrations in the combined sewer flows often include not just contributions from the urban surfaces and eroded sewer solids but also pollutants from the daily wastewater flow (cf. Section 5.4.1). Such reported concentration values are therefore typically larger than those shown in Table 5.2, however, still strongly depending on the design of the

TABLE 5.2
High and Median Values of Pollutant Concentrations, C_{runoff}, for Runoff Water in a Combined Sewer*

Pollutant (Unit)	Runoff Water in a Combined Sewer Network		Stormwater Runoff from Urban Surfaces and Roads (Median Values)	
	C_{runoff} Median	C_{runoff} High	C_{SWR} Urban	C_{SWR} Road
BOD_5 ($g\ m^{-3}$)	25	40	8	5
COD ($g\ m^{-3}$)	120	200	50	30
TSS ($g\ m^{-3}$)	150	250	100	100
Total Kjeldahl Nitrogen, TKN ($g\ m^{-3}$)	10	15	2	2
Total P ($g\ m^{-3}$)	2.5	4	0.5	0.3
Total Pb ($mg\ m^{-3}$)	30	100	30	30
Total Cu ($mg\ m^{-3}$)	30	50	20	20
Total Zn ($mg\ m^{-3}$)	300	500	300	200

Note: Corresponding concentrations of stormwater runoff, C_{SWR}, for pollutants originating from urban surfaces and roads are included for comparison. Data are based on several reported investigations originating from Northern Europe.

*Notice that contributions from the daily wastewater flow are not included (cf. Section 5.4.1).

sewer network and the concentration levels in the dry weather flow of wastewater. The data shown in Table 5.3 originate from Lager et al. (1977) and include contributions from the daily wastewater flow. Compared with the results reported in Table 5.2, these data are in general considerably larger.

A comprehensive study of the potential wet weather loads originating from combined sewers were carried out in six catchments in Paris (Kafi et al. 2008). Similar to the data shown in Table 5.3, contributions from pollutants originating from the daily wastewater flow are also included in the values reported in Table 5.4. A wet weather event is in this case pragmatically defined as the time span during which the water level at the outlet of a catchment exceeds the highest recorded value during a dry weather period. In Table 5.4, the pollutant concentrations of the wet weather flow are compared with corresponding values determined under dry weather conditions.

Based on the results shown in Table 5.4, it is interesting to notice that the flow of *runoff water* in the combined sewer (Paris) does not result in a *dilution* of the daily wastewater flow. It should be expected that suspended solids and organic matter primarily originate from eroded sewer solids whereas metals and PAHs may particularly originate from the urban surfaces.

The concentration levels of C_{runoff} (or C_{CSO}) and C_{SWR} cannot always be directly compared for a specific pollutant. As an example, COD values from a combined sewer include contributions of organic matter with a relatively higher biodegradability than those originating from a storm sewer. The reason is that organics from eroded sewer solids (i.e., settled wastewater components) may include easily degradable

TABLE 5.3

Average Pollutant Concentrations in CSOs Based on a Summary of Data from Several Studies

Pollutant (Unit)	C_{CSO}*	C_{SWR}	C_{sewage}
BOD_5 (g m^{-3})	115	20	200
COD (g m^{-3})	365	115	500
TSS (g m^{-3})	370	415	200
Total Kjeldahl Nitrogen, TKN (g m^{-3})	3.8	1.4	40
Total N (g m^{-3})	9	3	40
PO_4-P (g m^{-3})	1.9	0.6	10
Total Pb (mg m^{-3})	370	350	–

Source: Data from Lager, J. A., Smith, W. G., Lynard, W. G., Finn, R. M., and Finnemore. E. J., Urban stormwater management and technology: Update and users' guide. U.S. Environmental Protection Agency, Report EPA 600/8-77-014. 1977.

Note: Data from stormwater runoff and sanitary sewage are included for comparison.

*The concentration values include contributions from the daily flow of municipal wastewater. These concentrations are therefore referred to as C_{CSO} and not C_{runoff}.

TABLE 5.4

Pollutant Concentrations in the Wet Weather Flows and the Dry Weather Flows in Combined Sewers Originating from Measurements in Six Catchments in Paris

Pollutant	Wet Weather Flow Concentrations d_{10}–d_{50}–d_{90}	Dry Weather Flow Concentrations d_{10}–d_{50}–d_{90}
TSS (g m^{-3})	174–279–403	157–198–243
VSS (g m^{-3})	135–213–317	140–171–211
COD (g m^{-3})	286–432–633	315–388–528
BOD_5 (g m^{-3})	116–158–244	133–181–211
TKN (g m^{-3})	15–25–35	30–36–43
Cadmium, Cd (mg m^{-3})	0.64–1.20–2.03	0.28–0.50–0.70
Cupper, Cu (mg m^{-3})	66–130–231	60– 81–115
Lead, Pb (mg m^{-3})	55–98–289	16–22–34
Zinc, Zn (mg m^{-3})	760–1120–1832	131–172–388
PAHs (mg m^{-3})	1.04–2.12–4.81	0.37–0.80–1.12

Source: Data from Kafi, M., Gasperi, J., Moilleron, R., Gromaire, M. C., and Chebbo. G., *Water Research,* 42 (3), 539–49, 2008.

Note: The results are reported as median values (d_{50}) and as 10% (d_{10}) and 90% (d_{90}) percentiles, respectively.

TABLE 5.5
Typical Percentages of Pollutants Associated with Suspended Solids

Pollutant	Runoff Water in a Combined Sewer Network (%)	Stormwater Runoff from Urban Surfaces and Roads (%)
COD	70–80	—
Total P	40–60	60–80
Total Pb	—	70–80
Total Cu	—	30–40
Total Zn	—	30–40

Note: The constituents from the daily wastewater flow are not included in values given for the runoff water in a combined sewer network.

components whereas organics from stormwater flows originating from plant materials are low degradable.

In terms of both treatment and pollutant transport phenomena including accumulation, it is important to what extent the pollutants exist in soluble or particulate form. Based on a number of studies, Table 5.5 gives rough values with stormwater runoff data included for comparison.

5.6 CONTROL OF COMBINED SEWER OVERFLOWS

5.6.1 General Approaches for CSO Control

Control of CSOs means that measures are taken to reduce the load of pollutants discharged into adjacent receiving water bodies. *Control* is therefore observed if either the volume or the concentration of the discharge is controlled. As an example, *volume control* is possible by detention and diverting the flow via an interceptor to a treatment plant. The concept of *concentration control* can be accomplished by both treatment and flow control, the latter in an overflow structure by diverting a relatively high concentrated bed load to an interceptor. Methods for control of CSOs are therefore legion and include both structural and nonstructural measures or such measures in combination. Flow regulation is an integral part of such measures but will in the context of this text only be dealt with as a principle.

The stochastic nature of the rainfall, the variability in land use and the lack of solid knowledge on the efficiency of treatment measures are major obstacles for design of CSO control systems (Geiger 1998). The technology selected for treatment of CSOs often has its origin in systems applied for mechanical or chemical treatment of municipal wastewater. Adjustments of this technology are in general needed to perform in an efficient way under considerable varying flow and pollutant load conditions. Empirically based knowledge on the treatment performance of CSO controls is therefore central for design of such systems to observe specific receiving water quality objectives. Systems with a high flexibility—also after the construction is completed—are therefore of major importance for successful operation.

It is not possible, nor the objective of this text to describe all possible controls and treatment technologies, but commonly used methodologies and representative examples will be dealt with in the following sections.

5.6.2 Detention and Storage Facilities

The use of basins, tanks, and the network itself for storage purposes in combined sewer networks is probably the most common measure for hydraulic control of CSOs. Such "pure" detention measures that concern the hydraulic performance of sewer networks during runoff are, in this context, not a subject for this text. However, storage facilities and measures that reduce the volume of the overflows and divert part of the wet weather flow to a wastewater treatment plant or provide conditions for settling of solids also reduce the loads of pollutants from CSOs. Furthermore, a number of structural measures for treatment and reduction of pollutant loads from CSOs require volume for storage (e.g., for equalization). Storage of the wet weather runoff and reduction of pollutant loads on adjacent receiving waters are therefore closely related.

As already mentioned, establishment of storage capacity in a combined sewer network not only observes hydraulic objectives by reducing the volume of the overflows but also correspondingly affects the amount of pollutants discharged. The following examples will illustrate this. Further examples and construction details of CSO tanks and detention systems are found in WPCF (1989) and Mays (2001).

Figures 5.7 and 5.8 outline the two main types of storage basins, an off-line storage tank and an in-line tunnel, respectively (cf. Figure 5.2).

A great number of additional controls and treatment technologies can be applied to reduce CSO loads (Field, Sullivan, and Tafuri 2003). What is possible and feasible depends on numerous site-specific conditions and needs. Only an analysis that includes the local environmental, technical, and financial constraints will give an appropriate answer on what methods should be selected.

The choice of appropriate methods for reduction of CSO loads depends on the entire drainage system including the catchment, the sewer network, the wastewater treatment plant, and the receiving watercourse. The actual distribution of the rainfall at the catchment and the variability in rainfall pattern from event to event will affect the choice of an optimal control method. The objective for the CSO reduction must be stated, depending on what effects are important, what season of the year is of interest, and what points of discharge are the most important. Computer simulations of selected scenarios for reduction of CSO loads are therefore often relevant. It is important to assess the potential impacts of such measures on the performance of the sewer network during dry weather periods.

All sewer networks are unique and the more specific knowledge available, the better the general knowledge on wet weather phenomena is to apply. Therefore, monitoring and system analysis that support the characterization of the possible CSO control options and provide information on corresponding receiving water impacts are crucial.

Figure 5.9 shows the principle of a deep-tunnel sewerage system for storage of what otherwise would be discharged as CSOs. To avoid problems, such storage systems require careful consideration similar to what is known from other large underground

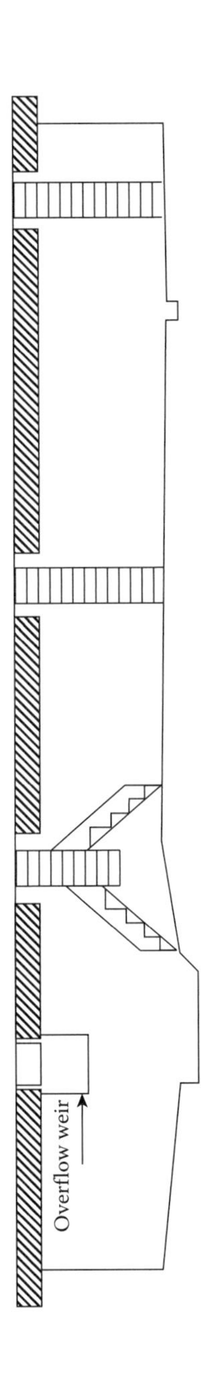

FIGURE 5.7 Outline of a large off-line storage tank for combined sewage placed under a parking area. The sewage in the tank is when transport capacity is available pumped to an interceptor.

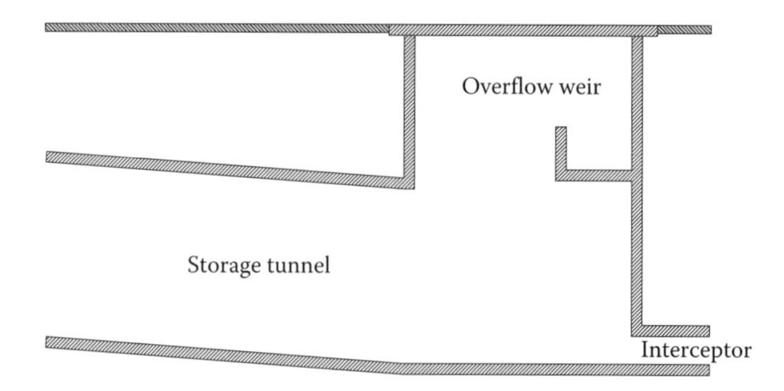

FIGURE 5.8 Outline of an in-line tunnel for storage of combined sewage.

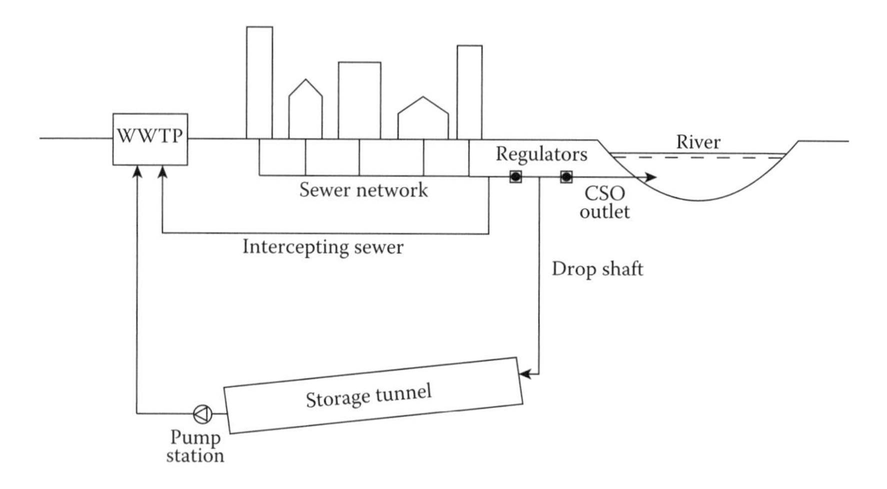

FIGURE 5.9 Principle of a deep-tunnel sewerage system for storage of CSOs.

civil projects. The geology and hydrogeology of the soil are therefore central. The use of corrosion resistant materials, proper hydraulic performance of the drop structures, self-cleansing systems, odor-control measures, and working shafts for safe entrance are examples of what are important to consider when constructing such facilities.

5.6.3 Structural Measures for Reduction of Pollutant Loads from CSOs

In addition to detention as dealt with in Section 5.6.2, a number of other structural measures exist. The following five main groups of methods are such options for reduction of CSO loads:

- Mechanical (physical unit operations)
- Chemical and physicochemical
- Biological

- In-sewer storage and flow management
- Reduction of surface runoff

When assessing the implementation of these measures, it is crucial to consider the conditions under which they should work. In this respect, they must comply with the fundamental characteristics of urban runoff. Therefore, any specific measure must ideally be able

> to manage and treat relatively large volumes of runoff with relatively low, but not negligible, concentrations of pollutants during a relatively short period of time—an event—that occurs stochastically.

When engineering a combined sewer network in terms of management and treatment, this combination of conditions comprises, when compared with the traditional management of the daily wastewater flow, a rather complex case to comply with.

In the following, different methods will be looked upon from this point of view. The five main methods will be separately described and exemplified, although some of them often operate in combination.

5.6.3.1 Mechanical Measures

Basically, mechanical measures, physical unit operations, should provide conditions for separation of particles from the combined sewer flow. There are two aspects of this separation: the solids are either concentrated in the flow diverted via an interceptor to a treatment plant or the solids should settle and the sludge being removed. The following methods belong to the group of mechanical measures:

- Basins and tanks that provide conditions for removal of solids by settling
- Lamella separators for settling of solids
- Sieves that remove gross solids
- Filters (e.g., sand filters that by high-rate filtration remove filterable solids)
- Screens for solid removal
- Vortex and swirl separators that remove particles subjected to an increased gravitational force

These measures can be located in separate ancillary works in the sewer network. Often, some measures provide volume for equalization of the flow to the treatment facility at the downstream end of a basin. Mechanical treatment units are designed following traditional approaches for such measures (cf. Tchobanoglous, Burton, and Stensel 2003). A few examples will illustrate this.

A simple methodology, the so-called Camp–Hazen equation, for the design of settlers was developed by Hazen (1904) and further improved by Camp (1946). This approach for removal of particles in tanks is based on the settling velocity for discrete particles, the inflow rate to the settler, and the horizontal settling surface. The Camp–Hazen equation, for the design of settling basins based on a limiting particle settling velocity is dealt with in Section 9.8 and applied in Example 9.6 for a catch basin removing sand particles.

Example 5.1: Removal of COD by Settling in Detention Basins

The basic theory for sedimentation is dealt with in Section 3.4.2.1. As discussed in this section, theoretical approaches for settling like Stokes's law have, however, a number of limitations for application under real conditions. As an example and based on simple empirically determined parameters, Figure 5.10 depicts how COD removal by settling can vary with time. A simple approach for this settling can be formulated in terms of a first-order removal rate:

$$COD_{rem} = COD_{sed}\,(1 - \exp(-k\,t)) \qquad (5.16)$$

where

COD_{rem} = COD fraction that is removed by settling at time t (g m^{-3})
COD_{sed} = COD fraction that can be removed by settling (g m^{-3})
k = first-order removal rate constant (h^{-1})
t = detention time in settling tank (h)

If it is assumed that 50% of COD_{sed} has settled within one hour of residence time in the settling tank, the first-order removal rate constant is $k = \ln(2)\ h^{-1} = 0.69\ h^{-1}$.

Equation 5.16 is in principle valuable for design and for assessment of settling performance in tanks. Locally performed measurements of settling characteristics for combined sewage are, however, needed to determine the parameters in Equation 5.16. Cumulative distribution curves for suspended solids determined on samples originating from the wet weather flow in combined sewers like those shown in Figure 5.11 are in this respect interesting. However, it must be realized that measurements performed under lab conditions do not necessarily reflect those observed in real systems. Curves like those shown in Figure 5.11 are just a starting point for determination of the parameters, COD_{sed} and k, needed for Equation 5.16 (cf. Ashley et al. 2004 for further details on the determination of the settling velocity for sewage). It is important that protocols have been developed to measure the settling velocity of urban wet weather flows (Gromaire et al. 2008).

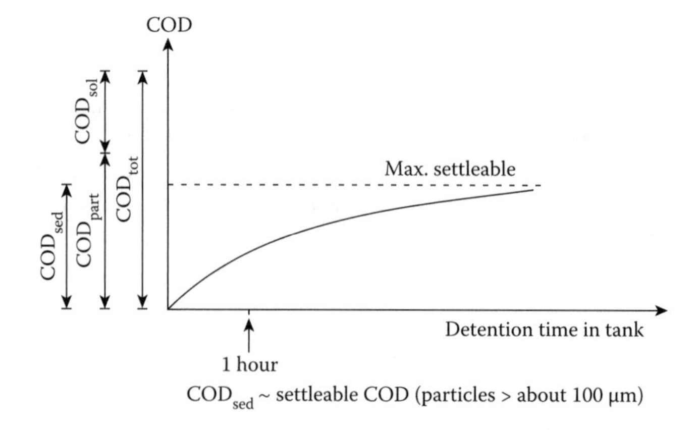

FIGURE 5.10 Outline of settling for COD in a detention tank.

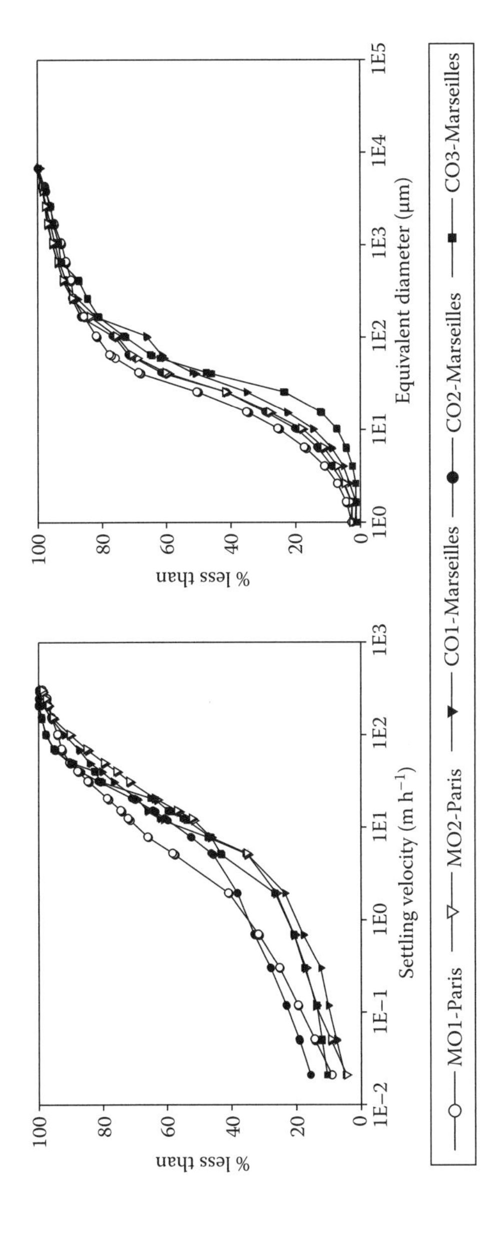

FIGURE 5.11 Cumulative distributions for particle size and settling velocities of particles in CSOs originating from combined sewer networks in Paris and Marseilles, France. (Modified from G. Chebbo, Solides des rejets urbains par temps de pluie: Caractérisation et traitabilité, PhD thesis, Ecole Nationale des Ponts et Chaussées, 1992).

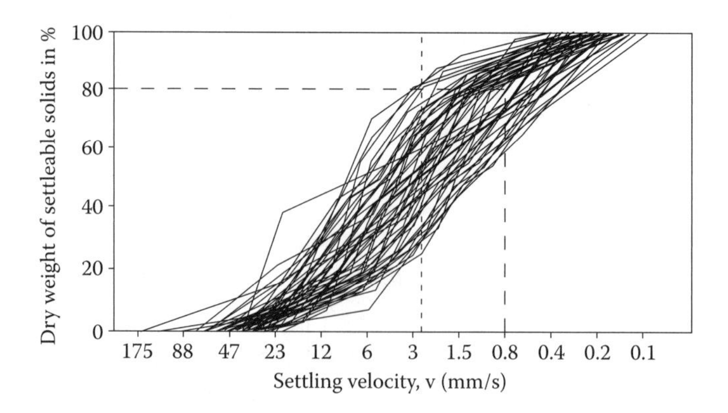

FIGURE 5.12 Cumulative distributions of settling velocities for solids in a mixture of wastewater and runoff water originating from a combined sewer. (Modified from S. Michelbach, *Water Science and Technology*, 31, 69–76, 1995).

The difficulties, and also the possibilities to apply results from studies on settling rates for sewage originating from combined sewers for the design of settling tanks, can be illustrated based on Figure 5.12 (Michelbach 1995). Each curve in this figure shows the result from one settling experiment performed on a sample of mixed wastewater and runoff water drawn from a combined sewer network. In this study, "settleable solids" refers to what will settle in two hours within an Imhoff-cone. There were 98 single experiments carried out of which about one third as representative curves are shown in Figure 5.12.

The surface loading rate is a rather simple criterion for design of settling tanks for combined sewage. As an example, this rate can be chosen equal to 10 $m^3\ m^{-2}\ h^{-1}$ (or 10 $m\ h^{-1}$) corresponding to a particle settling velocity $v_{design} = 2.8\ mm\ s^{-1}$. From Figure 5.12, and under the conditions selected, it is seen that between 30 and 80% of the solids were removed by settling (Michelbach 1995). From the figure it is also seen that solids removal with an average value of 80% requires a reduction in surface loading to about 0.8 $mm\ s^{-1}$ (2.9 $m\ h^{-1}$). Other investigations have shown that a settling rate of 0.5 $mm\ s^{-1}$ corresponds to 60–80% suspended solids removal and that these particles are larger than 50 μm. It seems clear that the variability in the different characteristics determining the conditions for settling may highly affect the removal efficiency.

Due to the weight distribution of particles from combined sewage as also shown in Figure 5.11, most of the curves in Figure 5.12 are *S*-shaped. The curves to the right-hand side mainly originate from nearly dry weather flows whereas the curves to the left mainly correspond to heavy storms. The higher settling velocities for samples corresponding to the latter group are probably caused by the better settling characteristics of the eroded sewer solids.

Inorganic particles settle faster than organic rich particles (cf. Figure 3.12) and the fastest settling organic particles in combined sewage are those with the lowest rate of biodegradation, Table 5.6. Therefore, inorganic materials and organic constituents with the lowest biodegradability are the compounds that tend to accumulate in the bottom materials of settling tanks. Compared with an average composition

TABLE 5.6
Typical Relative Distribution of Particulate Organic Matter of Different Biodegradability Expressed in Terms of COD

Settling Velocity (mm s^{-1})	$X_{s,fast}$: Fast Biodegradable Particles (% of COD)	$X_{s,med}$: Medium Biodegradable Particles (% of COD)	$X_{s,slow}$: Slowly Biodegradable Particles (% of COD)	X_{Bw}: Heterotrophic Biomass (% of COD)
>20	4	10	83	3
2–20	6	20	68	6
0.02–2	8	30	56	6

Source: Data from Vollertsen, J., Hvitved-Jacobsen, T., McGregor, I., and Ashley. R. M., *Water Science and Technology,* 39 (2), 231–41, 1999.

of the solids transported in a combined sewer, the organic and most biodegradable particle fractions are those expected to occur in the CSOs in the relatively highest concentrations. This fact may correspondingly have a negative impact on the receiving water quality (cf. Section 6.3).

In general, settling velocities for combined sewage recorded during wet weather will exceed what is measured for the wastewater itself (Gromaire et al. 2008). The contents of inorganic particulate matter originating from surface runoff and eroded materials from sewer deposits may be reasons for this observation.

Over time, a large number of empirical models have been developed for prediction of solids removal in different types of separators like tanks and basins. A major problem in this respect is the reliability of these models when applied on systems that are not sufficiently well defined. An overview of such models is presented in Kutzner, Brombach, and Geiger (2007).

5.6.3.2 Chemical and Physicochemical Measures

A physicochemical measure can be implemented in different ways. The following physicochemical processes are in several treatment systems for CSOs central unit operations (cf. Sections 3.2.4.1.2 and 3.5):

- Adsorption
- Coagulation
- Flocculation

As an example, addition of coagulants in terms of iron and aluminum salts or polymers may enhance the formation of larger particles with improved settling characteristics. Ballasted flocculation and flotation are specific treatment strategies for CSOs where the unit operations mentioned are important (cf. Examples 5.2 and 5.3).

In general, physicochemical treatment methods are appropriate technologies when dealing with stochastic occurring, short, and intensive runoff events because the treatment processes can be put into operation relatively faster and be designed to treat large volumes of low pollutant concentrations relatively faster. In addition to treatment of CSOs, these methods are also feasible for treatment of SWR.

Compared with mechanical treatment processes like settling, a chemical or physicochemical treatment technology might add a number of benefits in terms of treatment efficiency and footprint requirement. It is important that addition of chemicals enhances removal of pollutants like heavy metals and organic micropollutants that are predominantly associated with the small particles and colloids in the combined sewage. However, such treatment technologies are also typically expensive and require manpower for proper operation.

Example 5.2: Treatment of CSOs with the Ballasted Flocculation Method

The ballasted flocculation, high-rate treatment system combines chemical coagulation and flocculation of particles in the inflowing diluted wastewater (CSO) with settling of the flocs in a lamella separator after incorporation of microsand, Figure 5.13 (Plum et al. 1998; Young and Edwards 2003). The colloidal particles in the diluted wastewater are chemically destabilized to form a coagulated suspension and finally larger particles by adding polymers for formation of flocs (cf. Section 3.5). By incorporation of microsand into the flocs, as 0.1–0.3 mm diameter sand particles, the density of the flocs will be increased to enhance the settling rate in the lamellar separator. Settling rates ranging from about 100 m h^{-1} are thereby possible (Young and Edwards 2003). The coagulant is a metal salt, typically ferric chloride or aluminum sulfate (alum) also precipitating orthophosphate. The return flow from the lamellar separator passes a hydrocyclone where the sand is separated from the sludge (CSO solids) and returned to the process. The sludge from the hydrocyclone can be returned to the interceptor and thereby to a wastewater treatment plant.

Design of ballasted flocculation reactors (BFRs) are based on relatively simple empirical formulations and experience. The central design criterion is the hydraulic

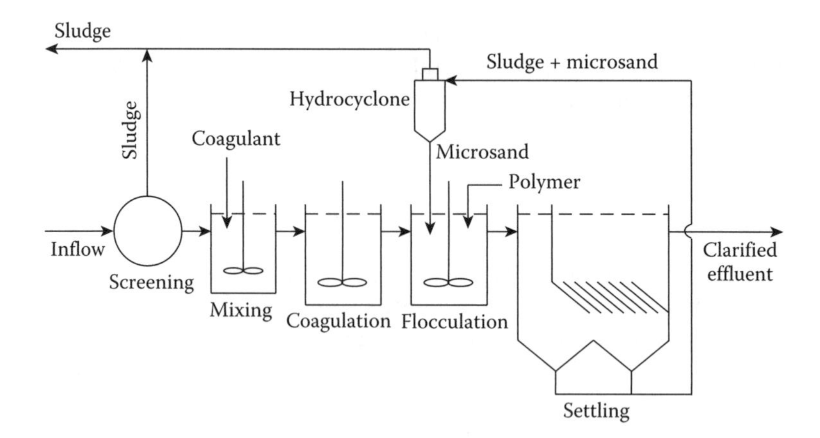

FIGURE 5.13 Outline of the ballasted flocculation treatment system.

load on the lamella separator taking into account the required treatment level for the CSO. For CSOs with TSS concentrations varying between 100 and 300 g m^{-3}, a recommended design value of the hydraulic load is in the order of 50–100 m^3 m^{-2} h^{-1} (m h^{-1}) corresponding to a removal efficiency for TSS > 70–80% although even higher hydraulic loads (100–150 m^3 m^{-2} h^{-1}) and corresponding removal efficiencies (85–90%) have been reported (Landon et al. 2006). The hydraulic load is calculated based on the wetted surface area of the lamella separator. With this design parameter, the following typical treatment efficiencies are reported for CSO (cf. Table 5.7).

The following Table 5.8 outlines rather high, however, typical levels of coagulants and polymers to observe the treatment efficiencies shown in Table 5.7.

Investigations have shown that the total alkalinity is a central parameter for determining an optimal dose of coagulants (El Samrani, Lartiges, and Williéras 2008).

In addition to the hydraulic load, a number of secondary criteria for design of the ballasted flocculation system must be observed. These criteria are related to the hydraulic residence time of the different parts of the treatment system in terms of the time needed for mixing, coagulation, and floc formation. These residence times are typically in the order of minutes (e.g., 2–3 minutes for mixing and coagulation and 3–8 minutes for both flocculation and settling). A total hydraulic residence time of 10–20 minutes for treatment of overflow water in a ballasted flocculation system is therefore typical.

TABLE 5.7
Typical Treatment (Removal) Efficiencies for CSOs in Percentage Obtained with the Ballasted Flocculation Method

	%
Suspended solids, TSS	70–90
COD	50–60
Total P	80–90

Source: Data from Plum, V., Dahl, C. P., Bentsen, L., Petersen, C. R., Napstjert, L., and Thomsen. N. B., *Water Science and Technology,* 37 (1), 269–75, 1998.

Note: The treatment efficiencies correspond to typical concentration levels in CSOs.

TABLE 5.8
Typical Concentration Levels of Coagulants and Polymers in Ballasted Flocculation Treatment Systems

Type of Chemical	Concentration Level (g m^{-3})
Ferric chloride	50–60
Aluminum sulfate (alum)	50–100
Polyaluminum hydroxychloride	20–25
Polymer	1.5–2.5

Addition of coagulant, polymers, and microsand can be controlled by flow regulation. In addition to the criteria mentioned, the volume of the inflow basin to the treatment system must be considered to ensure a stabile flow during runoff to the treatment processes.

A treatment method for CSOs like the ballasted flocculation method will generally reduce the load of pollutants on adjacent receiving water bodies. However, it can, depending on the local conditions, have further implications by keeping the water upstream in the catchment, by reducing needs for detention volume, and by reducing the hydraulic load on the wastewater treatment plant. Design and management of such a CSO treatment system must therefore be looked upon as a measure that might include considerations in terms of local rainfall (e.g., historical series), sewer capacity (e.g., intercepting capacity and volume for detention), receiving water impacts (e.g., including statistics like return periods for discharges), and wastewater treatment capacity (e.g., hydraulic loading and settling capacity). A hydraulic and pollutant load analysis that is based on appropriate models describing the performance of the entire wastewater system during a runoff event might therefore be appropriate when assessing and optimizing an option for treatment.

Example 5.3: Treatment of CSOs with Flotation

Flotation is a treatment methodology that combines settling and floating for removal of suspended particles. It is a technology that can be applied for treatment of CSOs prior to discharge into receiving waters or into an infiltration device. The suspended particles removed are normally solids, however, theoretically they may also be hydrophobic liquid particles like oil.

The central removal process for particles in dissolved air flotation is based on the injection of fine air bubbles into a reactor (flotation tank), Figure 5.14. Different procedures for the injection of air exist. However, injection of air dissolved in water under high pressure (dissolved-air flotation) is typically applied when dealing with wastewater and CSOs. During the injection, the release of the pressure causes the formation of fine air bubbles that, depending on the characteristics of the particles, can be attached to the particle surface. Addition of various chemicals (e.g., aluminum and ferric salts) to the suspension can create a particle surface structure that enhances the adsorption of the air bubbles. Particles with attached

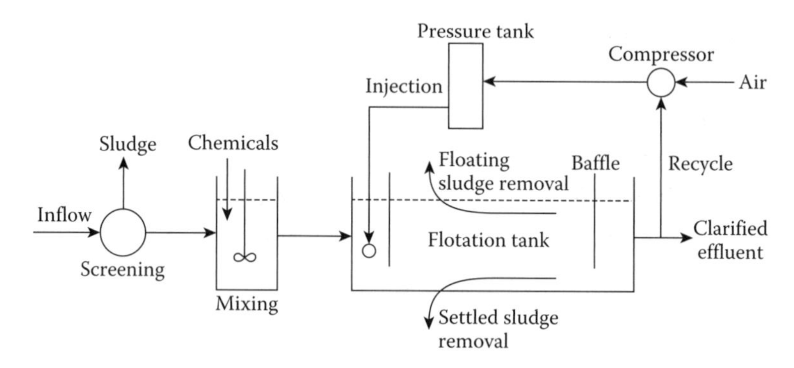

FIGURE 5.14 Principle of the flotation process with pressurized recycle flow.

air and with a bulk density lower than the suspension will rise to the surface and float whereas the more heavy particles settle. Floating materials are removed from the water surface of the flotation tank by skimming and settled materials are collected and removed.

The design of a flotation system largely depends on the type of the particles and the adsorption characteristics of the air bubbles onto the particle surface. Flotation is a complex methodology that requires laboratory or pilot scale experiments be performed prior to the design of a flotation system. A rather pragmatic design criterion in terms of the mass ratio of air injected to the mass of solids is normally applied. Empirical expressions for determination of this ratio including the operating pressure, removal efficiency characteristics, and influent concentration of suspended solids have been developed (Tchobanoglous et al. 2003). A screening design parameter for the air/solids mass ratio for CSOs may vary considerably (i.e., between 0.05 and 0.35; Field and Sullivan 2003). Other important design parameters for a flotation system for efficient solids removal are the solids loading rate, sedimentation rates for those particles that settle and rise-velocities for particles that will float. Typical values of design parameters for high and low overflow rate systems are in the order of 20 and 5 $m^3\ m^{-2}\ h^{-1}$, respectively. The hydraulic detention time of the water in the flotation tank may typically vary between 10 minutes and 1 hour (Field and Sullivan 2003). In terms of space and volume requirement for the treatment system, these design parameters are attractive compared with those for conventional settling.

5.6.3.3 Biological Measures

As mentioned in Section 5.1.2.1, a combined sewer network is typically designed with a value of the mixing ratio for the start of an overflow event, m_s, varying between 2 and 20. Depending on this design value and the available storage volume, a portion, often an important portion, of the wet weather flow in the network is diverted to a downstream treatment plant where it undergoes biological treatment. Biological treatment of the runoff is therefore to some extent indirectly taken into account by the design of the combined sewer system.

In addition to biological treatment of the flow entering the interceptor, biological treatment of the CSO is an option. However, it is in contrast to physicochemical treatment associated with basic problems:

- The transformation rates for the biological processes are, in general, rather low compared with the detention time of the wet weather flow that is typically realistic. These processes are therefore negatively influenced when treating CSOs at relatively low concentrations, at high flow rates, and during a relatively short period of time.
- Biological processes may require maintenance during interevent dry periods.

Biological treatment of CSOs requires process and design principles that can overcome such constraints (e.g., in combination with more conventional treatment). As an example, constructed wetlands where the CSO discharge is temporarily stored on top of a sand filter planted with reed can reduce the COD and solid contents considerably and at the same time nitrify ammonia (Gervin and Brix 2001; Uhl and Dittmer 2005; Dittmer, Meyer, and Langergraber 2005). To prevent clogging of the filter, pretreatment of the flow in a settling tank or another mechanical device prior to the inflow of the filter is needed. Outflow from the filter takes place through drainage pipes at the bottom.

5.6.3.4 In-Sewer Storage and Flow Management

The following two main interacting objectives of a combined sewer network are crucial to observe during wet weather periods:

- The frequency and the extent of CSOs should be reduced to a level acceptable for the receiving water, particularly in the case of overflows of high pollutant concentrations.
- The diluted wastewater, at least the most polluted part of it, should be diverted to a downstream treatment plant depending on its capacity to treat the wet weather flow.

Such integrated objectives can be accomplished, applying sound criteria and methods by optimal design and operation of a sewer network, its overflow structures, and the treatment plant located downstream. The following management options of the wet weather flow and use of an in-line storage capacity are closely related to the two points mentioned:

- Volume available for storage (detention) in the network itself will reduce overflows and increase the possibility for diverting the diluted wastewater to an interceptor and to treatment.
- Low-turbulent flow conditions in the sewer will enhance discharge of diluted wastewater with low concentration of particulates and corresponding high concentrations in the near-bed flow that can be diverted to treatment.
- Capture and diversion of a first flush, if it exists, or at least part of it will reduce the pollutant level in the overflow.
- Permanent settling of solids should be avoided and solids diverted to the treatment plant.

In principle, the simple point is to design and operate a combined sewer network that observes the two objectives and that results in a well-functioning system under both dry and wet weather conditions. The complex interactions can basically be analyzed by the use of hydraulic simulation models. Although in-sewer storage to some extent is comparable to detention in tanks and basins, it may enhance settling, which is not intended in a network.

5.6.3.5 Reduction of Surface Runoff

Reduced runoff flows from a catchment into the combined sewer system will correspondingly reduce the wet weather flow in the network and thereby the frequency and extent of CSOs. In general, space available at the catchment for detention and diversion of the flow is crucial. The following are examples that may also improve the possibility of the beneficial use of the flow:

- Infiltration of roof runoff (cf. Section 9.4).
- Collection of roof runoff in tanks and used for garden watering or toilet flushing following secondary treatment (cf. Section 8.2).
- Use of porous pavement for parking lots.

Such measures can be valuable by reducing peak flows.

5.6.4 Disinfection

Disinfection of CSO serves the purpose of controlling discharges of pathogens (cf. Section 3.2.1.6). The use of disinfection varies depending on the local regulations and legislation. In some countries and for some receiving waters, disinfection will not be required.

In principle, methods for disinfection of CSOs follow what might be used for the daily flow of wastewater from conventional treatment facilities (Tchobanoglous et al. 2003). Central design criteria for disinfection are the contact time and in the case of a chemical agent, its concentration and kinetic characteristics. Parameters like temperature and pH are for some methods also central.

Disinfection of CSOs operates under different conditions compared with disinfection of the effluent from a wastewater treatment plant. Such specific conditions are determined by the intermittent operation and a highly variable flow rate, a typical high concentration of particles in the overflow, and the fact that an overflow structure normally operates without regular supervision. The following three methods are those mostly used:

- *Chlorination*
  The chemical agent is chlorine (Cl_2) and its compounds: chlorine dioxide (ClO_2) and hypochlorite (OCl^-). One of the problems with the use of chlorine and hypochlorite is a potential reaction with ammonia and organic compounds producing substances that are toxic for aquatic life.
- *Ozone*
  Ozone (O_3) is a strong oxidizing agent. It destroys not only pathogens but to some extent organics and odor-causing compounds. Ozone is rather unstable and must therefore be produced on-site.
- *UV-light*
  The use of ultraviolet light is flexible and overcomes some of the problems associated with the nature of CSOs. To be effective, the overflow must, however, be pretreated to remove particles that otherwise might shade the microorganisms against the UV-light. Turbulent flow around the UV lamps can therefore also enhance the efficiency of the disinfection.

Disinfection is typically applied to protect bathing water against fecal pollution. A number of indicator microorganisms are used for assessment of the water quality (e.g., 6.6 fecal coliforms (FC), intestinal enterococci (IE), and escherichia coli (EC); cf. Section 6.6).

5.6.5 Measures for Improvement of Existing Overflow Structures

In addition to the structural measures, Sections 5.6.2 and 5.6.3, a number of controls exist that require little investment. The following methods are recommended by the U.S. Environmental Protection Agency (EPA) and are considered useful in this respect. Some of these controls have been included in the previous sections and are considered structural measures:

- *Sewer maintenance*
  Proper operation and regular maintenance of the sewer network can reduce the potential load of pollutants. As an example, flushing of a sewer line and CSO structures during a dry weather period can divert accumulated sediments to a point where the flow velocity is sufficiently high for its transport. Ultimately, the solids can be diverted to a storage sump (trap) from where it is regularly collected or to the treatment plant located downstream thereby reducing one of the important sources for the CSO loads.
- *In-sewer storage of runoff water*
  As already mentioned in Sections 5.6.2 and 5.6.3.1, the use of detention tanks in combined sewer networks can reduce CSO loads. Similarly, already existing capacity in a sewer network can be applied. The implementation of in-sewer storage may require that a device (e.g., an adjustable weir, a structural measure) be implemented at the downstream end of the collection system used for storage.
- *Increased wet weather flow to a downstream located treatment plant*
  Changes of flow regulation devices in a sewer network can affect the flow in interceptors to the treatment plant. Attention must be paid to the function of the treatment unit during wet weather periods.

REFERENCES

Ashley, R. M., J.-L. Bertrand-Krajewski, T. Hvitved-Jacobsen, and M. Verbanck, eds. 2004. *Solids in sewers: Characteristics, effects and control of sewer solids and associated pollutants.* Scientific and Technical Report No. 14. London: IWA Publishing.

Ashley, R. M., and T. Hvitved-Jacobsen. 2003. Management of sewer sediments. In *Wet-weather flow in the urban watershed: Technology and management*, eds. R. Field and D. Sullivan, 187–223. Boca Raton, FL: Lewis Publishers.

Banasiak, R., and S. Tait. 2008. The reliability of sediment transport predictions in sewers: Influence of hydraulic and morphological uncertainties. *Water Science and Technology* 57 (9): 1317–27.

Banasiak, R., R. Verhoeven, R. D. Sutter, and S. Tait. 2005. The erosion behaviour of biologically active sewer sediment deposits: Observations from a laboratory study. *Water Research* 39:5221–31.

Brombach, H. 1992. Solids removal from CSOs with vortex separators. Proceedings from Novatech 92, International Conference on Innovative Technologies in the Domain of Urban Water Drainage, Lyon, France, November, 447–59.

Camp, T. R. 1946. Sedimentation and the design of settling tanks. *ASCE (American Society of Civil Engineers) Transactions* 58: 895–936.

Chebbo, G. 1992. Urban runoff solids: Characterization and management. PhD thesis, Ecole Nationale des Ponts et Chaussées.

Dittmer, U., D. Meyer, and G. Langergraber. 2005. Simulation of a subsurface vertical flow constructed wetland for CSO treatment. *Water Science and Technology* 51 (9): 225–32.

El Samrani, A. G., B. S. Lartiges, and F. Williéras. 2008. Chemical coagulation of combined sewer overflow: Heavy metal removal and treatment optimization. *Water Research* 42 (4–5): 951–60.

Field, R., and D. Sullivan, eds. 2003. Management of wet weather flow in the urban watershed. In *Wet-weather flow in the urban watershed: Technology and management*, 1–41. Boca Raton, FL: CRC Press/Lewis Publishers.

Field, R., D. Sullivan, and A. N. Tafuri, eds. 2003. *Management of combined sewer overflows.* Boca Raton, FL: CRC Press/Lewis Publishers.

Geiger, W. F. 1998. Combined sewer overflow treatment: Knowledge or speculation. *Water Science and Technology* 38 (10): 1–8.

Gervin, L., and H. Brix. 2001. Removal of nutrients from combined sewer overflows and lake water in a vertical-flow constructed wetland system. *Water Science and Technology* 44 (11–12): 171–76.

Gromaire, M. C., M. Kafi-Benyahia, J. Gasperi, M. Saad, R. Moilleron, and G. Chebbo. 2008. Settling velocity of particulate pollutants from combined sewer wet weather discharges. *Water Science and Technology* 58 (12): 2453–65.

Hazen, A. 1904. On sedimentation. *ASCE (American Society of Civil Engineers) Transactions* 53: 45–71.

Kafi, M., J. Gasperi, R. Moilleron, M. C. Gromaire, and G. Chebbo. 2008. Spatial variability of the characteristics of combined wet weather pollutant loads in Paris. *Water Research* 42 (3): 539–49.

Kutzner, R., H. Brombach, and W. F. Geiger. 2007. Sewer solids separation by sedimentation: The problem of modeling, validation and transferability. *Water Science and Technology* 55 (4): 113–23.

Lager, J. A., W. G. Smith, W. G. Lynard, R. M. Finn, and E. J. Finnemore. 1977. Urban stormwater management and technology: Update and users' guide. U.S. Environmental Protection Agency, Report EPA 600/8-77-014.

Landon, S., C. Donahue, S. Jeyanayagam, and D. Cruden. 2006. Rain check. *Water Environment and Technology* 18 (7): 30–35.

Mays, L. W., ed. 2001. *Stormwater collection systems design handbook.* New York: McGraw-Hill.

Michelbach, S. 1995. Origin, resuspension and settling characteristics of solids transported in combined sewage. *Water Science and Technology* 31 (7): 69–76.

Plum, V., C. P. Dahl, L. Bentsen, C. R. Petersen, L. Napstjert, and N. B. Thomsen. 1998. The actiflo method. *Water Science and Technology* 37 (1): 269–75.

Schellart, A. N. A., F. A. Buijs, S. J. Tait, and R. M. Ashley. 2007. Estimation of uncertainty in long term combined sewer sediment behaviour predictions, a UK case study. Proceedings of the 5th International Conference on Sewer Processes and Networks, Delft, the Netherlands, August 29–31, 191–98.

Skipworth, P., S. Tait, and A. Saul. 1999. Erosion of sediment beds in sewers: Model development. *Journal of Environmental Engineering, ASCE (American Society of Civil Engineers)* 125 (6): 566–73.

Tchobanoglous, G., F. L. Burton, and H. D. Stensel. 2003. *Wastewater engineering: Treatment and reuse.* New York: McGraw-Hill.

Torfs, H. 1995. Erosion of mud/sand mixtures. PhD thesis, Katholieke University of Leuven, Belgium.

Uhl, M., and U. Dittmer. 2005. Constructed wetlands for CSO treatment: An overview of practice and research in Germany. *Water Science and Technology* 51 (9): 23–30.

Vollertsen, J., and T. Hvitved-Jacobsen. 2000. Resuspension and oxygen uptake of sediments in combined sewers. *Urban Water* 2: 21–27.

Vollertsen, J., T. Hvitved-Jacobsen, I. McGregor, and R. M. Ashley. 1999. Aerobic microbial transformations of pipe and silt trap sediments from combined sewers. *Water Science and Technology* 39 (2): 231–41.

Wotherspoon, D. J. J. 1994. The movement of cohesive sediment in a large combined sewer. PhD thesis, Dundee Institute of Technology, UK.

WPCF. 1989. *Combined sewer overflow pollution abatement, No. FD-17.* Alexandria, VA: Water Pollution Control Federation Manual of Practice.

Young, J. C., and F. G. Edwards. 2003. Factors affecting ballasted flocculation reactions. *Water Environment Research* 75 (3): 263–72.

6 Effects of Combined Sewer Overflows and Runoff from Urban Areas and Roads

The discharges of pollutants originating from stormwater runoff (SWR) and combined sewer overflows (CSOs) were dealt with in Chapters 4 and 5, respectively. In Chapter 6, the effects of these discharges will be focused on. In very general terms, the urban wet weather effects are briefly described in Section 1.3 and outlined in Table 1.1. In particular, the general characteristics in terms of time and spatial effects of pollutants like acute and accumulative effects are dealt with in Section 1.3.2. The different types and characteristics as well as the extent of these effects are central for why and how to manage and control the wet weather discharges. This chapter is, in this respect, a prerequisite for the following chapters focusing on engineering solutions to the wet weather discharges.

Each receiving water system has its very specific characteristics. The discharges are also specific and the effects originating from a discharge of pollutants vary in principle from location to location and from time to time. The mode of wet weather induced discharges is a central phenomenon in this respect. This mode of discharge can briefly be expressed as either an acute or an accumulative effect, formulated in terms of the time and the spatial scales of the effect. A general formula that describes the effect of a pollutant load taking into account its mode does, however, not exist when it comes to the detailed level. In spite of that, such difficulties do not alter the fact that prediction of the effects from urban and road wet weather discharges is a central task that must be dealt with. The authors are aware of the difficulties. It is a challenge to approach these difficulties in a conceptual as well as a pragmatic way. In this respect it is considered relevant to include examples, not only for illustration of a theory or a concept, but also to highlight general characteristics of a phenomenon.

It is not always possible to identify an effect caused by the wet weather discharges as a direct harmful impact onto humans or the ecosystem. The load-effect relations are very complex. Particularly, the lack of detailed knowledge on how the different pollutants and their actual speciation cause an effect, the relative importance of and interactions between the dry and wet weather loads, and the lack of appropriate methods for analysis and monitoring are examples that are obstacles for a deeper insight. A pragmatic measure and substitute of an effect is therefore often needed. It might sometimes be expressed in terms of a load or a concentration below which no harmful impact should be detected.

As an example, fish species can negatively be affected by CSOs but it is not easy to determine what is the cause of that. Is the harmful effect caused by DO depletion, a high concentration of nondissociated ammonia or toxic micropollutants in the overflow and is it a synergistic effect? It is therefore difficult to establish a sound and generally expressed criterion for the harmful effect of such wet weather impact for such criterion to manage the discharges in a meaningful way. A critical DO concentration caused by degradation of biodegradable organic matter depending on the return period for the critical event is, maybe, a possible pragmatic criterion for the harmful effect caused by CSO discharges (cf. Section 6.3.4).

6.1 GENERAL CHARACTERISTICS OF EFFECTS

6.1.1 General Overview

Table 1.1 outlines in general terms the main effects related to wet weather discharges. In a slightly modified and somewhat extended form, these effects are

- Physical habitat changes of downstream receiving waters caused by flooding, erosion, and sediment deposition.
- Receiving water quality changes and toxic effects onto the aquatic life caused by a large number of pollutants associated with the runoff.
- Groundwater recharge deficits caused by diversion of the runoff.
- Soil and groundwater deterioration caused by infiltration of polluted runoff.
- Public health risks directly or indirectly caused by pollutants, bacteria, and viruses in the runoff.
- Aesthetic deterioration and public perception.

Although several of these effects will be included in this chapter, the focal point is the quality changes of surface waters, directly or indirectly related to the environment and humans. The reason is that surface waters typically are receiving systems for both SWR and CSOs. Among these surface water systems, the flowing waters, creeks, channels, and rivers, and the lakes and reservoirs are commonly discharged to, although the coastal zone for cities is relevant.

Having this in mind, Table 6.1 outlines what is typically considered important effects caused by pollutants. In addition, flooding and erosion are specific hydraulic induced effects.

Table 6.1 is a first estimate of what pollutants should be addressed in case of wet weather effects. The table is a starting point for a deeper analysis. It can also be relevant in selecting means to reduce the pollutant load and when comparing the wet weather pollutant loads with those from other sources.

The following specific comments relate to Table 6.1:

- *Sediments*
  Dust from urban and road surfaces and deposits and sediments from sewer networks can be eroded and suspended under high flow conditions. By nature, a major part of such solids therefore show potential for settling under

TABLE 6.1
Outline of Potential Effects of Wet Weather Discharges That Are Considered Both Typical and Central

Type of Pollutant	CSO	SWR
Sediments	Deposition in flowing waters and close to the inlet in stagnant waters	Deposition in flowing waters and close to the inlet in stagnant waters
Biodegradable organic matter	Dissolved oxygen (DO) depletion and excessive microbial growth; particularly important in flowing waters	In general of minor importance
Nutrients, N and P	Eutrophication; particularly important in stagnant waters	Eutrophication; particularly important in stagnant waters
Heavy metals and organic micropollutants*	Acute and accumulative toxic effects in both the water phase and in the aquatic sediments	Acute and accumulative toxic effects in both the water phase and in the aquatic sediments
Pathogenic microorganisms	Effects of pathogenic bacteria, viruses, and protozoa	In general of minor importance
Gross solids	Aesthetic deterioration	In general of minor importance

*Including hydrocarbons (oil and grease).

conditions where the flow rate decreases (i.e., in the receiving water and often close to the point of discharge). In principle, such deposits show accumulative effect by changing and deteriorating the bottom of the receiving water system.

- *Biodegradable organic matter*
  Two major effects are related to the discharge of biodegradable organic matter. One type of effect is related to the aerobic degradation and corresponding dissolved oxygen (DO) depletion. The other type of effect is as substrate for heterotrophic microorganisms causing excessive microbial growth, particularly by microorganisms attached to fixed surfaces like biofilms. The first type is an acute effect whereas excessive biomass growth in principle is accumulative of nature, at least when the discharges occur rather frequently. The effect of biodegradable organic matter is relevant in case of CSOs whereas the organic matter in SWR (e.g., originating from leaves and other types of vegetation), is typically less biodegradable.
- *Nutrients*
  Growth of aquatic primary producers (e.g., rooted plants and free living algae in the water body) requires both nitrogen and phosphorus. The N/P stoichiometric mass ratio between these two nutrients is about 7, the so-called Redfield ratio. In urban runoff, the ratio between N and P is typically larger and consequently is phosphorus compared with nitrogen, often the limiting nutrient for algal growth in lakes and reservoirs. To some extent, it is also the case in larger rivers and estuarine waters. Control of eutrophication therefore

typically concerns the control of phosphorus inflow to surface waters. Compared with the runoff from rural areas, the urban impervious surfaces play a central role (cf. Example 4.2). The contribution of P per unit area of an urban catchment that is served by a separate sewer system is often higher than from a similar area with combined sewers although the concentration of phosphorus is generally higher in CSOs than in SWR (cf. e.g., Tables 5.2 and 5.3). The reason for a lower contribution from a combined sewer system is that a relative large portion of the runoff, on an annual basis, is diverted to a wastewater treatment plant and thereby also typically to a different, and more robust, surface water body. It is important that nutrients (phosphorus) exhibit an accumulative and not an acute effect (cf. Figure 1.2).

- *Heavy metals and organic micropollutants*
  A great number of heavy metals and organic micropollutants that occur in both CSOs and SWR are potentially toxic. These components are to a large extent associated with particles and might therefore be accumulated in the sediments of surface waters. In general, the level of concentration and the speciation of the components in both CSOs and SWR can potentially result in both acute and accumulative effects. Both heavy metals and organic micropollutants might concentrate in living organisms via the food chain and thereby result in long-term impacts onto the ecosystem including humans.
- *Pathogenic microorganisms*
  The CSOs constitute a risk for contamination of receiving waters with pathogenic bacteria and viruses affecting human health. The effect of such discharges is basically acute. To some extent animals like cats, dogs, ducks, and pigeons are also a potential source of fecal organisms and can thereby contaminate SWR. Furthermore, wrong connections of foul into storm sewers may also contaminate SWR.
- *Gross solids*
  In contrast to the effect of sediment discharges, gross solids from CSOs are particularly undesirable for aesthetic reasons.

Referring to Table 6.1 and the corresponding comments, it is important to note that the nature of the effects related to pollutants discharged from SWR are mainly accumulative, although it is the acute effects that have been monitored and identified for the toxic pollutants—heavy metals and organic micropollutants. It is, however, likely that accumulative effects will occur at concentration levels lower than those causing acute effects (cf. Section 6.5). In case of SWR, it is therefore the total load of pollutants during a season (year) that is important to determine. When dealing with CSOs, the potential for DO depletion and for the effects related to the discharge of pathogenic microorganisms and ammonia can result in acute effects. Under such conditions, it is therefore important to determine the effects from extreme single events.

The complex nature of urban drainage quality impacts and the rather dilute knowledge that exists makes it difficult to establish a comprehensive review. An attempt in this direction is found in House et al. (1993) and in Ellis and Hvitved-Jacobsen (1996).

6.1.2 Classification of Wet Weather Discharges

As described in the introduction to this chapter, it is in principle expected that both the variability of the pollutant discharges during wet weather and the specific characteristics of the receiving water systems will greatly affect the nature and the extent of the impact. A conceptually based description of the wet weather effects is complex and problematic. It is, however, considered an important goal of this text to establish predicting tools that, as much as possible, relies on basic principles and a conceptual understanding of the underlying phenomena and processes.

The overview shown in Section 6.1.1 is a first systematic approach. This overview has the pollutants and the mode of discharge as central elements influencing the effect onto the environment. In addition to that, the receiving water, being the system impacted, and the time scale of the impact play central roles. The first approach for establishing an overview and a systematic and central approach to effects of urban wet weather and road runoff pollution rests on the following characteristics:

- The type of pollutant
- The type, or mode, of the wet weather discharges (i.e., SWR and CSOs)
- The type of receiving water system, the spatial scale
- The time scale of the effect

It is in the organization of this chapter that it was decided that the type of pollutant is superior to the other (important) factors. There are several reasons for that. First of all, the description can be organized more straightforward and with a minimum of repetitions. Secondly, it follows the main concept of the entire text with the wet weather pollutants and their associated processes being central. The decision to let the type of pollutant be the superior characteristic means that the organization of the subject shown in Table 6.1, at least to some extent, can be followed.

An organization of the text according to the type of pollutant, however, also means that a systematic approach of the discharge mode, the receiving system, and to some extent the time scale of the effect will be possible. As an example, it appears from Table 6.1 that the discharge of sediments and biodegradable organics are expected to exert direct negative effects onto flowing waters. Nutrients and toxic pollutants that potentially accumulate in stagnant waters during extended periods may result in negative effects in lakes and reservoirs.

6.2 HYDRAULIC RELATED EFFECTS ON RECEIVING WATERS

Both SWR and CSOs may cause physical changes of receiving waters. The morphology of the receiving water can thereby be altered in two ways: either by deposition of mainly fine particulate matter (i.e., silting), or deteriorated by erosion of bottom materials. In both cases, the original (natural) system is physically changed resulting in a potential deterioration of the receiving water as a habitat.

In the case of deposition of fine materials in receiving waters, it is clear that particulate matter that, at either urban surfaces or in the combined sewer network, was eroded and resuspended in the runoff water further downstream may have a potential for settling. Conditions for deposition may occur when the runoff is discharged into channels, streams, and lakes where the flow velocity, and the bottom shear stress, typically is significantly reduced. Such changes often occur far downstream from the site where the erosion took place.

The opposite phenomenon (i.e., erosion of the receiving water system) is potentially possible because of the increased flow during the runoff event. Erosion is a problem in the small upstream receiving water systems and detention of the runoff therefore often is required to reduce peak flows.

The basic phenomena related to deposition and erosion of particulate matter are described in Section 3.4.2. The shear stress, τ, at the bottom is the central parameter that determines the conditions for a hydraulic impact. The phenomenon is dealt with in Section 4.4 and Section 5.3 in case of stormwater systems and combined sewers, respectively. Example 4.1 concerns scouring of an urban channel and is central as an example of a hydraulic related effect and therefore referred to in this respect.

6.3 EFFECTS OF BIODEGRADABLE ORGANIC MATTER

The basic characteristics of biodegradable organic matter are outlined in Section 3.2.1.1 in terms of parameter estimation, fractionation, biodegradability, transport, and mass balance determination.

The effects of wet weather discharges of organic matter onto receiving waters particularly relate to a potential DO depletion in case of degradation and to the fact that organic matter is a substrate for invertebrates, bacteria, fungi, and so forth that may cause excessive growth of such microorganisms. These negative effects of organic matter concern both the dissolved and colloidal fractions in the water body and the particulate fractions that might accumulate in the sediments. In particular, the fractionation according to the biodegradability of organic matter is in this respect important (cf. Table 5.5).

In principle, these negative effects can occur in any receiving water system, however, they are often more pronounced expressed in flowing waters where DO depletion is exerted onto the same volume of water over a relative long time span. A specific water volume in a stream or channel is affected by settled organic matter in the sediments and dissolved and suspended matter in the water body. These receiving water impacts are relevant in the case of CSOs and in general not for SWR. The reason is that CSOs include contributions from the daily wastewater flow and sewer solids. A typically lower content and concentration of organic matter in SWR mainly originate from less biodegradable plant materials.

6.3.1 BIODEGRADATION OF ORGANIC MATTER AND DISSOLVED OXYGEN DEPLETION IN STREAMS

Biodegradation of organic matter and a corresponding consumption of oxygen can lead to DO depletion in receiving waters and a negative effect onto the higher animal life like fish. The DO depletion, relative to saturation, is a phenomenon that occurs naturally in surface waters or caused by continuous inflow of wastewater or by CSO discharges. In addition, plants cause DO fluctuations during day and night by photosynthesis and respiration. In this section particular focus is on the wet weather impacts that, however, occur on the top of the dry weather variability.

Although DO depletion caused by wet weather discharges is of general relevance, as previously mentioned it is particularly important in streams receiving CSOs. There are two reasons for that:

- Degradation in the polluted water volume will in a stream—except for the diluting effect of dispersion—affect the same volume of water when it is transported downstream
- Biodegradable organics that have settled during the discharge will exert an oxygen demand onto the water passing by

In case of CSO discharges into a stream there are two different types of impacts onto the DO concentration:

- A direct effect caused by biodegradation of discharged organic matter—in the water phase or at the bottom—followed by DO depletion
- An indirect and hydraulically induced effect caused by scouring of the stream bottom resulting in an increased sediment oxygen demand (SOD)

Both types of impacts are acute effects. In the first case the DO depletion can be defined as either immediate or delayed:

- *Immediate oxygen demand*
  The immediate effect is caused by degradation of (soluble and colloidal) organic matter in the water phase and by absorption of organic matter by benthic organisms, mainly bacteria and fungi. In case of CSOs, these processes take place in the polluted water volume moving downstream, resulting in an immediate consumption of dissolved oxygen.
- *Delayed oxygen demand*
  The delayed effect is caused by organic matter that is removed from the water phase by sedimentation or adsorption of solids followed by accumulation at the bottom without being degraded and thereby resulting in immediate consumption of oxygen. The organic matter is attached to sediments, stones, and plants. In this position the organic matter is degraded and a corresponding delayed oxygen demand onto the water passing by is the result. In case of CSOs, this type of DO depletion takes place during and after the discharge event (i.e., degradation will also occur in the stream after the polluted water volume has passed the station in question).

The relation between the removal processes for organic matter (OM) and the consumption of oxygen is illustrated in Figure 6.1. It should be noted that the distinction between immediate and delayed oxygen consumption is important in the case of intermittent discharges like CSOs. The reason is that the immediate consumption is dominating for the DO depletion in the plug of mixed river water and overflow water passing downstream in the river. The delayed oxygen

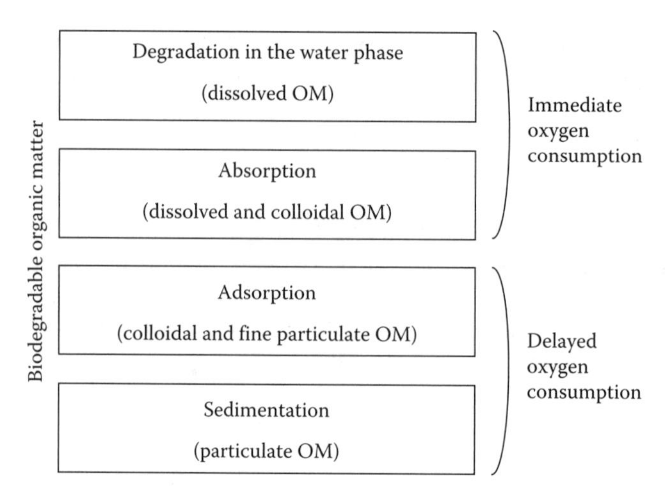

FIGURE 6.1 Relations between removal processes for biodegradable organic matter (OM) from the water phase and the corresponding immediate and delayed oxygen consumption.

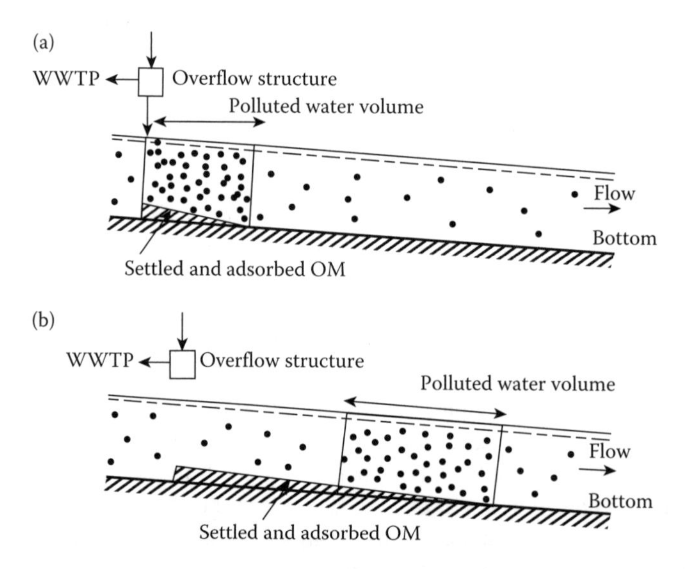

FIGURE 6.2 Figure (a) shows the principle of the formation of a polluted plug of water in a river during a CSO event and the corresponding accumulation of organic matter (OM) at the bottom. Figure (b) shows the transport of the entire discharged polluted water volume after the cessation of the overflow event and the continuation of formation of deposits.

consumption, however, can affect the DO concentration after the polluted plug has passed the actual position in the river, hours or even days later (cf. Figure 6.2). Both types of effects, the immediate and the delayed, will also occur for more or less continuous discharges, but they will not under such conditions be identified as separated phenomena.

A quantification of the relations shown in Figure 6.1 requires that the organic matter be fractionated according to the removal processes and that a mass balance can be established. Referring to Section 3.2.1.1, COD and its fractionation in a soluble and a particulate part is a first approach to take into account different characteristics of the organic matter in terms of mass transport, removal processes, and biodegradability:

$$COD_{tot} = COD_{sol} + COD_{part} \tag{6.1}$$

where
COD_{tot} = total concentration of COD (g m^{-3})
COD_{sol} = concentration of soluble COD (g m^{-3})
COD_{part} = concentration of particulate COD (g m^{-3})

In relation to this fractionation of COD, the immediate oxygen consumption can be associated with COD_{sol} and the delayed effect with COD_{part}. Following this rather simple approach and assuming a first-order reaction for both the immediate and the delayed oxygen consumption, these two contributions to the DO depletion in rivers receiving CSOs can be formulated.

The immediate oxygen consumption is

$$s_{imm} = K_1\, COD_{sol}\, A \tag{6.2}$$

where
s_{imm} = sink term for COD_{sol} per unit length of the stream (g h^{-1} m^{-1})
K_1 = first-order degradation constant for COD_{sol} (in the water phase; h^{-1})
A = cross-sectional area of the river water phase (m^2)

It should be noted that s_{imm} be interpreted both as a sink term for COD and a corresponding DO consumption.

The delayed oxygen consumption: the removal rate for the transfer of COD_{part} from the water phase to the bottom by adsorption or sedimentation is

$$s_{rem} = k \frac{COD_{part}}{h} A \tag{6.3}$$

where
s_{rem} = sink term for removal of COD_{part} from the water phase to the bottom per unit length of the stream (g h^{-1} m^{-1})
k = first-order rate of a combined adsorption and sedimentation of COD_{part} (m h^{-1})
h = river water depth (m)

It is important that Equation 6.3 only expresses a transfer of organic matter from the water phase to the bottom surfaces and not as was the case for the immediate effect (cf. Equation 6.2), a corresponding consumption of oxygen. A portion of the particulate fraction of COD (COD_{part}) in the water phase of the river is during the transport of the polluted water plug continuously transferred to the river bottom by adsorption and sedimentation where it is degraded causing delayed oxygen consumption. To combine the removal of COD_{part} with its degradation at the bottom, the mass balance for the accumulated particulate COD at the bottom, COD_{acc}, must be formulated (cf. Equation 6.4). The two terms at the right side of this equation express what is added to the bottom from the water phase and what is degraded, respectively. Equation 6.4 thereby combines the CSO load of particulate COD with its following degradation at the river bottom:

$$\frac{d(COD_{acc})}{dt} = k\ COD_{part} - K_3\ COD_{acc} \tag{6.4}$$

where

COD_{acc} = amount of adsorbed or settled particulate COD at the bottom originating from the CSO event (g m^{-2})

t = time (h)

K_3 = first-order degradation constant for COD_{acc} (h^{-1})

The impact on the DO concentration of the river—the delayed oxygen consumption—by degradation of COD_{acc} is therefore expressed as follows:

$$s_{del} = K_3 \frac{COD_{acc}}{h} A \tag{6.5}$$

where

s_{del} = sink term for degradation of COD_{acc} at the stream bottom (g h^{-1} m^{-1})

Equations 6.2 and 6.5 for s_{imm} and s_{del}, respectively, are the basis for the formulation of a DO mass balance for a river receiving CSO discharges (cf. Section 6.3.3). The contribution to the DO mass balance for the river caused by degradation of biodegradable organic matter discharged during the CSO event is therefore

$$\frac{dC}{dt} = -K_1 COD_{sol} - K_3 \frac{COD_{acc}}{h} \tag{6.6}$$

where

C = DO concentration in the water body of the river (g m^{-3})

6.3.2 Dry Weather Variability of Dissolved Oxygen in Flowing Waters

The contribution to the DO mass balance in a stream from CSOs is as expressed by Equation 6.6, in principle a superposition of the dry weather DO concentration. The formulation of the DO mass balance for a river during a period with no wet weather discharges is the basis for modeling the DO variability during a CSO event.

The following are the main DO consuming and DO producing processes affecting the variability of the DO concentration of a stream:

- Degradation of organic matter in the water phase and at the bottom resulting in consumption of oxygen.
- Nitrification of ammonia caused by nitrifying bacteria at the stream bottom and resulting in consumption of oxygen.
- Photosynthesis and respiration of plants. The photosynthesis affected by the daily fluctuations of the influx of sunlight results in production of oxygen, whereas the respiration of the plants during both day and night consumes oxygen.
- The reaeration from the atmosphere.

Taking these processes into account, a DO mass balance (i.e., a DO stream model) can briefly be formulated:

$$\frac{dC}{dt} = K_2(C_s - C) + P(t) - \sum R(t) \quad \text{for } \frac{dx}{dt} = u \tag{6.7}$$

where

K_2 = reaeration constant (h^{-1}); cf. Section 3.5.1 and Equation 3.58
C_S = DO saturation concentration ($g\ m^{-3}$)
$P(t)$ = photosynthesis; DO production rate of aquatic plants ($g\ m^{-3}\ h^{-1}$)
$\Sigma R(t)$ = total respiration rate, DO (dry weather) consumption rate for all DO consuming processes ($g\ m^{-3}\ h^{-1}$)
x = space coordinate, length (m)
u = cross-sectional average stream velocity ($m\ h^{-1}$)

Equation 6.7 is expressed in the case of uniform and steady-state flow conditions and neglecting the effect of dispersion (cf. Section 3.4.1). Equation 6.7 is the basis for any kind of dissolved oxygen model formulation. It is basically an extension of the classical Streeter–Phelps DO model that did not include the DO fluctuation caused by the plant activity (Streeter and Phelps 1925; Schnoor 1996). According to Simonsen and Harremoës (1978), the simply expressed solution to Equation 6.7 is

$$C = C_m + 0.5DC\frac{b}{b'} \tag{6.8}$$

where

C_m = mean DO concentration in a stream during dry weather conditions ($g\ m^{-3}$)
DC = total daily DO amplitude during dry weather conditions ($g\ m^{-3}$)
b = dimensionless term for DO fluctuation during day and night (–)
$b' = |b_{min}|$ for $b < 0$ and b_{max} for $b > 0$.

The dimensionless daily DO fluctuation, b, in a stream can be formulated in different ways. In Simonsen and Harremoës (1978) and Schaarup-Jensen and

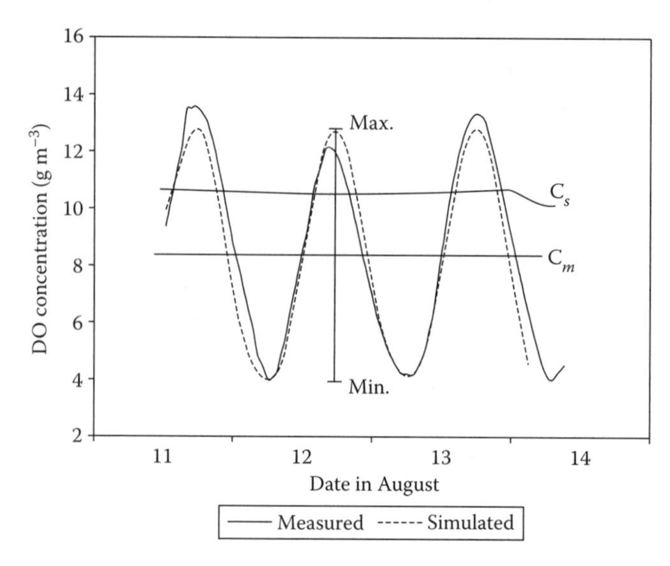

FIGURE 6.3 DO fluctuation in a small river in the temperate climate region (Denmark) during a dry weather period, August 11–14, 1993. The DO concentration is monitored continuously and simulated with a numerical model that includes Equation 6.8.

Hvitved-Jacobsen (1991) it is formulated as a cosine function including the following three parameters:

- The relative length of the daytime at the location in question
- The relative time of the day
- The reaeration coefficient, K_2

Figure 6.3 exemplifies, by the simulated curve, how Equation 6.8 fits to measured values of the DO concentration in a stream.

6.3.3 CSO Impacts on Dissolved Oxygen in Streams

The basic theoretical background for the impact of CSOs on the DO concentration in streams and the dry weather DO fluctuations were focused on in Sections 6.3.1 and 6.3.2, respectively. Before these phenomena are further integrated to assess the effect of CSO discharges, a couple of examples will illustrate how the DO concentration of a river might change from a dry weather period to a period affected by CSOs (cf. Figures 6.4 and 6.5).

The two streams used for illustration, Figures 6.4 and 6.5, were both affected by CSO discharges. Although a detailed monitoring of the two streams and the discharges to these are not available, and although a CSO impact is very complex, the two cases indicate both immediate and delayed oxygen depletion as detailed in Section 6.3.1. It is therefore considered a pragmatic approach to formulate the effect of a CSO event in terms of DO depletion in a river by Equation 6.6. The complex interactions between the immediate effect and the delayed effect will be described next based on Figure 6.6.

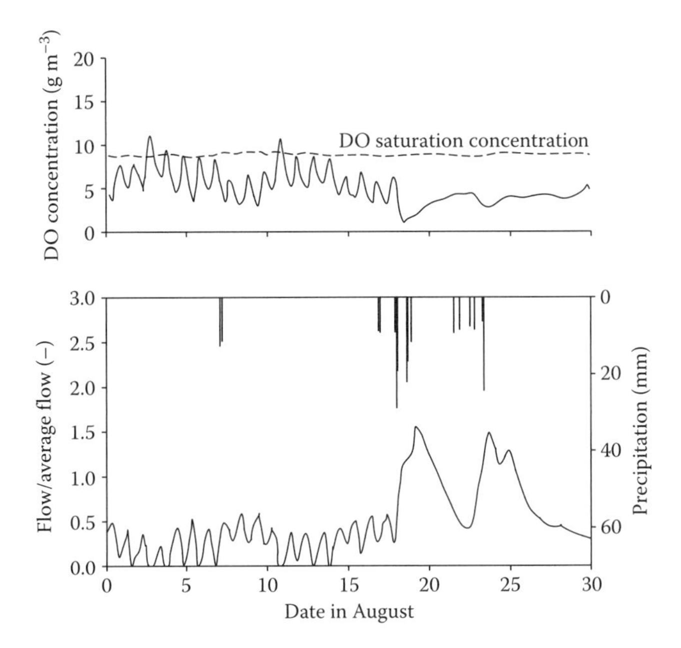

FIGURE 6.4 Dry and wet weather DO concentrations and ratios of flow to yearly average flow in Scioto River, Ohio during the period August 1–30, 1972.

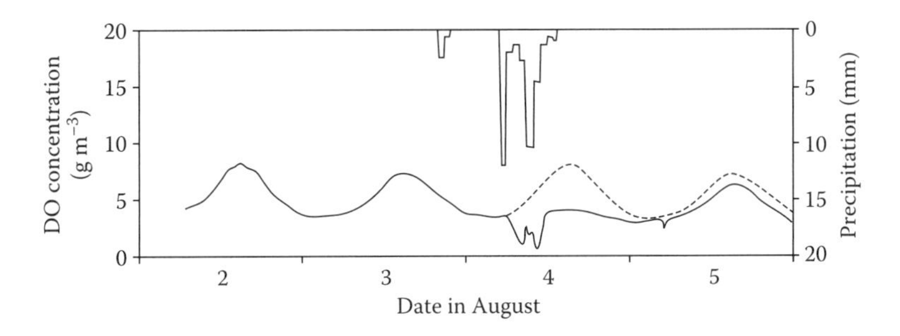

FIGURE 6.5 Dry and wet weather DO concentration in a small Danish stream during the period August 2–5, 2004. The dashed curve shows the estimated DO concentration in case of no CSO discharge.

Referring to Figure 6.6a, the impact on the DO concentration of a CSO discharge at point 0 of a stream is illustrated. The effect is separated in terms of the immediate effect (curve #1) and the delayed effect (curves #2) with the DO concentration versus the transport time of the flow of water, $t_h = x/u$, from point of discharge. The CSO discharge starts at time $t = -t_{CSO}$ and stops at $t = 0$ (i.e., corresponding to a duration of the discharge event equal to t_{CSO}). The immediate effect that takes place in the polluted water volume transported downstream is a DO sag curve that is basically identical with the one described by the concept of Streeter–Phelps in case of continuous discharge of wastewater (Streeter and Phelps 1925; Nazaroff and Alvarez-Cohen 2001). Corresponding to this concept the DO concentration has its minimum

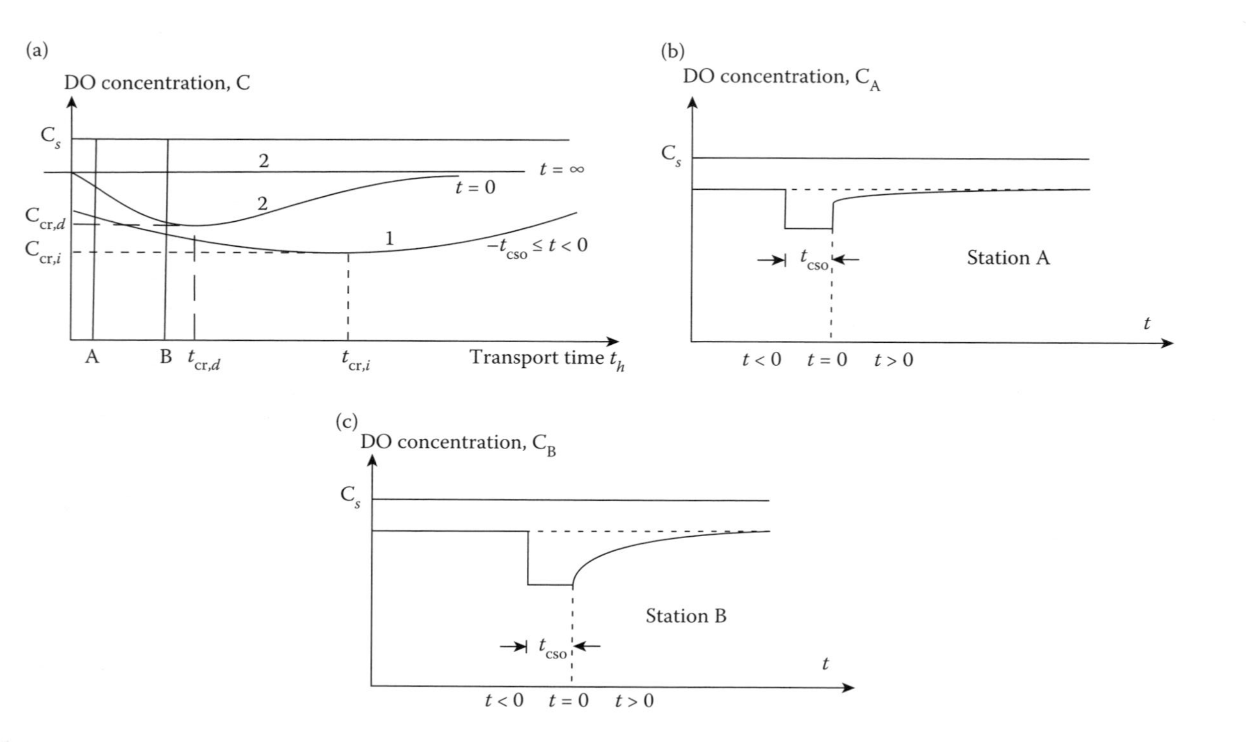

FIGURE 6.6 Principle of immediate effect, curve #1 in (a) and delayed effect, curves #2 in (a), versus the residence time, t_h, in a stream affected by a CSO event. The critical (minimum) DO concentrations, $C_{cr,i}$ and $C_{cr,d}$, for the immediate and the delayed effect, respectively, are indicated. The curves in (b) and (c) show the principle of the DO concentration versus time t for two different positions in the stream, A and B, respectively.

value, $C_{cr,i}$, at the so-called critical point, $t_h = t_{cr,i}$. The delayed effect has its maximum impact on the DO concentration at $t = 0$ (i.e., when the polluted plug of water has just passed the point of the stream in question). The delayed effect is hereafter reduced and approaches the DO concentration upstream at the point of discharge faster. In this respect it should be noted that the time t varies downstream of the point of discharge with a start ($t = 0$) when the polluted water volume has just passed the point in question and the maximum amount of particulate COD has been transferred to the river bottom. Pragmatically, the delayed effect is considered to start at $t = 0$ although particulate organics has been continuously transferred to the river bottom from $t = -t_{CSO}$ to $t = 0$. The distance from point of discharge to the critical point for the immediate effect, $t_{cr,i}$, is typically several km (e.g., 5–10 km). The corresponding distance for the delayed effect, $t_{cr,d}$, is normally considerable shorter because the DO response is caused by degradation at the stream bottom and not occurring in the water phase.

Contrary to the curves in Figure 6.6a that show the DO concentration versus the distance downstream (shown as transport time, t_h), the curves in Figure 6.6b and c show the DO concentration versus the real time, t, for two arbitrarily selected positions (stations A and B) in the stream. The curves in Figure 6.6b and c thereby show the principle of the time dependent development of the DO concentration at a fixed position of a stream during passage of the polluted plug and after this event (i.e., the immediate as well as the delayed effect). As seen from the two examples, Figures 6.4 and 6.5, the delayed effect may last for a few days until the river returns to its steady state. It should be noted that the change in DO concentration that takes place at $t = 0$ corresponds to the pragmatic and simple understanding that the delayed effect starts at $t = 0$ whereas the real variation of the DO concentration is more continuously occurring.

The principle for DO depletion in streams caused by CSOs as illustrated in Figure 6.6 can be mathematically expressed as shown in Equation 6.9 by combining Equations 6.6 and 6.7. The assumption thereby made is that the DO deficits originating from the dry weather processes, the immediate effect, and the delayed effect can be superposed.

$$\frac{dC}{dt} = K_2(C_s - C) + P(t) - \sum R(t) - K_1 COD_{sol} - K_3 \frac{COD_{acc}}{h} \tag{6.9}$$

An analytical solution to Equation 6.9 is shown in Equation 6.10 (Hvitved-Jacobsen 1986; Schaarup-Jensen and Hvitved-Jacobsen 1991). Equation 6.10 is simple assuming a constant flow and a constant COD concentration of the overflow. The river is considered having uniform flow and exposed to constant external conditions during the period where the immediate and the delayed effect are effective.

$$C = C_m + 0.5DC\frac{b}{b'} - L_0(t)\left(\frac{hK_2}{k}\right)^{-1}\left(\exp\left(\frac{-k}{h}t_h\right) - \exp(-K_2 t_h)\right) \tag{6.10}$$

where

$L_0(t) = \dfrac{K_3 P}{Q_{river} + Q_{CSO}} \exp(-K_3 t)$ (gm^{-3})

t = real time (s); cf. Figure 6.6 and related text

TABLE 6.2
Central Parameters Related to a DO Stream Model Receiving CSO Discharges

Parameter	Typical Value (Unit); Temperature 20°C
k = first-order rate of a combined adsorption and sedimentation of COD_{part}	1–2 (m d^{-1})
K_1 = first-order degradation constant for COD_{sol} (in the water phase)	0.2–0.5 (d^{-1})
K_3 = first-order degradation constant for COD_{acc}	1–2.5 (d^{-1})

Source: Data from Hvitved-Jacobsen, T., and Harremoës. P., Impact of combined sewer overflows on dissolved oxygen in receiving streams. In *Urban stormwater quality, management and planning*, Littleton, CO: Water Resources Publication. 226–35, 1982; Schaarup-Jensen, K., and Hvitved-Jacobsen. T., Simulation of dissolved oxygen depletion in streams receiving combined sewer overflows. In *New Technologies in Urban Drainage, Elsevier Applied Science*, 273–82, 1991.

Note: The parameters refer to relatively small streams.

t_h = transport time of water from point of discharge (s); cf. Figure 6.6 and related text
P = total amount of COD_{part} discharged during the CSO event (g)
Q_{river} = constant baseflow in the river (m^3 s^{-1})
Q_{CSO} = mean flow of the CSO event (m^3 s^{-1})

$L_0(t)$ is a time dependant and fictitious COD concentration in the water phase. It is fictitious in the sense that it exerts exactly the same effect onto the DO concentration as the real occurring COD_{acc} at the river bottom. The effect of the discharged amount of COD_{part} (P) thereby decreases as time t increases.

Contrary to the analytical solution shown in Figure 6.10, Equation 6.9 can more generally be used as a basis for development of computer models for the prediction of the DO depletion caused by CSOs (Schaarup-Jensen and Hvitved-Jacobsen 1991). Equation 6.4 describing the transfer of COD_{part} from the water phase to the river bottom is an integrated part of such model descriptions.

The parameters needed for a DO model are expected to vary according to river and CSO discharge characteristics. The central parameters related to the effect of a CSO discharge to a river are shown in Table 6.2. These values are just shown to give an impression of their level of magnitude.

6.3.4 Criteria for Assessment of the Wet Weather Effect of Discharged Biodegradable Organic Matter

It is crucial to understand—and accept—that if discharges from an overflow structure to a receiving water body are accepted, it must also be accepted that extreme overflow events can cause fatal effects. The dry weather water quality criteria and standards are therefore not applicable under such conditions.

In case of CSO discharges into flowing waters, and following the course of Section 6.3, the discharge of biodegradable organic matter and its effect on the DO concentration is a potential critical phenomenon. In general, a water quality criterion in terms of a DO concentration refers to the activity level of the fish species (e.g., trout and carp). Such criteria may vary from one region to another. For a trout river, a daily dry weather median and minimum DO concentration is 9 and 6 g m^{-3}, respectively. A criterion formulated in this way is expected to be violated in the case of an extreme CSO event.

Water quality criteria related to wet weather events have been formulated by applying the use of cumulative frequency curves for dissolved oxygen versus the DO concentration (Medina 1980). However, this criterion does not observe the fact that a fish kill caused by DO depletion has exterminated the fish population for the period it takes to establish a new population or to reestablish it by migration. The concern is how infrequent the CSO event occurs or how frequent it is expected to occur. It is judged by extreme event statistics and not through a percentage time of occurrence.

A water quality criterion for the DO concentration of a stream that takes into account the nature of intermittent CSO events can be formulated in terms of a concept including the DO concentration, the duration of the event, and the frequency of the event. The following five points are in this respect central:

- A statistics of the magnitude of each single CSO event.
- The duration of low DO concentrations and the corresponding effect on the fish population.
- A statistical, formulated acceptance that critical, and infrequent, – CSO events might affect the survival of the fish population.
- That frequently occurring CSO events should observe the general formulated dry weather water quality criterion.
- That different fish species, corresponding to the intended water quality, set different requirements for the DO concentration.

Figure 6.7 shows a concept for a DO water quality criterion taking into account these five points (Ellis and Hvitved-Jacobsen 1996). It is included in the criterion that half the fish population will be killed caused by a low DO concentration at events with return periods from eight to 16 years. The criterion is in Figure 6.7 shown for two durations of exposure time, one and 12 hours.

The concept depicted in Figure 6.7 is further discussed in Section 11.3. It is shown in Figure 6.8 a Danish and a United Kingdom version based on information from DWPCC (1985) and WRC (1990), respectively. Water quality criteria for wet weather conditions applying a similar type of statistics for the impact of heavy metals is formulated by U.S. EPA (1983).

6.3.5 Biodegradable Organic Matter and DO Depletion: Complex Interactions and Effects

The effect on receiving waters of biodegradable organic matter discharged with the CSOs was, in the preceding sections, focused on in terms of DO depletion caused by microbial degradation. It must, however, be noticed that discharge of biodegradable

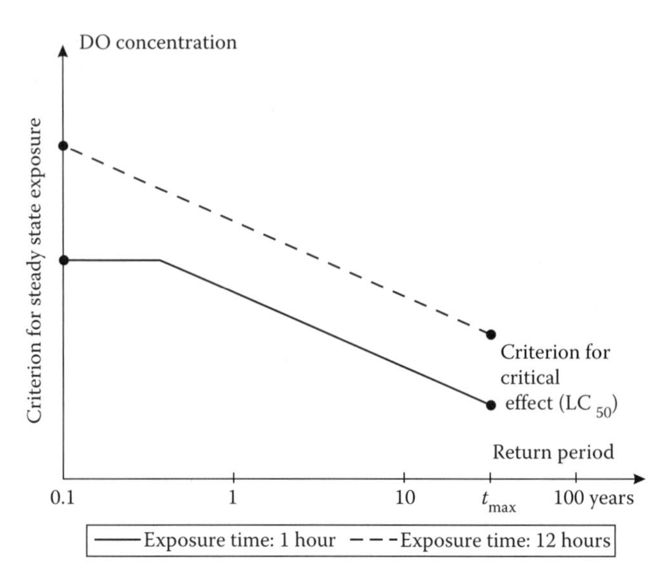

FIGURE 6.7 Principle of a DO water quality criterion for CSO discharges into rivers based on the potential impact on a fish population.

organic matter can result in other types of effects. It is important to understand that although the DO concentration is selected for assessment, other complex phenomena can also be important, siltation of particulate organic matter, deterioration of habitats for the sediment associated microorganisms, and excessive growth of such species. These facts add to the complexity of effects from CSO discharges.

In principle, there are similarities between the effects caused by the discharge of biodegradable organic matter and the discharge of nutrients. Both types of discharges may directly or indirectly cause DO depletion and excessive growth of biomass. Organic matter is a substrate for growth of microorganisms and nutrients are substrates for plants that can also affect the DO concentration (cf. Section 6.4).

Although the interactions between the different types of pollutants and their effects make the assessment of the wet weather impacts difficult, the nature of the pollutants and their importance in different types of receiving waters is a help to classify type of impact and determine load-effect relationships in concrete situations (cf. Section 6.1.2). As an example, an effect in terms of DO depletion must therefore be assessed according to its cause, being either biodegradable organic matter discharged to streams (cf. Section 6.3.4), or nutrients discharged to lakes and ponds (cf. Sections 6.4.2 and 9.3.1.8).

6.4 EFFECTS OF NUTRIENT DISCHARGES

The characteristics of nutrients, nitrogen and phosphorus, in the case of eutrophication and contributions from wet weather sources are briefly outlined in Section 6.1.1. In the following, a number of rather simple empirical models for assessment of eutrophication of surface waters will be presented. These models can also be applied

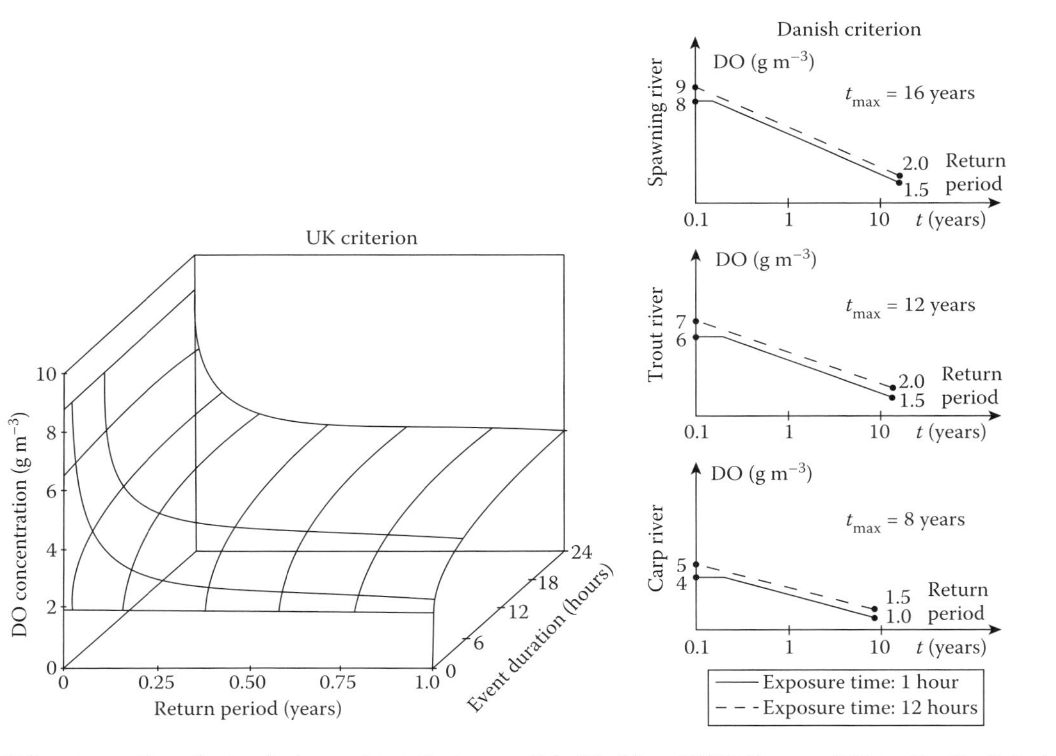

FIGURE 6.8 The UK DO water quality criterion for intermittent discharges. (Modified from WRC, Proposed Water Quality Criteria for the Protection of Aquatic Life from Intermittent Pollution, Report PRS2498-NM, WRC, Medmenham, UK, 1990; Clifforde, I. T., Crabtree, R. W., and Andrews, H. O., 10 Years Experience of CSO Management in the United Kingdom, Proceedings from the WEFTEC 2006 Conference, Water Environment Federation, 3744–56, 2006.) The Danish water quality criterion for the DO concentration in rivers receiving CSO discharges. (Modified from DWPCC, Pollution of Rivers from Overflow Structures, Report No. 22 from Danish Water Pollution Control Committee, 1985.)

in case the nutrients originate from wet weather sources. The more complex process models for evaluation of eutrophication that require data in terms of a rather large number of process parameters are not dealt with in this context.

The following aspects of nutrient discharges from wet weather sources are considered important to note:

- Nutrients cause accumulative effects. A seasonal or a yearly discharge, and not the event-based discharge, of nutrients is therefore the relevant time scale when assessing the effect (cf. Section 1.3.2).
- Stormwater or road runoff will be focused on although the models can also be applied in case of CSOs.
- It is normally the case that phosphorus is the limiting nutrient for eutrophication of stagnant waters. In this respect, it is noticed that phosphorus loads from impervious surfaces compared with rural contributions, per unit area of contributing catchment, in general, are relatively more important than nitrogen (cf. Example 4.2).
- Eutrophication models that are particularly relevant in case of lakes or similar types of stagnant surface waters will be focused on.

6.4.1 Nutrient Loads on Surface Waters

Nutrient (phosphorus) loads from both dry weather and wet weather sources can be determined based on sampling of the relevant flows. In addition to that, the contribution from diffuse sources must be estimated. In principle, a seasonal or annual load can simply be estimated from the following main sources:

- *Rural contributions*
  The area load of phosphorus is based on estimated values of export coefficients (i.e., the contribution per unit area and unit time; kg ha^{-1} yr^{-1}).
- *Wet weather urban and road runoff*
  Contributions from urban and road runoff are dealt with in Sections 4.5 and 4.6. In the case of CSOs, corresponding contributions are subjects of Sections 5.4 and 5.5.
- *Point source contributions*
  Specific contributions from sources like wastewater treatment plants.

6.4.2 Eutrophication

Eutrophication of surface waters is caused by nutrient discharges and results in a number of complex effects. In general, the impacts refer to the biological productivity characterized by an increased (excessive) growth of plants (algae). In brief, the main characteristics and effects of eutrophication are

- *Ecosystem changes*
  Ecosystem changes refer to increased biomass production and often a changed composition and diversity of the plant community (e.g., a change

from rooted plants to algae that are free-living in the water body. A high amount of algae is therefore characteristic for eutrophic surface waters and often quantified in terms of the chlorophyll-a concentration, the active component for photosynthesis of plants.

- *Penetration of light*
  Reduced penetration of sunlight into surface waters is typically due to the free-living algae (the phytoplankton biomass). The reduced light penetration causes changes of the ecosystem for both plants and animals (e.g., the carnivorous fish).
- *Increased oxygen consumption*
  Degradation of algae in the water body and at the bottom may cause increased DO consumption. Under extreme conditions, the deeper parts of the water body and the bottom sediments turn anaerobic.
- *Alkalinity changes*
  A change in the alkalinity of the water is caused by plant utilization of inorganic carbon (i.e., causing a change in the carbonate system). The reduced alkalinity may, in highly eutrophic surface waters, increase the pH value to a level of 9 to 10.

When omitting the deterministic formulated eutrophication process models, the assessment of eutrophication is basically restricted to empirical knowledge and often performed by a comparison with cases that have been investigated more. Criteria for the extent of eutrophication are discussed in the following. Such examples are a critical phosphorus concentration level, an analysis based on the biological productivity of the water body or a penetration depth for sunlight.

The models dealt with in the following are at different levels based on a mass balance for phosphorus. Processes important for a P-balance of a lake where the complex reactions of P-speciation and plant growth are excluded is shown in Figure 6.9.

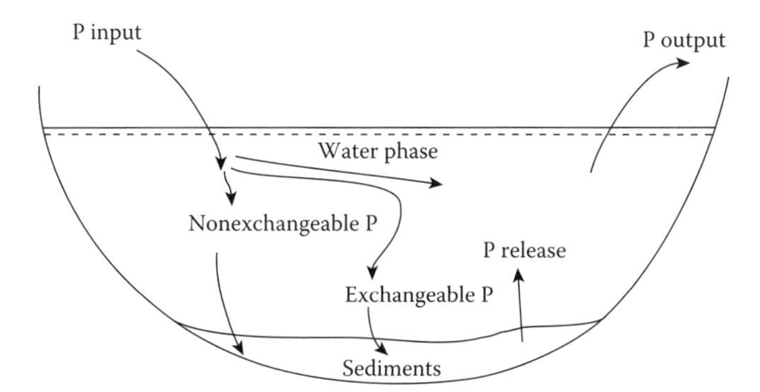

FIGURE 6.9 Processes related to a simple P-balance for a lake. A number of details in terms of different P-species and P-processes in both the water phase and the sediments are omitted. Processes like algal growth (P-uptake in the algae), grazing, excretion, predation, and mineralization are therefore not shown.

The trophic state (biological productivity level) of a lake is central for assessment of the level of eutrophication. The trophic state of lakes and reservoirs can be expressed in the following terms:

- *Oligotrophic lakes*
  Lakes with low biological productivity and therefore characterized by low levels of nutrients to support growth of algae. The chlorophyll-a concentration in an oligotrophic lake is typically < 10 mg m^{-3}.
- *Mesotrophic lakes*
  Lakes with a limited potential for biological production.
- *Eutrophic lakes*
  Lakes with a high biological productivity and a relatively large amount of nutrients. A high inflow rate of the nutrients is often the cause. The chlorophyll-a concentration in eutrophic lakes is typically > 50–100 mg m^{-3}.

6.4.3 Trophic State Models for Lakes and Reservoirs

Eutrophication is a phenomenon that is an environmental problem of general relevance for surface waters (i.e., lakes, reservoirs, flowing waters, estuarine waters, and marine waters). However, the following focus is on models for lakes and reservoirs. First of all, the wet weather flows of phosphorus are of particular relevance in the case of urban lakes. Secondly, the mass balance can be easily illustrated for a lake or reservoir compared with open surface waters. Thirdly, the residence time for nutrients is typically relatively high in stagnant waters.

The group of trophic state models for lakes is based on the fact that the level of eutrophication depends on the surface water load of nutrients (phosphorus). The models express the potential for biological productivity based on a trophic state analysis of a number of lakes and a following comparison with the lake in question. A fundamental requirement is that the lake in question, in terms of its ecology and geology, belongs to the group of lakes with which it is compared.

The specific external surface load of phosphorus, L_P, in units of gP m^{-2} yr^{-1} is a central parameter for the assessment.

6.4.3.1 Vollenweider's Model

The Vollenweider model is probably the simplest example of an empirical-statistical eutrophication model (Vollenweider 1968). The model is based on an analysis of a large number of lakes and reservoirs of different trophic state and located in the temperate region. Vollenweider found that when applying a double logarithmic plot of the specific annual load of phosphorus, L_P, versus the average water depth of the lake, z_{av}, the lakes would be placed in different parts of the diagram according to their trophic state (cf. Figure 6.10).

If a lake in terms of its trophic state belongs to the group of lakes analyzed by Vollenweider, the specific annual load of phosphorus and the average water depth are parameters for estimation of the degree of its trophic state. It is estimated by the distance of the lake's position to two lines indicated by "critical loading" and "acceptable loading." By selecting values for reduced phosphorus load from wet weather

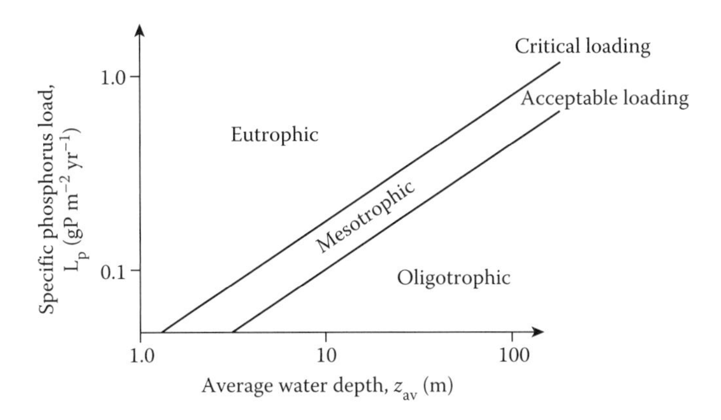

FIGURE 6.10 The plot used by the Vollenweider model.

sources, it is possible to give a first estimate of the importance of the corresponding control measure. It is also, by means of the model, possible to estimate the trophic state of the lake under natural conditions.

6.4.3.2 Dillon's Model

Vollenweider's model is a simplification by assuming a direct flow of phosphorus through the lake. The model does not take into account that part of the phosphorus inflow might undergo temporary or permanent inactivation. Orthophosphate may form precipitates with metals or via incorporation in algae settle and accumulate in the bottom sediments (cf. Figure 6.9). The model of Dillon (1975) takes into account this phenomenon by introducing a phosphorus retention coefficient, R, defined as follows:

$$R = \frac{J_i - J_o}{J_i} \tag{6.11}$$

where

R = phosphorus retention coefficient for a lake (–)
J_i = yearly inflow of phosphorus to the lake (gP yr^{-1})
J_o = yearly outflow of phosphorus from the lake (gP yr^{-1})

$$J_i = L_P A_O \tag{6.12}$$

where

A_O = surface area of the lake (m^2).

The average outflow concentration of phosphorus is

$$C_{av} = \frac{L_P T_w (1-R)}{z_{av}} = \frac{L_P (1-R)}{q_s}, \tag{6.13}$$

where

C_{av} = average outflow concentration of phosphorus from the lake (g m^{-3})
$T_w = V/Q_i$ = hydraulic residence time for the lake (yr)

V = lake volume (m^3)
Q_i = annual average hydraulic load of the lake ($m^3\ yr^{-1}$)
$q_s = z_{av}/T_w$ = annual average inflow of water depth to the lake ($m\ yr^{-1}$)
z_{av} = average water depth (m)

As a substitute for the hydraulic residence time, T_w, the following parameter is often used:

$$r = \frac{1}{T_w} \tag{6.14}$$

where
r = annual flushing rate (yr^{-1})

Equations 6.11 through 6.14 can be included in an extended version of Vollenweider's model referred to as Dillon's model (Dillon 1975; cf. Figure 6.11). By taking into consideration the retention of phosphorus and the hydraulic residence time of the lake, the validity of the model is improved compared with Vollenweider's model. However, the data requirement is correspondingly extended. The use of Dillon's model follows what was described for Vollenweider's model.

From Figure 6.11 it is seen that the slope of the two lines is equal to the average outflow concentration, C_{av}. In case of a completely mixed water body, this concentration is equal to the concentration of phosphorus in the lake.

6.4.3.3 Larsen–Mercier's Model

It is possible to produce a large number of different versions of Vollenweider's model concept. It is possible to use a variety of different information that may affect the trophic state of surface water systems.

Larsen–Mercier's eutrophication model is basically identical to Dillon's model because it uses the same type of information (Larsen and Mercier 1976). However, it is differently expressed in terms of showing the inflow concentration to a lake, C_i,

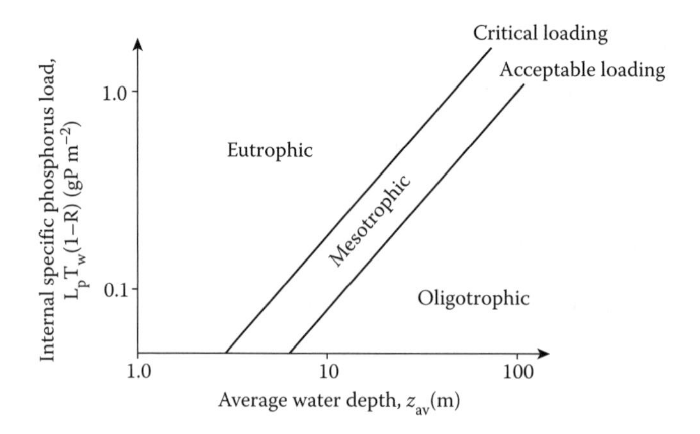

FIGURE 6.11 Dillon's eutrophication model.

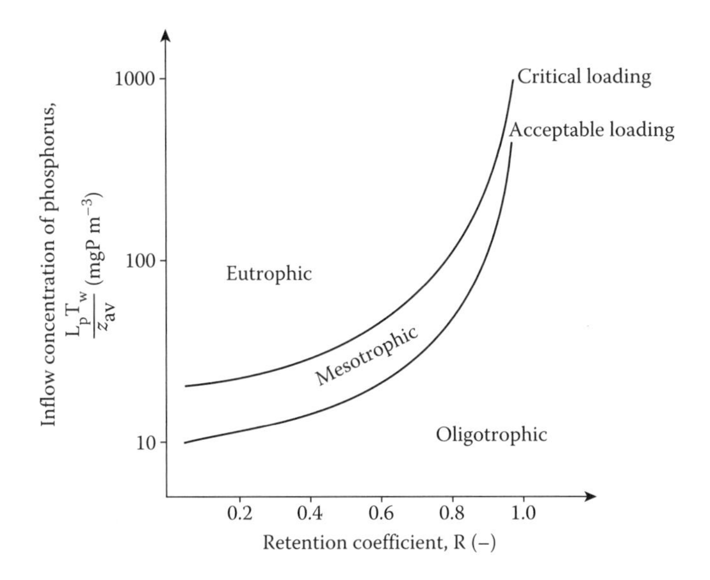

FIGURE 6.12 Larsen–Mercier's eutrophication model.

versus the retention coefficient, R (cf. Figure 6.12). Larsen–Mercier's model thereby expresses the capability of the lake to "absorb" this inflow depending on the retention of phosphorus in the lake. For R = 0 it is from Figure 6.12 seen that the limit between the eutrophic and mesotrophic states corresponds to $C_i = 20$ mgP m^{-3} whereas it is 10 mgP m^{-3} in case of the mesotrophic and oligotrophic states.

$$C_i = \frac{L_P T_w}{z_{av}}, \tag{6.15}$$

where

C_i = average inflow concentration of phosphorus to the lake (notice: mg m^{-3}).

6.4.4 Empirical–Statistical Eutrophication Models for Lakes

The empirical–statistical eutrophication lake models comprise a concept where statistical and empirical knowledge on eutrophication in lakes is combined and applied for prediction of the effect. Quite a number of model types and levels of details can be established on this basis. In the following, a stepwise model type will be described. It combines a phosphorus mass balance for a lake with the effect in terms of eutrophication. In the following, a rather simple model that comprises the stepwise application of three submodels will be described:

1. A phosphorus balance model for a lake
2. A model for estimation of the chlorophyll-a concentration in the lake
3. A model for light penetration into the water body of the lake

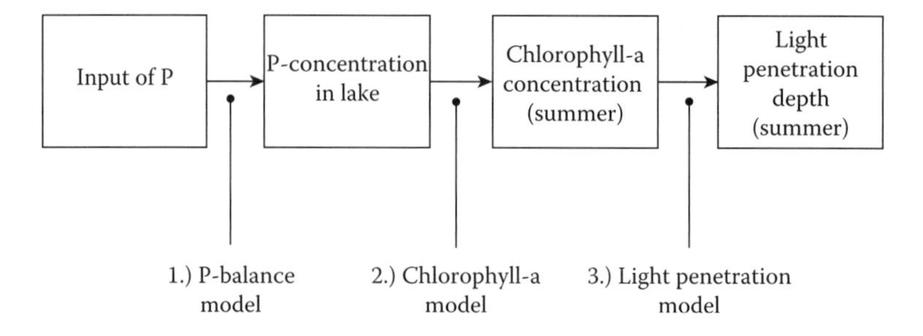

FIGURE 6.13 Example of an empirical–statistically formulated eutrophication model complex for a lake.

In Figure 6.13, the concept of this model is outlined.

The basic objective of the model complex is to establish a relation between the load of phosphorus onto the lake—the cause of eutrophication—and a rather simple but quantitatively expressed measure of the related effect. The three steps of the model are more detailed and exemplified in the next section.

6.4.4.1 Phosphorus Mass Balance Model

It is assumed that phosphorus is the limiting nutrient for growth of both rooted plants and algae and that phosphorus is the governing nutrient for eutrophication.

A P-balance can be established at different levels depending on what details are available. Typically, the P-balance is performed rather simply as exemplified in Figure 6.9. In general, an annual average total phosphorus concentration is therefore the result of this model.

The trophic state models described in Section 6.4.3 are examples that at a simple level can provide the information required, Dillon's and Larsen–Mercier's models are relevant. These two models basically express the P-balance for a lake in a steady-state situation as follows:

$$\text{P-inflow rate} = \text{P-accumulation rate} + \text{P-outflow rate}. \tag{6.16}$$

6.4.4.2 Chlorophyll-A Model

Chlorophyll-a is the active substance for the photosynthesis of plants by absorbing energy from sunlight used for production of organic matter in terms of new biomass. Chlorophyll-a is therefore controlling the biological production and it plays a central role in the eutrophication of lake waters. In the water phase of a lake it is a measure of the activity of the free-flowing algae. The link between the concentration of phosphorus in the water phase and the concentration of chlorophyll-a in the algae is the central characteristic of the chlorophyll-a model.

A number of chlorophyll-a models have been developed for different types of lakes and valid for different climate regions. The mathematical form of these models is typically expressed in terms of an exponential equation:

$$C_{\text{chl}} = a\,(C_{\text{av}})^{b} \tag{6.17}$$

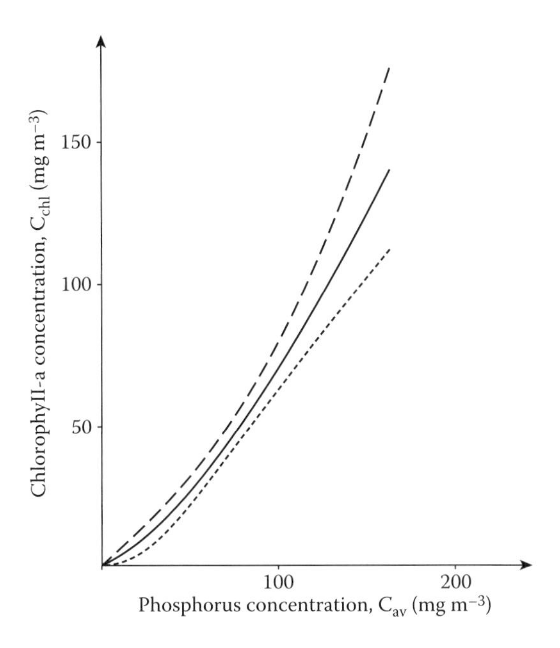

FIGURE 6.14 Correlation between the average chlorophyll-a concentration in lake waters during the period July through August and the annual average total concentration of phosphorus. (Modified from Broegger, J., and Heintzelmann, F., Lake Restoration: The Application of Simple Mass Balance and Eutrophication Models in Water Quality Planning, The National Environmental Protection Agency, Environmental Project No. 16, 1979).

where

C_{chl} = summer concentration of chlorophyll-a (notice: mg m^{-3})
C_{av} = annual average total concentration of phosphorus (notice: mg m^{-3})
a and b = constants

Figure 6.14 is an example illustrating Equation 6.17 with $a = 0.11$ and $b = 1.41$. The curve is based on investigations of 10 lakes located in the temperate climate zone (Denmark). The regression line and the 95 percentiles are shown.

6.4.4.3 Light Penetration Model

The degree of penetration of sunlight onto a surface water body affects the water depth to which the development of rooted plants is possible. Furthermore, the penetration also affects where activities for carnivorous fish in the water body are possible. A penetration depth for light, z_{Sec}, can pragmatically be defined as a water depth, the Secchi disc depth, where it is just possible to identify a white colored circular plate as a contrast to the surrounding water body. At this depth about 90% of the sunlight at the water surface has been absorbed (i.e., a water depth to which approximately 10% of the light will penetrate). Twice the value of z_{Sec} corresponds to a depth where approximately 1% of the light has penetrated. The average net primary production of the phytoplankton during day and night is, at this depth, approximately equal to 0.

According to Lambert-Beer's law, light penetration in water decreases exponentially with increasing water depth. The decrease in light penetration in surface

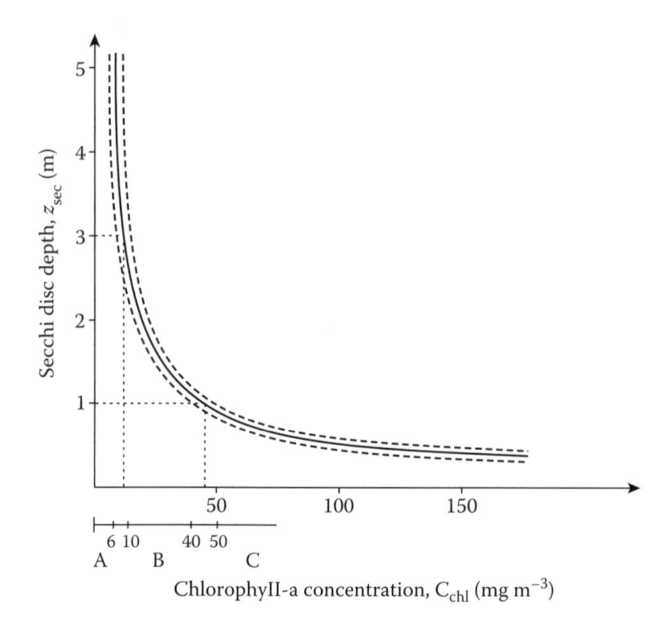

FIGURE 6.15 The light penetration model depicted as a correlation between average values for the period July through August of the chlorophyll-a concentration and the Secchi disc depth for 41 lakes. The regression line and the 95 percentiles are shown. A lake classification is illustrated.

waters is in addition to the absorption of the water molecules and particulate matter mainly influenced by chlorophyll-a in the free-floating algae. A simple model that only takes into account the light absorption of chlorophyll-a can be expressed as an exponential equation:

$$z_{Sec} = c\ (C_{chl})^d \tag{6.18}$$

where

z_{Sec} = Secchi disc depth (m)
c and d = constants

In Figure 6.15, Equation 6.18 is exemplified based on data from 41 lakes located in the temperate region (Denmark) with $c = 10.8$ and $d = -0.622$.

The results according to Figure 6.15, the light penetration model, can be used for a simple classification of lakes according to the degree of eutrophication (cf. Table 6.3).

6.5 POLLUTANTS WITH TOXIC EFFECTS

Toxic effects or toxicity refers to a harmful, damaging, poisoning, or injurious impact of a substance or a mixture of substances onto living organisms (i.e., animals, plants, or humans). A toxic impact can lead to loss of biodiversity, ultimately to the elimination of organisms.

In general, all pollutants that occur in CSOs or SWR may exert toxic effects but the word toxic effect—or toxicity—is normally used when a negative effect occurs

TABLE 6.3
Lake Classification According to the Secchi Disc Depth

Group	Overall Lake Characteristics	Total Primary Production (of Algae) ($gC\ m^{-2}\ yr^{-1}$)	Secchi Disc Depth (m)
A	Unpolluted	< 100	> 3
B	Slightly polluted	100–500	1–3
C	Strongly polluted	> 500	< 1

at relatively low concentration levels. Toxic effects are identified or measured as changed (reduced) activity of the organisms or ultimately their destruction (i.e., sublethal and lethal effects), respectively.

Toxic effects are complex and highly dependent on the organism that is exposed to the pollutant and the substance that causes the toxicity. Furthermore, and in particular for heavy metals, the species in question play a central role. The distribution and exchange of those species between phases, a complex dynamic interaction referred to as partitioning, is therefore important (cf. Section 3.2.4). In general, an effect is associated with a breakdown of some part of the physiological system of the organism. External factors like temperature, salinity, and dissolved oxygen that affect the physiology of an organism, as well as the partitioning, normally affect the toxic effect exerted by a pollutant.

Both acute toxic effects and accumulative (chronic) toxic effects can occur in receiving systems as a result of SWR and CSOs. In the context of wet weather related pollution, it is not well defined what should be considered an acute and a chronic toxic effect, respectively. Although a chronic effect will occur as a long-term effect, there is a gradual transition between these two types of effects. Typically, a chronic effect is associated with an accumulation of a pollutant via the food chain. In case of an acute effect, the median lethal concentration to kill 50% of a test population, the LC_{50} value, for a set period of exposure (e.g., one hour to five days) can be used as a measure of a toxic effect. Often an acute toxicity test lasts two or four days. A range of different species, fish and invertebrate crustaceans, can be selected for such tests. In addition to this test, a great number of bioassays and toxicological methods can be applied to assess the effects but will not be dealt with in this context (Dutka 1988; Pitt et al. 1995; Herricks 1995; Rochfort et al. 1997; Marsalek et al. 1999a, 1999b; Burton and Pitt 2002; Hoffman et al. 2003).

Toxic effects caused by CSOs and SWR are normally considered in relation to the occurrence of the following groups of pollutants:

- Heavy metals
- Organic micropollutants
- Specific substances like chloride (Cl^-) and ammonia (NH_3)

The pollutants belonging to the two first mentioned groups are briefly dealt with in Sections 3.2.1.3, 3.2.1.4, and 3.2.1.6, respectively. The number of potentially occurring substances in SWR and CSOs for each of these groups is very large and in the case of organic micropollutants subject to considerable changes in terms of both new substances being introduced and others being banned. The technological development and the improved knowledge on the potential impacts of the substances affect their use in society and thereby their occurrence in both wastewater and runoff water. Furthermore, it must be realized that the number of substances within these groupings and in addition to their speciation make it, based on field investigations, unrealistic to identify what type of effects are caused by what pollutants. It is, however, in some cases, possible to identify both lethal and sublethal effects in surface water systems where the dominating pollutant load is caused by wet weather discharges. This aspect will be illustrated in Example 6.1.

A potential occurrence of a toxic effect is typically determined by tests on runoff samples and not by field observations. Marsalek and colleagues (1999a) refer to toxicity tests of urban runoff samples from 14 sites showing about 20% of the samples indicated severe toxicity and another 20% confirmed toxicity. Because of relatively high concentrations of both metals and organic micropollutants, snowmelt may exert toxic effects (Marsalek et al. 2003; cf. Sections 4.2 and 4.6.2).

The majority of tests for toxicity are designed for relatively short periods of duration. In general, such tests express a potential acute, and not an accumulative, effect. However, it is likely that accumulative effects occur at a lower concentration level of a toxic species than an acute effect. This fact—and other aspects too—therefore call for taking into account the concept of risk of toxic pollutants. This approach is not only based on scientific knowledge but also political of nature.

Although not generally correct, it is to some extent expected that SWR is more toxic than CSOs. The reason is that a number of substances in the CSO, different organics, can adsorb toxic substances or via complex reactions detoxify such substances and thereby reduce a potential effect.

6.5.1 Effects of Heavy Metals

Selected heavy metals that may occur in the wet weather urban and road runoff are briefly outlined in Section 3.2.1.3.

Each heavy metal has its specific characteristics and the potential harmful effect on humans and the environment depends, in a complex way, on these metal characteristics, the biological system that is affected, and a number of external conditions. Particularly the chemical environment in terms of the metal solubility—including metal association with particles—and the potential for speciation of the heavy metals are such governing external conditions, often determined by the following characteristics:

- Complexing agents for the heavy metals
- Hardness
- pH and alkalinity
- Dissolved oxygen

Important aspects of these characteristics are dealt with and exemplified in Sections 3.2.4 and 3.2.5, respectively.

Furthermore, and closely related to these characteristics, the following parameters are central for the toxicity and potential for detoxification of heavy metals (cf. Section 3.2.5):

- Acidity: pH
- Concentration: pC = –logC
- Redox potential: pe = –logE

As an example Figure 6.16 shows the solubility diagram for lead (Pb), pC versus pH under aerobic conditions, at a relatively high value of pe (cf. Section 3.6.1). This diagram is constructed based on the equilibrium solubility equations that exist when chloride, sulfate, phosphate, and carbonate occur in the water phase, as is the case for

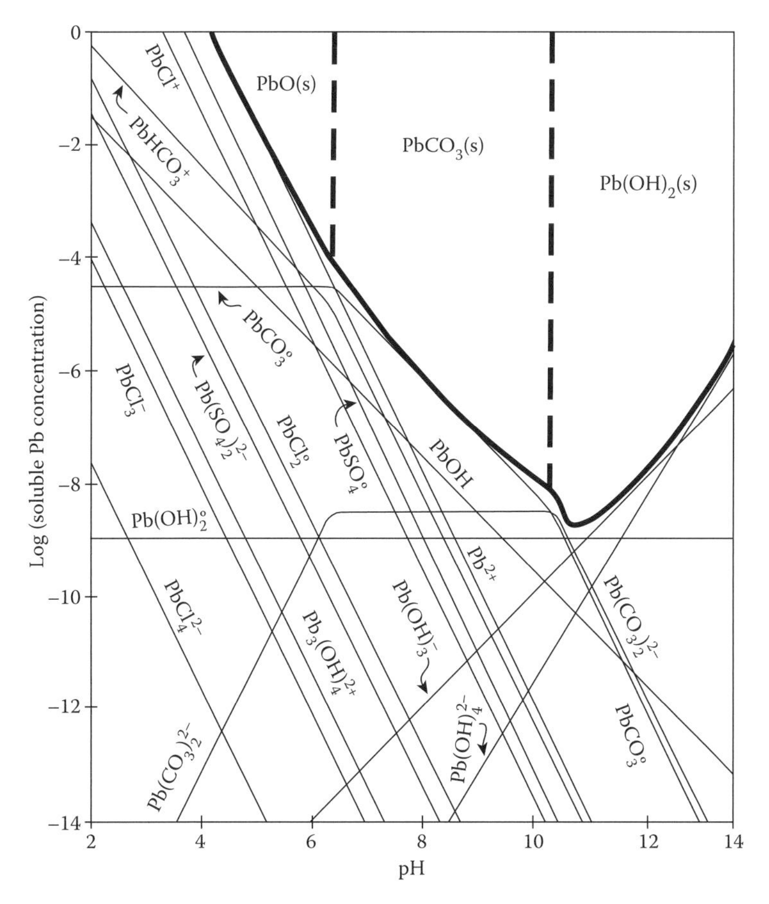

FIGURE 6.16 Solubility of inorganic Pb-ions in surface waters characterized by concentration of total inorganic carbon, $C_T = 5\ 10^{-4}$ M; chloride (Cl^-) concentration = $5\ 10^{-4}$ M; sulfate (SO_4^{2-}) concentration = 10^{-4} M; phosphate (PO_4^{3-}) concentration = 10^{-4} M. (Modified from Yousef, Y. A., Wanielista, M. P., Harper, H. H., Pearce, D. B., and Tolbert, R. D., Best Management Practices: Removal of Highway Contaminants by Roadside Swales, Report No. FL/DOT/BMR-84-274, Tallahassee: Florida Department of Transportation, 1985.)

natural surface waters, also those receiving wet weather discharges (cf. Section 3.2.5). It is seen that the solubility of lead below a pH value of about 6.5 is determined by the hydrated Pb^{2+} ion, whereas above this value is mainly governed by the carbonate ion, CO_3^{2+}, and the hydroxyl ion, OH^-, both inorganic ions being ligands for the formation of metal complexes. In terms of toxicity it is important that free metal ions, metal ions that are hydrated (i.e., exist surrounded by the dipole-forming water molecules), in general show higher level of toxic effects compared with those corresponding metal ions where complexes with both inorganic and organic ligands have been formed. In other words, the formation of complex metal ions will, to some extent, detoxify heavy metals. In the actual case it is therefore likely that Pb in stormwater is more toxic below pH = 6.5 than above this value. However, an exception to this statement is complexes that are soluble in fat and therefore exert a potential for accumulation in the fat tissue of an organism. An example is a number of complexes with mercury (e.g., CH_3HgCl).

Corresponding to the inorganic complex formation exemplified in Figure 6.16, metal complexes with organic ligands can also stabilize and detoxify metals in solutions. In natural waters the occurrence of humic substances, fulvic acids and tannic acids, will form complexes with copper and to some extent also cadmium. These organics are complex aromatic polymers with a carboxylic group (–COOH), a hydroxyl group (–OH), and an amine group ($-NH_2$) attached to the molecules.

Based on solubility diagrams for the different heavy metals as the one shown in Figure 6.16, the dominating inorganic complexes for selected heavy metals occurring in the wet weather flows and at different pH ranges are summarized in Table 6.4.

TABLE 6.4
Dominating Inorganic Speciation of Selected Heavy Metals at Various pH Ranges in Typical Natural Surface Waters

Heavy Metal	pH < 6.5	6.5 < pH < 7.5	pH > 7.5
Copper, Cu	Cu^{2+}	Cu^{2+}	$Cu(OH)_2^0$
		$CuCO_3^0$	
		$Cu(OH)_2^0$	
Lead, Pb	Pb^{2+}	Pb^{2+}	$PbCO_3^0$
		$PbCO_3^0$	$Pb(CO_3)_2^{2-}$
			$Pb(OH)_3^-$
			$Pb(OH)_4^{2-}$
Zinc, Zn	Zn^{2+}	Zn^{2+}	$ZnCO_3^0$
		$ZnCO_3^0$	$Zn(OH)_2^0$
			$Zn(OH)_4^{2-}$
Cadmium, Cd	Cd^{2+}	Cd^{2+}	Cd^{2+}
			$CdCO_3^0$
			$Cd(OH)_2^0$

Source: Data from Yousef, Y. A., Wanielista, M. P., Harper, H. H., Pearce, D. B., and Tolbert. R. D., Best management practices: Removal of highway contaminants by roadside swales. Report No. FL/DOT/BMR-84-274, Tallahassee: Florida Department of Transportation, 1985.

It is readily seen that most toxic species of the metals, the hydrated ions, in general are dominating at the relatively low pH ranges. Further analysis of metal speciation and interactions can be done by the MINTEQ-model (cf. Section 3.2.5).

The formation of metal complexes as referred to in Table 6.4 depends on the occurrence and level of concentration of the ligands. Chloride, carbonate, and sulfate ions are typically present in urban and road runoff water and in most surface waters. These ions are therefore important for the formation of metal complexes.

Heavy metals can accumulate in biomass, plants, bacteria, and invertebrates (insect larvae), and be further concentrated through the food chain. The complex effects of this metal accumulation may result in inhibition of the enzyme systems that manifests itself in a reduced or changed activity level. Structural deformities, loss of cell membrane stability and breakdown of cell materials, are other possible effects of heavy metals. As shown in Table 6.4, a number of metals form complexes with the charge 0. These complexes have low tendency for adsorption to particles and will mainly occur in the water phase.

Several heavy metals are essential micronutrients for living organisms and the toxic effects are associated with water phase concentrations exceeding certain background levels. The bioaccumulation process in the cell biomass is an active process that requires energy from the organism. The process is therefore a relatively slow process compared with adsorption of heavy metals.

The toxic effect exerted by heavy metals occurring in typical concentration levels found in SWR and CSOs cannot be exemplified by giving corresponding typical levels of toxicity. The nature of the problem is too complex to give a meaningful impression. However, the following will be concluded:

- The level of heavy metal concentrations that typically occur in surface waters receiving considerable inflows of SWR or CSOs can—with reference to results from laboratory studies with different test organisms—result in toxic effects. Generally, it is the acute effects and not the accumulative effects that have been tested under such conditions.
- A number of external factors dealt with in this section will affect the toxicity of the heavy metals (e.g., in terms of detoxification).

6.5.2 Effects of Organic Micropollutants

Organic micropollutants—also called organic priority pollutants or organic xenobiotic compounds—that potentially occur in the wet weather flows are briefly outlined in Section 3.2.1.4. Further details are found in Burton and Pitt (2002). Numerous organic micropollutants can be found in both SWR and CSOs exerting complex effects onto humans and the environment.

In general terms, the following characteristics of a component are important for its effect:

- *Transport characteristics and mobility*
  In general, the transport characteristics and mobility of a component determine where it is found in the environment. Particularly, the solubility (solubility in water and sorption on particles, cf. Section 3.2.4.3.2) and the

volatilization of the organic micropollutants are central in this respect. Several organic micropollutants are hydrophobic and adsorb to particles (cf. Section 3.2.4.3.3). Such species can therefore accumulate in sediments of surface waters.

- *Bioaccumulation*
  Bioaccumulation refers to the uptake characteristics for a component in an organism and its following fate in the food chain. Particularly, hydrophobic substances may accumulate in the fat tissue of an organism.
- *Degradation*
  Organic micropollutants are potential substrates for living organisms and might therefore be metabolized (i.e., undergo chemical and biological transformation). The metabolism will typically proceed in steps and often rather stable intermediates are produced. Compared with the original compound, such intermediates may show either reduced or increased solubility and toxicity. The rate of degradation is a parameter dependant on the actual substance, the organism, and a number of external conditions. The redox potential in terms of aerobic and anaerobic conditions in the environment is in this respect often central. Also photolysis—breakdown caused by sunlight—is in some cases important.
- *Persistency*
  The persistency of a xenobiotic compound refers to a low degradability and corresponding prolonged occurrence in the environment. A persistent compound accumulates in animals, sediments, or snow. Persistency also includes the formation of intermediate toxic products with low degradability. Examples of persistent compounds are PCBs that are still found in the environment although they were banned in the 1970s.

In general, organic micropollutants typically occur in the runoff water in concentrations below 1 mg m^{-3} or even below detection limit. However, and in addition to total hydrocarbon, the total concentration of PAHs can typically reach values up to 10 mg m^{-3}. Also micropollutants like DEHP (di(2-ethylhexyl phthalate) and nonylphenol often occur in relatively high concentrations.

The PAHs may have adverse ecological effects on receiving water systems and it is well known that PAHs can exert human genotoxic effects and cancer. Because of that and because PAHs occur in relatively high concentrations in the runoff water, they are often selected as indicators for wet weather pollutant loads of organic micropollutants. Furthermore, the hydrophobic characteristics of the PAHs mean that they can accumulate in the sediments. In particular, the low-molecular PAHs are bioavailable (Nakajima et al. 2005).

The toxic effects of both heavy metals and organic micropollutants are complex and can, in general, not be assessed in receiving waters by direct experimental studies in terms of a simple cause-effect relation. The following Example 6.1 will illustrate this fact and also exemplify levels of potentially toxic substances originating from SWR.

Example 6.1: Toxic Effects Observed in Lakes Receiving Urban Stormwater and Highway Runoff

The example concerns toxic effects that are identified for biota in a shallow lake system receiving urban stormwater and highway runoff (Harremoës 1982). The lake is located in the temperate region (Denmark) and is, in principle, an artificially constructed twin-lake, here named lake A and lake B, Figure 6.17. Lake A is located upstream of lake B and receives runoff water from a highway and a mixed residential and industrial catchment with a total contributing impervious area of about 140 ha.

Measurements of the accumulation of different heavy metals and selected organic micropollutants have been performed. Tables 6.5 and 6.6 exemplify typical values for lead and PAH, respectively, which are accumulated in the two

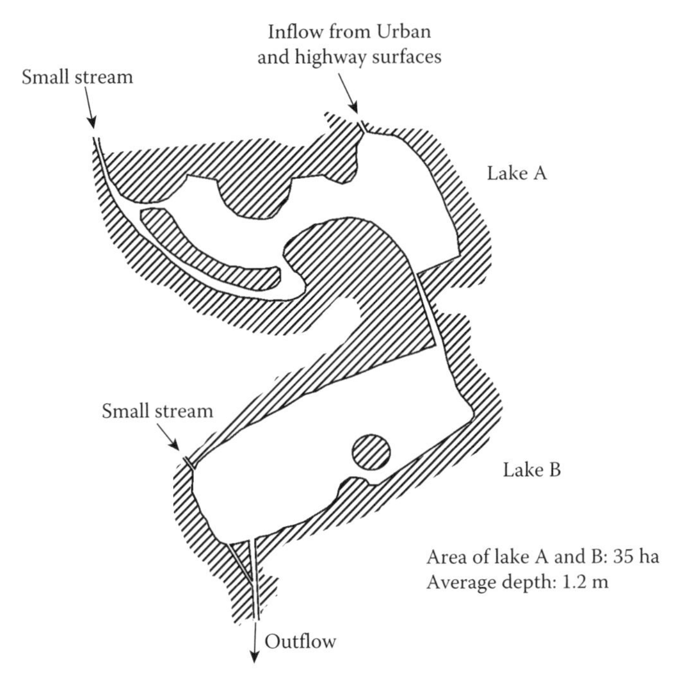

FIGURE 6.17 The artificial twin-lake system receiving urban stormwater and highway runoff.

TABLE 6.5
Typical Values for Lead (Pb) Accumulated in Lake A and Lake B

Pb (mg (kg DM)$^{-1}$)	Lake A	Lake B
Sediments	640	102
Invertebrates	33	24
Bivalves	2.4	0.75
Plants	82	6

TABLE 6.6
Typical Values for PAH Accumulated in Lake A and Lake B

PAH (mg (kg DM)$^{-1}$)	Lake A	Lake B
Total PAH in sediments	8.0	1.5
Benzopyrene in sediments	0.45	0.055
Benzopyrene in fat tissue of bivalves	0.40	0.20

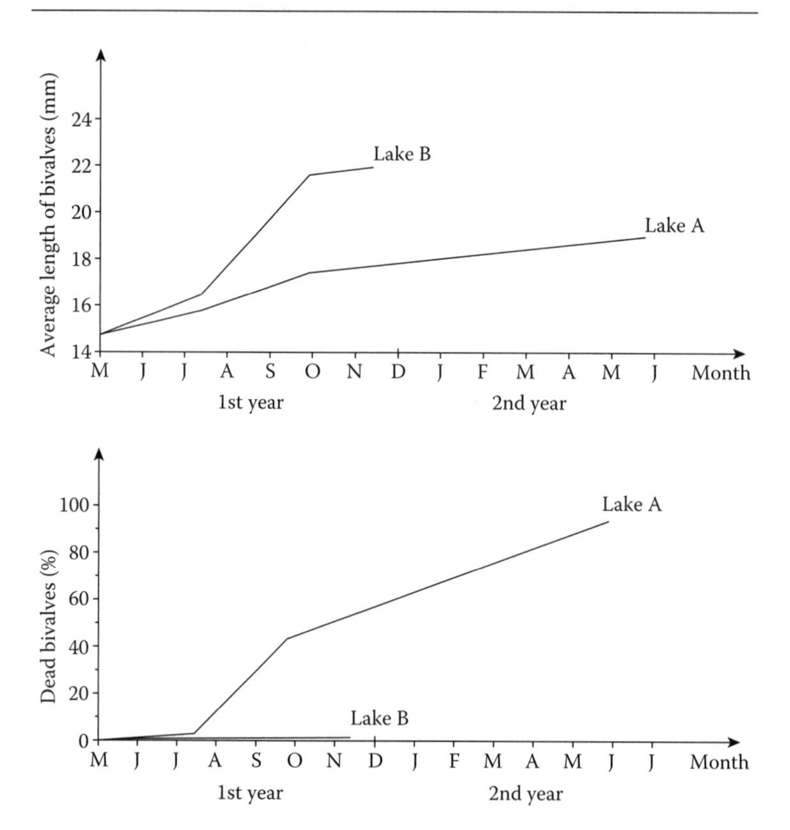

FIGURE 6.18 Growth (shown as changes in length) and survival of bivalves used as test organisms in lake A and lake B. The bivalves in lake B were unintentionally removed from lake B after about six months!

compartments of the twin-lake. The values refer to what was measured after receiving stormwater discharges during a period of eight years.

Tables 6.5 and 6.6 show that there is a substantially higher concentration of lead and PAH accumulated in the sediments and in the biota of lake A compared with lake B. This fact is assessed as a result of adsorption, sedimentation, and bioaccumulation that are considered central pollutant accumulation processes in the two compartments. It is crucial that there is also different effects observed onto the ecosystem in the two compartments of the twin-lake. Figure 6.18 shows how growth and survival of bivalves that are used as test organisms in the two compartments develop over time. From this figure it is readily seen that both growth and survival of the test organisms are subject to the largest stress factors in the upstream and most heavily pollutant loaded compartment.

Although the bivalves are exposed to the most severe toxic conditions in the upstream compartment of the twin-lake, there exists no proof of the cause of their reaction. However, because the two artificially constructed lake compartments are uniform in several ways, it is a correct hypothesis that some pollutants, probably among the heavy metals and the organic micropollutants, originating from SWR result in toxic effects. This hypothesis is basically twofold: The biota in lake A is negatively affected but the cleaning processes in this compartment in terms of pollutant removal from the water phase also reduces the effect onto the biota in lake B. It is therefore an important—and probably also correct—statement that natural occurring processes can improve the quality of waters contaminated with pollutants originating from SWR. This aspect of treatment will be dealt with in detail in Chapter 9.

6.5.3 Effects of Specific Substances

A number of specific substances not included in the groups of heavy metals or organic micropollutants and occurring in SWR or CSOs can, in different ways, cause unintended impacts onto the environment. Chloride and ammonia should be mentioned as important examples. Other examples of (toxic) substances like sulfide can occur in CSOs under certain conditions, however, in general not in concentrations that affect receiving water quality.

6.5.3.1 Chloride

Chloride is widely used as a deicing agent and therefore found in snowmelt runoff. Basically, it is not considered a toxic substance, however, and often together with sodium, it can cause a number of adverse effects that are related to its occurrence in concentrations found in the snowmelt runoff:

- Chloride affects the speciation of heavy metals and by formation of soluble complex species it can transfer particle bound metals to a soluble form (cf. Sections 3.2.5, 4.2, and 6.5.1). The bioavailability of heavy metals can thereby be increased.
- When discharged to lakes and reservoirs, chloride may result in chemical stratification. Under extreme conditions, this chemically initiated stratification can hinder or even prevent spring overturn and thereby the transport of dissolved oxygen to the bottom layer.
- High chloride concentrations may exert negative effects onto the freshwater biota.
- Deicing salts (NaCl) can change (destruct) the structure of soils and thereby lower the soil fertility caused by cation replacement (i.e., Na^+ for Ca^{2+} and Mg^{2+}, cf. Section 6.7). Correspondingly, leaching of heavy metals can take place.

Chlorides can furthermore exert negative impacts onto the built environment in terms of corrosion.

6.5.3.2 Ammonia

In general, ammonia will occur in CSOs originating from the daily wastewater flow and eroded deposits. Compared with the potential contribution from combined sewers, its occurrence in SWR is negligible. The molecular form of ammonia (NH_3) and not NH_4^+ exerts acute toxic effects to fish because occurrence of NH_3 in the water phase obstructs the diffusion of that species across the gills of the fish. The following momentary established equilibrium between these two forms is therefore crucial:

$$NH_4^+ \leftrightarrows NH_3 + H^+. \tag{6.19}$$

The pK_a value for this equilibrium is at 20°C about 9.2 (cf. Section 3.2.3). According to Figure 6.19, the toxic effect increases with both pH and temperature.

The influence of pH and temperature on the relative content of molecular ammonia means that a given total (analytically determined) concentration of ammonia in a surface water system receiving CSOs may result in varying toxic effect. The most severe potential effect will occur during hot summer periods at the time of a day (shortly after noon) when the primary production of the aquatic plants is high and the alkalinity of the water is correspondingly low (cf. Section 3.2.4.2).

Both lethal and sublethal acute effects on the fish population are possible. As previously mentioned in Section 6.5, the lethal concentration to kill 50% of a test population is formulated in the so-called LC_{50} value. The LC_{50} for ammonia in the case of fish is subject to considerable variability depending on exposure time, temperature, and DO concentration. For rather short periods of exposure ($<$ 10 minutes), a standard value proposed by Whitelaw and Solbé (1989) for salmon and carp is 2 and 10 g NH_3-N m^{-3}, respectively. Increase in exposure time decreases these values considerably with lethal effects for salmon observed at 0.2 to 0.5 NH_3-N m^{-3}. To avoid sublethal effects for salmon, concentrations are recommended below 0.1 NH_3-N m^{-3}.

In case of CSO discharges, it is important to notice that the effect is often acute. Therefore, it must be evaluated on the basis of single events—particularly the extreme

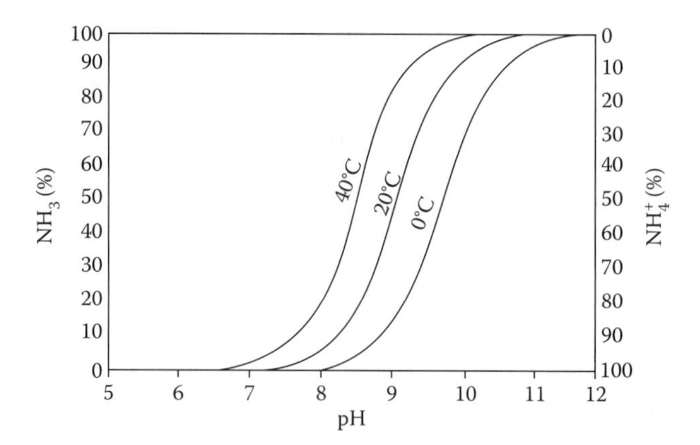

FIGURE 6.19 The influence of pH and temperature on the equilibrium between NH_4^+ and NH_3. (Modified from U.S. EPA, *Process Design Manual for Nitrogen Control, Technology Transfer*, Washington, DC, 1975.)

ones—and not as is the case for the accumulative effects in terms of a number of events occurring during a rather long period of time. A criterion for CSO discharges of ammonia that in principle follows the five points formulated in Section 6.3.4 for the DO concentration in a stream receiving CSO discharges is therefore a basic approach. A statistics for salmon formulated similar Figures 6.7 and 6.8 might result in 0.1 and 0.5 g NH_3-N m^{-3} being acceptable for frequent events and return periods 5–10 years, respectively. If the water temperature is 20°C and pH = 8.0, 0.5 g NH_3-N m^{-3} will, according to Figure 6.19, correspond to a total ammonia concentration of about 14 g N m^{-3}. However, if the pH value is 8.5, the total ammonia concentration equivalent to 0.5 g NH_3-N m^{-3} is just about 4.5 g N m^{-3}. In addition to the surface water characteristics in terms of pH, temperature, and the DO concentration, it is the total ammonia concentration in the CSO outlet and the following dilution in the receiving water that determines if an ammonia concentration is critical for the fish population. In House et al. (1993) a corresponding, statistically based water quality criterion for ammonia has been formulated for United Kingdom surface waters in the case of return periods < 1 year.

6.5.4 CSO versus SWR: Comments on Toxic Effects from Wet Weather Sources

In Section 6.5 on toxic effects caused by pollutants in the wet weather flows it is shown that the effect related phenomena are very complex. It is not surprising that the large number of pollutants with different characteristics occurring under highly varying external conditions makes it very difficult to draw clear conclusions concerning toxicity. Site-specific investigations based on a number of test procedures are therefore typically needed to assess if toxic effects occur.

However, it is interesting to discuss in general terms potential toxic effects from CSOs versus SWR. In several countries it was 40–70 years ago an important issue that SWR was clean compared with CSOs and therefore motivated a shift away from constructing combined sewer systems. It is well known today that the runoff water from both urban and industrial areas and from roads is polluted. As also shown in this text, there are situations where the discharges from storm sewers are more problematic than from overflows (e.g., compared per unit area of contributing catchment). There are reasons to believe that SWR in general will exert more serious toxic effects than CSOs. This statement is particularly due to the following four reasons:

- In most cases the volumes discharged from CSO structures are, per unit area of contributing catchment, rather limited compared with the volumes from a corresponding storm sewer.
- For several potentially toxic pollutants there are, in general, not significant differences in concentrations between runoff water and the wet weather flows in combined sewers.
- Wastewater in combined sewers has a detoxifying effect on several toxic pollutants including those originating from urban surfaces and roads (Marsalek et al. 1999b). There are several reasons why it is the case, adsorption processes, different immobilization reactions, and formation of complexes due to the

contents of organic matter like humic substances. Furthermore, the hardness and alkalinity may also contribute positively to the detoxification.

- As a consequence of the detoxifying effect of the combined sewage, spills, and accidents that occur in areas with storm sewers may result in the most serious environmental impacts.

6.6 EFFECTS OF PATHOGENIC MICROORGANISMS

The grouping and occurrence of pathogenic microorganisms and examples of their disease-causing impacts on humans are dealt with in Section 3.2.1.6. Furthermore, microorganisms as indicators for their potential harmful effect are also outlined. The pathogenic microorganisms originate from both dry and wet weather sources. Concerning the wet weather flows, particularly the flows of untreated wastewater in CSOs are problematic (Soyeux, Blanchet, and Tisserand 2007). Further details concerning pathogens can be found in Burton and Pitt (2002).

The concern of pathogenic microorganisms is particularly related to their occurrence in surface waters that are classified as bathing waters (i.e., waters for recreational purposes). Furthermore, microbial contamination in case of rainwater harvesting is problematic. It is a potential pollution with fecal matter that is assessed by the number of occurring indicator organisms. Because of that, it is a human health risk that is the pragmatic basis for a microbial water quality standard and not some level for real occurring pathogens. In terms of wet weather discharges, it is important to notice that the bathing water quality is assessed by regularly sampling during the bathing season and therefore represents a mix of both dry and wet weather impacts.

Indicator microorganisms for fecal pollution that are typically used for assessment of bathing water quality are outlined in Section 3.2.1.6. As an example, the proposed European Union bathing water directive distinguishs between three levels of bathing water quality based on intestinal enterococci (IE) and escherichia coli (EC) (EU 2002):

- Excellent quality
- Good quality
- Poor (i.e., a quality that is not acceptable)

Taking this proposal of a directive as an example, Table 6.7 shows the maximum acceptable levels of the indicator microorganisms. Statistically, the assessment is

TABLE 6.7
EU Proposal for Upper 95 Percentiles, X_{95}, for Excellent and Good Bathing Water Quality

Indicator Microorganism	Excellent Quality (per 100 ml)	Good Quality (per 100 ml)
Intestinal enterococci (IE)	100	200
Escherichia coli (EC)	250	500

based on the upper 95-percentile value based on a set of n values of the measured number of microorganisms, x_i:

$$X_{95} = \exp(u + 1.65\, s), \tag{6.20}$$

where

X_{95} = upper 95-percentile value (per 100 ml)
u = arithmetic mean value of n values of $\log_{10}(x_i)$ (–)
s = standard deviation of n values of $\log_{10}(x_i)$ (–).

The risk levels of fecal pollution shown in Table 6.7 are based on epidemiological studies. A number of such studies have revealed that when the effect is defined as respiratory illness, the 95-percentile values for excellent and good bathing water quality correspond to about 1% and 2% risk for illness, respectively, when comparing bathers with nonbathers. The World Health Organization (WHO) considers such risk levels acceptable.

6.7 CONTAMINATION OF SOILS AND GROUNDWATER

In this section, the impact of pollutants—including those originating from the wet weather sources—related to soil, groundwater, and sediments will be outlined in general terms. Further details related to infiltration of stormwater are dealt with in Section 9.4.

The previous sections of this chapter have focused on the effects of SWR and CSOs discharged to receiving waters. However, discharge of stormwater into soil systems also occurs. Examples are runoff from impervious urban areas and roads into adjacent vegetated areas, runoff directly through porous pavements, filter strips and swales or infiltration from infiltration ponds (cf. Sections 9.4 and 9.7).

The concern of pollutants in soils is basically identical with that dealt with in the previous sections, although often differently expressed. The concern of groundwater contamination is particularly crucial in case of its use for drinking water purposes and irrigation. Contaminated soil is directly a problem in the case of growth in plants for human or animal food supply or due to its ecological value. Because of such impacts, the concern of soil contamination is particularly important along rural roads and highways where the adjacent land is sensitive to contamination. Contamination caused by pollutants originating from the runoff is therefore site-specific. A risk assessment should be included in any project for construction or development of roads and highways.

Similar to what is referred to in Section 6.5.2 for the organic micropollutants, important characteristics for the effect of pollutants in soil systems are closely related to transport characteristics and mobility, bioaccumulation, and degradation in the soil system.

The mobility of a pollutant will determine whether a pollutant will occur associated with soil particles, often accumulated in the top soil, or it will be transported with the interstitial water and potentially end up in the groundwater zone. The transport of water through the soil medium is therefore a prerequisite and driving force for this movement (cf. Section 3.4.3). Main characteristics of soils related to the movement of pollutants, like heavy metal ions, is the adsorption characteristics including the cation exchange capacity, CEC (cf. Section 9.4.2.3). For the degradable substances like the organic micropollutants, the rate of biodegradation is central.

In several regions there is a need for groundwater recharge by infiltration of urban and road runoff. The risk of soil and groundwater contamination, however, calls for a feasible methodology to assess a potential risk of groundwater contamination and to find appropriate treatment technologies (Pitt, Clark, and Field 1999). The following pragmatic procedure by Clark and Pitt (2007) is subdivided in three steps:

- *Determination of loadings for selected pollutants and their chemical forms*
  Pollutants should be selected among those with a potential impact on the groundwater quality. Pollutant characteristics in terms of their mobility (adsorption characteristics and association with particles) and their degradability should be determined.
- *Determination of soil characteristics*
  Soil characteristics in terms of organic matter content, pH, infiltration rate, and contents of microorganisms determine to what extent pollutants are retained or degraded in the soil medium.
- *Determination of pretreatment requirement*
  Pretreatment before infiltration can be needed to reduce loads of pollutants. A high concentration, a high soluble fraction, and a high mobility are all pollutant characteristics that indicate a need for pretreatment.

Depending on what details are included in the three steps, the procedure can be more or less complex.

REFERENCES

Broegger, J., and F. Heintzelmann. 1979. Lake restoration: The application of simple mass balance and eutrophication models in water quality planning (in Danish). The National Environmental Protection Agency, Environmental Project No. 16.

Burton, G. A., and R. E. Pitt. 2002. *Stormwater effects handbook: A toolbox for watershed managers, scientists and engineers*. Boca Raton, FL: Lewis Publishers.

Clark, S. E., and R. Pitt. 2007. Influencing factors and a proposed evaluation methodology for predicting groundwater contamination potential from stormwater infiltration activities. *Water Environment Research* 79 (1): 29–36.

Clifforde, I. T., R. W. Crabtree, and H. O. Andrews. 2006. 10 years experience of CSO management in the United Kingdom. Proceedings from the WEFTEC 2006 Conference, Water Environment Federation, 3744–56.

Dillon, P. J. 1975. The phosphorus budget of Cameron Lake, Ontario: The importance of flushing rate to the degree of eutrophy of lakes. *Limnology and Oceanography* 20: 28–38.

Dutka, B. J. 1988. Priority setting of hazards in waters and sediments by proposed ranking scheme and battery of tests approach. *Zeitschrift für Angewandte Zoologie* 75: 303–16.

DWPCC (Danish Water Pollution Control Committee). 1985. *Forurening af vandloeb fra overloebsbygvaerker* [Pollution of rivers from overflow structures]. Report No. 22 from DWPCC.

Ellis, J. B., and T. Hvitved-Jacobsen. 1996. Urban drainage impacts on receiving waters. *Journal of Hydraulic Research* 34 (6): 771–83.

EU. 2002. Proposal for a directive of the European Parliament and of the Council concerning the quality of bathing water, Directive 2002/0254(COD). *Official Journal of the European Union* C 45 E.

Harremoës, P. 1982. Urban storm drainage and water pollution. In *Urban stormwater quality, management and planning*, ed. B. C. Yen. Proceedings of the 2nd International Conference on Urban Storm Drainage, Urbana, IL. June 15–19, 1981, 469–94.

Herricks, E., ed. 1995. *Stormwater runoff and receiving systems: Impact, monitoring and assessment*. Boca Raton, FL: CRC Press.

Hoffman, D. J., B. A. Rattner, G. A. Burton, and J. Cairns, eds. 2003. *Handbook of ecotoxicology*, 2nd ed. Boca Raton, FL: CRC Press/Lewis Publishers.

House, M. A., J. B. Ellis, E. E. Herricks, T. Hvitved-Jacobsen, J. Seager, L. Lijklema, H. Aalderink, and I. T. Clifforde. 1993. Urban drainage: Impacts on receiving water quality. *Water Science and Technology* 27 (12): 117–58.

Hvitved-Jacobsen, T. 1986. Conventional pollutant impacts on receiving waters. In *Urban runoff pollution*, eds. H. C. Torno, J. Marsalek, and M. Desbordes, 345–78. Berlin, Heidelberg, Germany: Springer Verlag.

Hvitved-Jacobsen, T., and P. Harremoës. 1982. Impact of combined sewer overflows on dissolved oxygen in receiving streams. In *Urban stormwater quality, management and planning*, ed. B. C. Yen, 226–35. Littleton, CO: Water Resources Publication.

Larsen, D. P., and H. T. Mercier. 1976. Phosphorus retention capacity of lakes. *Journal of Fisheries Research Board of Canada* 33 (8): 1742–50.

Marsalek, J., G. Oberts, K. Exall, and M. Viklander. 2003. Review of operation of urban drainage systems in cold weather: Water quality considerations. *Water Science and Technology* 48 (9): 11–20.

Marsalek, J., Q. Rochfort, B. Brownlee, T. Mayer, and M. Servos. 1999a. An exploratory study of urban runoff toxicity. *Water Science and Technology* 39 (12): 33–39.

Marsalek, J., Q. Rochfort, T. Mayer, M. Servos, B. Dutka, and B. Brownlee. 1999b. Toxicity testing for controlling urban wet-weather pollution: Advantages and limitations. *Urban Water* 1 (1): 91–103.

Medina, M. A. 1980. Continuous receiving water quality modeling for urban stormwater management. In *Urban stormwater and combined sewer overflow impact on receiving water bodies*, eds. Y. A. Yousef, M. P. Wanielista, W. M. McLellon, and J. S. Taylor, 466–501. Washington, DC: U.S. EPA 600/9-80-056.

Nakajima, F., A. Baun, A. Ledin, and P. S. Mikkelsen. 2005. A novel method for evaluating bioavailability of polycyclic aromatic hydrocarbons in sediments of an urban stream. *Water Science and Technology* 51 (3–4): 275–81.

Nazaroff, W. W., and L. Alvarez-Cohen. 2001. *Environmental engineering science*. New York: John Wiley & Sons, Inc.

Pitt, R., S. Clark, and R. Field. 1999. Groundwater contamination potential from stormwater infiltration practices. *Urban Water* 1 (3): 217–36.

Pitt, R., R. Field, M. Lalor, and M. Brown. 1995. Urban stormwater toxic pollutants: Assessment, sources and treatability. *Water Environment Research* 67 (3): 260–75.

Rochfort, Q., J. Marsalek, J. Shaw, B. J. Dutka, B. Brownlee, A. Jukovic, R. McInnis, and G. McInnis. 1997. Acute toxicity of combined sewer overflows and stormwater discharges. Report No. 97-190, National Water Research Institute, Burlington, Ontario, Canada.

Schaarup-Jensen, K., and T. Hvitved-Jacobsen. 1991. Simulation of dissolved oxygen depletion in streams receiving combined sewer overflows. In *New Technologies in Urban Drainage, Elsevier Applied Science*, ed. C. Maksimovic, 273–82, London and New York.

Schnoor, J. L. 1996. Environmental modeling: Fate and transport of pollutants in water, air and soil. New York: John Wiley & Sons, Inc.

Simonsen, J. F., and P. Harremoës. 1978. Oxygen and pH fluctuations in rivers. *Water Research* 12 (7): 477–89.

Soyeux, E., F. Blanchet, and B. Tisserand. 2007. Stormwater overflow impacts on the sanitary quality of bathing waters. *Water Science and Technology* 56 (11): 43–50.

Streeter, H. W., and E. B. Phelps. 1925. *A study of the pollution and natural purification of the Ohio River,* Vol. III, Public Health Bulletin, No. 146. Washington, DC: U.S. Public Health Service.

U.S. EPA. 1975. *Process design manual for nitrogen control.* Washington, DC: Technology Transfer.

U.S. EPA. 1983. Final report of the nationwide urban runoff program, U.S. EPA Planning Division, Vol. 1, September 30.

Vollenweider, R. A. 1968. Scientific fundamentals of the eutrophication of lakes and flowing waters, with particular reference to nitrogen and phosphorus as factors in eutrophication, OECD, Technical Report. DAS/CSJ/68.27, Paris.

Whitelaw, K., and J. F. de L. G. Solbé. 1989. River catchment management: An approach to the derivation of quality standards for farm pollution and storm sewage discharges. In *Urban discharges and receiving water quality impacts*, ed. J. B. Ellis, 145–58, Oxford: Pergamon Press.

WRC (Water Research Centre). 1990. Proposed water quality criteria for the protection of aquatic life from intermittent pollution. Report PRS2498-NM, WRC, Medmenham, UK.

Yousef, Y. A., M. P. Wanielista, H. H. Harper, D. B. Pearce, and R. D. Tolbert. 1985. Best management practices: Removal of highway contaminants by roadside swales. Report No. FL/DOT/BMR-84-274, Tallahassee: Florida Department of Transportation.

7 Experimental Methods and Data Acquisition within Urban Drainage

In principle, this chapter has its focus on how to establish a "quality information system" and to outline what elements are important in this respect. The design and operational details of a measurement program is therefore central. However, it should be realized that the design and performance of such a measurement program to a great extent depends on the actual objective in addition to numerous site characteristics. Focus in this chapter is therefore on general aspects of the pieces that form the entire program, particularly defined in terms of sampling, monitoring, and analysis. Procedures for these elements are considered basic for all types of measurement programs and they are central for information on the performance of the urban drainage system. Literature is rich in examples on how measurement programs and data acquisition in specific cases have been designed and managed for urban and road drainage systems. Based on the pieces dealt with in this chapter, it is up to the reader to form the entire puzzle leading to a measurement program in a specific case.

As briefly mentioned, sampling, monitoring, and analyses are the key elements of a measurement program. Treatment, assessment, and use of data related to the quality of urban wet weather pollution require that these elements be well performed. The "nature" and the "empirical basis" focusing on the way of achieving information in terms of data, often quite different compared with similar considerations in case of dry weather, will be emphasized. It is not an objective to deal with details like specific methods for monitoring and analysis, types of equipment, or different statistical methods. Such specific details can be found elsewhere in texts and technical publications (Wiersma 2004). In case of chemical analyses, APHA-AWWA-WEF (2005) is a central source for information. A number of specific methods and technologies in terms of stormwater sampling, monitoring, and analysis are found in Burton and Pitt (2002) and principles for data acquisition within urban water management are focused on in Fletcher and Deletic (2007). Numerous textbooks on statistical methods exist, for example, Miller and Freund (1965) and several procedures are available via the internet.

Although wet weather conditions may require specific methods for sampling, monitoring, and analysis, there are a number of similarities with the methods applied in case of dry weather measurement programs. In general, laboratory methods for chemical analysis are identical and it is often the case for the sampling equipment. Such aspects are therefore, in this brief overview, given less attention. It is, however, important to focus on those general aspects that must be considered, particularly in case of monitoring and sampling in the field. The objective of this chapter is to provide the reader with general formulated information that is central for the production of data that as much as is practically possible reflect the true nature of urban wet weather quality phenomena. Basic knowledge on sampling, monitoring, and analysis is important for those producing the data but also for the user of the data.

7.1 GENERAL CHARACTERISTICS OF SAMPLING, MONITORING, AND ANALYSIS

Prior to the discussion of details on the experimental methods used for water quality assessment, it is important to understand the terminology used. The term water quality is basically understandable. However, the term is described in this text in relation to legislation and regulation (cf. Section 11.3). The meaning of the terms sampling, monitoring, and analysis are, however, often expressed in different ways. The common meaning of these words cannot always be defined clearly, however, it is in this text the intention to use the words as follows:

- *Sampling* is ideally the collection of a representative part of a system. The sample collected is subject to "handling" prior to a more detailed investigation that typically takes place under laboratory or pilot scale conditions where a number of methods for this investigation are available and can be controlled. Handling means, in this respect, appropriate pretreatment or transport that will protect the sample safely until it has reached its final destination (e.g., a laboratory).
- *Monitoring* is a continuous or regular registration of the performance of a system in terms of selected characteristics. Monitoring takes place under field, pilot scale, or laboratory conditions and is often on-line by use of sensors and equipment for transmission of the monitored data to a computer.
- *Analysis* is, in contrast to monitoring, a procedure that is performed on a sample, typically under laboratory conditions. Standardized methods are recommended, however, analysis also covers the use of appropriate experimental investigations that are selected according to a specific objective.

A specific experimental procedure with the purpose of producing information for quality assessment is not always unambiguously described by one of the terms mentioned. As an example, the term "biological monitoring" might include elements of both sampling and analysis.

To a large extent, the basis for design and management of urban drainage systems depends on measured data. In principle, such data determine characteristics in terms of water and pollutant transport as well as the changes in the quality of the pollutants. It is clear that sampling and monitoring for quality purposes is meaningless without monitoring flow and transport of the pollutants. Logically, sampling and monitoring follows the routes and pathways of the water and associated pollutants as depicted in Figure 7.1.

In Section 1.2, a number of fundamental wet weather phenomena and characteristics, often in contrast to the pollution originating from dry weather sources, are

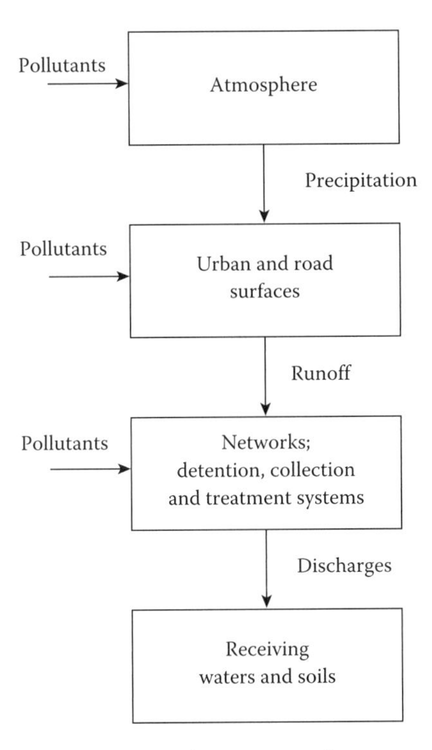

FIGURE 7.1 Overview of the urban drainage system in terms of possible locations and flows for sampling and monitoring.

depicted. Such aspects also specify, in general terms, the basis on which we must design and perform experiments and measurement programs for procuring information on the wet weather pollution. According to that, there are, at the conceptual level, three basic aspects that must be taken into account in the case of sampling and monitoring for assessment of the wet weather characteristics:

- *The event-based nature of the pollution*
  A main characteristic of urban wet weather aspects concerns their event-based nature related to precipitation and runoff. This basic aspect is central and must always be observed. In case of sampling and monitoring, it affects a number of practical aspects (when and where to perform the measurements, for how long of time to sample and monitor, choice of equipment for monitoring, and use of statistical methods for analysis). The overall nature of the wet weather phenomena determines what experimental methods and techniques are relevant. The objective of the measurement program is central (determined by acquisition of either short- or long-term information).
- *The variability of the phenomena*
  Compared with dry weather, the variability of the wet weather phenomena and data is in most cases to be characterized as very large. Due to that, it is important to consider what phenomena and conditions can affect the variability. Prior to sampling and monitoring and as a part of the planning of a

measurement program, it is therefore relevant to consider the importance of the variability for choice of sampling location, sampling period, and number of samples.

- *The interaction between wet and dry weather periods*
  The wet weather phenomena will in principle occur on the top of the dry weather phenomena. Sampling and monitoring should for that reason, in principle, take place before the event starts and continue after it has terminated. As examples, it is important in the case of monitoring and sampling of flows in combined sewer networks and in the receiving environment. Although the wet weather pollution is event-based and follows the runoff pattern, the effect following a discharge—a delayed effect—is basically a part of the event and sampling and monitoring during a period of time after the runoff event can be very relevant. Similarly, it might as well be needed to determine the dry weather quality of the system prior to the wet weather event.

7.2 THE NATURE OF DATA

Throughout the entire text, it has been both directly and indirectly expressed that the understanding of the wet weather phenomena to a great extent depends on experimentally based data. These data are often used systematically in terms of mathematical formulations and models. To a great extent, the theoretical and conceptual understanding included in such mathematical expressions originates from experimental investigations in the field as well as from laboratory experiments. It is therefore evident that sampling, monitoring, and analysis play a central role, not just for the quality and nature of the data that are directly used, but also indirectly via models and corresponding procedures for prediction.

7.2.1 Ideal Data

Ideal data are data with a variability that is only a result of the nature of the system investigated. The statistically determined distribution of the data is therefore a basic characteristic of the system or the population dealt with. The pollutant variability dealt with in Section 2.4 in terms of the variability within an event, the variability between events at a specific site, and the variability between sites is basically represented by the term *ideal*.

In this way, ideal data originate from a population characterized by a statistical probability distribution having a well-defined mean value and variance. It is clear that the nature of the statistical distribution must be known (e.g., being normal or log-normal). Ideal data require that the conditions for sampling from the population are defined and stochastically carried out. A sample that observes the requirements for producing ideal data is therefore representative for the population from which it originates.

As a consequence, methods for sampling, monitoring, and analysis must be well defined. It is clear that different methods for these procedures can be applied. However, it also means that data—describing the same characteristic of a system but originating from different methods—in principle cannot be compared. Unfortunately, it is often seen that this basic requirement has not been observed. If data produced by different methods are compared, one has to be very careful.

7.2.2 Nonideal Data

In contrast to ideal data, the nonideal data are affected by the choice and use of the methods for sampling, monitoring, and analysis. The limitation of interpreting specific data (e.g., to which extent a BOD value reflects the biodegradability of a sample) has nothing to do with its possible nonideal state. It is basically an inappropriate selected procedure or a wrong or insufficient use of a procedure that is the cause of why data can be characterized as nonideal. It must be stressed that all data in principle are nonideal. However, it must be a clear objective to approach as much as possible the ideal state and the true values, which for practical reasons can be the case if the procedures are carefully selected and correspondingly used. In this respect it is important for nonideal data to distinguish between systematic errors and stochastic variability.

- *Systematic error*
  A systematic error, also referred to as bias, occurs because one or more of the procedures (sampling, monitoring, and analysis) is inappropriate in terms of the objective or the procedure is not calibrated according to its specific use. A systematic error is thereby defined as the deviation between the mean value of the measured data and the *true* value. It is important to understand that it is not the system characteristics that cause the systematic errors. As an example, inhomogeneity is a characteristic of a system. On the other hand such characteristics may require specific attention in the case of sampling, for example, not to result in a systematic error. It is therefore important to understand the nature of the system as well as the basic characteristics and limitations of the procedures applied.
- *Stochastic variability*
  As mentioned in Sections 7.2.1 and 2.4, ideal data characterizing urban drainage systems and phenomena normally show stochastic variability. However, stochastic variability also concerns the procedures applied in terms of sampling, monitoring, and analysis. It is important to use procedures with low stochastic variability. However, if the number of data is sufficiently high, a procedure with a relatively high stochastic variability can produce valuable results.

7.2.3 Quality Assurance of Data

Quality assurance and quality control of data are subjects of importance. In the case of data characterizing urban wet weather phenomena where the data typically are subject to considerable stochastic variability, assessment of data is crucial. Furthermore, the event-based nature and the associated variability of the wet weather phenomena require specific attention when planning and performing sampling and monitoring.

A number of statistical methods are generally available for quality assurance and quality control of data. Such methods and procedures are dealt with in several textbooks, Miller and Freund (1965) and Walpole and Myers (1993). The choice of statistical methods and procedures for the assessment depend on the specific purpose and will not be dealt with in this text.

A number of central concepts that relate to quality assurance of data will be presented. The definition of these concepts may vary. In the following, they are defined and explained according to what is convenient for data characterizing urban drainage phenomena. The terms accuracy and precision are graphically depicted in Figure 7.2.

- *Accuracy*
  The term accuracy represents the total error associated with data values (i.e., accuracy includes both systematic errors and stochastic variability). The accuracy of a data value (or a number of measured values) is thereby defined as a measure of the proximity of that value (or a number of values) to a value that is considered the true value.
- *Precision*
  The precision—also known as reproducibility—of data values refers to the stochastic variability, however, only the variability that is a result of sampling, monitoring, and analysis. The precision is thereby characterized by the statistically determined standard deviation. Since data (ideal data) on urban drainage phenomena as previously mentioned by nature includes considerable variability, it is in general and based on field investigations not possible to distinguish between precision and the naturally occurring variability. Only for controlled experiments (e.g., performed under laboratory or pilot scale conditions), will it be possible to determine the precision of a procedure for measurement.
- *Selectivity*
  The selectivity of a method (sampling, monitoring, or analysis) refers to its response in terms of the ability to measure a specific characteristic value (e.g., pollutant concentration) or to collect a representative sample. Selectivity must be judged for the matrix where the specific characteristic should be determined and a potential interference of other substances is therefore important for the selectivity. Selectivity is a central characteristic for any analytic method or sensor for monitoring but also in case of

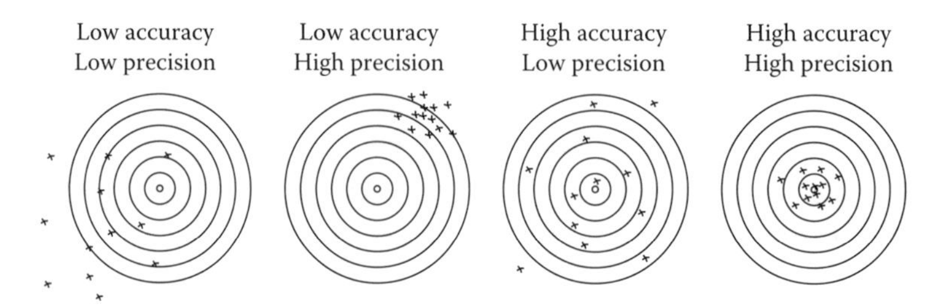

FIGURE 7.2 Graphical representation of accuracy and precision.

sampling. Selectivity related to sampling is often expressed by a sample or a collection of samples being representative for a system (e.g., determined by the ability of a water sampler to collect a representative sample with suspended particles).

- *Sensitivity*
  Sensitivity is defined as the change in a measured value per unit of change in that value. Sensitivity is particularly relevant in the case of substances determined in concentration units. Sensitivity can be considered an extended description of the term selectivity.

The terms *accuracy* and *precision* define corresponding characteristics of data. The same is the case for the terms *selectivity* and *sensitivity*.

In addition to the terms described for data quality, a number of other definitions are relevant. The limit of detection defines, in the case of a chemical analytical method, the lowest concentration of a component that can be determined with a specified probability, typically not lower than 95%. In contrast to the limit of detection, the criterion of detection of an analytical method is defined as a limit that determines if a component is present or not. If the measured value is higher than the criterion of detection, there is a low probability (typically < 5%) that the true value is zero.

7.3 MEASUREMENT PROGRAMS

The specific objective of a measurement program for quality assessment and performance of a system is the starting point for its design. The objective of a program often concerns determination of wet weather pollutant loads, environmental impacts, effects, and needs for control and treatment. The loads of pollutants, their transport and transformations, and following the establishment of mass balances for central pollutants are therefore typically central. The effects in terms of lethal or sublethal responses on biological communities and humans are often an ultimate goal and are therefore either directly or indirectly important parts of a measurement program. A specific objective might also be to provide data for models and to assess if water quality standards are observed.

To fulfill the objective of a measurement program in practice, the contributing catchment, the drainage system, and the receiving water system might be equipped with installations and devices for sampling and monitoring. In the following, central elements and steps are, in this respect, described in general terms. Major elements of the design process for a measurement program are shown in Figure 7.3.

By establishing an overall structure for a measurement program, Figure 7.3 depicts the starting point of a protocol for such a program. In this respect, there are several approaches that can be followed. A typical approach is to establish a system for input/output measurements as a basis for a mass balance. Another approach is to compare system performance under different external conditions (e.g., by measurement before/after a change in the system has been made).

The event-based nature and the variability of the parameters of the urban wet weather phenomena require that measurements typically be performed over a relatively

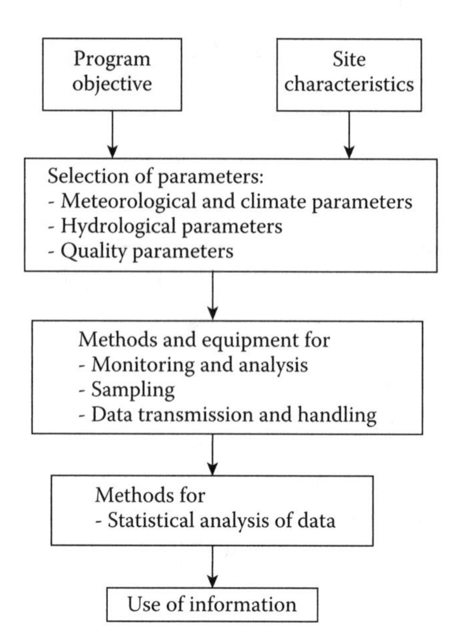

FIGURE 7.3 Elements of the design process for a measurement program.

long period of time. Furthermore, a program requires regular and time-consuming inspection from skilled personnel for proper function and extensive use of manpower is needed for handling and analysis of the samples and for a final statistical analysis. A wet weather measurement program is therefore expensive to perform and it becomes important that automated and electronic equipment for sampling, monitoring, and data collection and for their transmission are used as much as possible.

As already mentioned in the introduction to this chapter, literature on urban drainage quality is rich in examples of how measurement programs in specific cases have been designed and performed. In this respect, a number of both general and specific recommendations for the design of rather extensive measurement programs are listed and discussed in U.S. EPA (1983) and Strecker et al. (2001).

7.3.1 Catchment Characteristics

A number of site characteristics must be considered prior to the design of a measurement program. First of all, the size and the structure of the contributing catchment determine what capacity of the flow meters is required. Site characteristics that are important to estimate are

- The impervious area of the catchment
- The structure of the catchment (e.g., in terms of slope and detention of the runoff water)
- Periods with snowfall and snowmelt (i.e., statistics of the winter period)
- Estimation of the inflow of pollutants in terms of concentration levels that can determine what sampling volumes for analysis are needed

7.3.2 Meteorological and Climate Parameters

A number of meteorological parameters are potentially central for a measurement program. First of all, the precipitation pattern is important and it should be often recorded with a rather small time resolution (e.g., by means of a tipping-bucket rain gauge, cf. Section 2.1.2.1). Other meteorological parameters of potential importance are air temperature, evaporation rate, wind speed, wind direction, and solar radiation. Basically, meteorological data should be monitored at the catchment in question. However, for practical reasons they may partly originate from permanently established meteorological stations in the vicinity of the site.

7.3.3 Hydraulic Parameters

Hydraulic parameters in terms of flow and water level parameters are needed to calculate pollutant loads and to establish mass balances. Flow parameters can be recorded in several ways (e.g., by using calibrated overflow weirs or magnetic flow meters). The flow entering a system, pipe, and channel flows, the flow out of a system, and overflows are those of interest. Often the extreme variation in flow that characterize runoff events requires that more than one meter is installed to record both low and high flow rates with sufficient precision.

7.3.4 Parameter Monitoring

A number of parameters can be directly monitored by using sensors from where the signals can be transmitted (e.g., via internet connection to a server). Examples of parameters for continuous or intermittent monitoring are

- Flow
- Water table
- Conductivity
- Turbidity
- Temperature
- Solar radiation
- pH
- Dissolved oxygen

Sensors for monitoring are under continuous development and parameters that previously needed sampling and analysis might be considered for being monitored.

7.3.5 Quality Parameters Determined by Analysis of Samples

Analysis of quality parameters often concerns water samples, however, it is important to stress that pollutants accumulated in the sediments are also of interest for assessment. One of the advantages of sediment analyses is that sediments can provide central information on the pollutant history of the site. The central reference book for analysis is the so-called Standard Methods, APHA-AWWA-WEF (2005), which is used worldwide and regularly updated.

The noncontinuous measured water quality parameters are mainly those parameters (pollutants) that are analyzed based on flow or time proportional sampling. Sampling therefore constitutes a central part of a measurement program, normally performed by automated samplers. Timing for manual collection of samples must be planned. Due to the stochastically occurring variability of the generation and transport of pollutants, it is a central task to select time scale and frequency of sampling in the planning phase of the measurement program.

Sequential extraction procedures refer to analytical techniques for determination of the speciation and availability of components that occur in solid materials like sediments, biofilms, plants, and animals. Within the area of urban wet weather pollution, the procedure typically aims to provide information exceeding results on the total composition (e.g., as site-specific bioavailability of different substances and the potential occurrence of such species). The procedure is typically carried out stepwise by the use of different extraction agents like weak and strong acids and different oxidizing reagents. As an example, a sequential extraction procedure can be applied to determine the exchangeability of heavy metals occurring in a solid state. Although such procedures do not necessarily provide specific information on which species are available as shown in Table 6.4, the extractability may, under well-defined conditions, indicate a potential fate and harmful impact of a specific component. Several protocols exist for sequential extraction. Tessier, Campbell, and Bisson (1979) describes the classical procedures, which later have been modified by other researchers (e.g., Osuna et al. 2004).

The number of possible quality parameters analyzed for is basically legion. It is a specific task based on the objective of the measuring program to determine the extent of this part of the program. Information on potentially relevant pollutants is found in Section 3.2.1. Both chemical constituents and microorganisms are relevant.

7.3.6 Biological Methods

Methods for biological measurements in quality assessment of wet weather discharges can be based on single species as well as biological communities and analysis of both plants and animals are relevant. The methods applied can focus on pollutant uptake and accumulation as well as impact in terms of lethal or sublethal responses.

Biological measurements (ecotoxicological techniques) are useful in the assessment of the effects of the wet weather discharges into receiving waters. Biological communities take time to recover and their status reflect the pollution history at the site that is the combined, and complex, result of both dry and wet weather discharges. A number of different biological methods and tests exist for assessment of urban runoff impacts caused by both CSOs and SWR on receiving waters (Ellis and Hvitved-Jacobsen 1996):

- *In situ acute toxicity methods*
  This technique applies different animal species, particularly macroinvertebrates. As an example, the impact of CSOs can be assessed by measurement of the mortality of freshwater shrimps exposed to such discharges.

- *In situ bioaccumulation methods*
  In contrast to the acute toxicity methods, this type of method is based on accumulation of pollutants following an initial uptake (e.g., by accumulation in the soft tissue).
- *Laboratory experiments*
  Both fish species and invertebrates can be used to measure lethal or sublethal responses.
- *Community assessments*
  Several procedures exist for the assessment of the impact on the biological community. Such methods often use an index of diversity and species richness for macroinvertebrate communities as measures. These methods are particularly valuable if a degraded site can be referenced against a clean site.

Chapter 6 deals with effects of both CSOs and SWR. A number of experimental methods dealt with in this respect are therefore useful for assessment of water quality impacts. Just a single example: the measurement of chlorophyll-a and the penetration of sunlight (Secci depth) are central parameters for the assessment of lake eutrophication (cf. Sections 6.4.2 and 6.4.4).

7.4 MEASUREMENT PROGRAMS IN PRACTICE

In the introduction to this chapter it was mentioned that literature on urban drainage quality is rich in examples of how measurement programs have been designed and performed. However, such specific information is in general only valuable as an inspiration. The bridging between the concepts discussed in this chapter—particularly outlined in Section 7.3—and the specific design of a measurement program is crucial.

In addition to the objective and required outcome of a program, a large number of practical aspects must be considered when designing and implementing a measurement program. The following are important to consider:

- Financial resources available
- Manpower
- Equipment
- Construction details of the measurement system including transmission of data
- Accessibility of the measurement system

Last but not least it is important to stress that integration of scientific knowledge and practical skills is crucial. Theoretical concepts relevant for the objective of a measurement program and skills for handling the technology of sampling and monitoring must go hand in hand.

REFERENCES

APHA-AWWA-WEF. 2005. *Standard methods for the examination of water and wastewater*, 21st ed. Washington, DC: APHA (American Public Health Association), AWWA (American Water Works Association), WEF (Water Environment Federation).

Burton, G. A., and R. E. Pitt. 2002. *Stormwater effects handbook: A toolbox for watershed managers, scientists and engineers.* Boca Raton, FL: Lewis Publishers.

Ellis, J. B., and T. Hvitved-Jacobsen. 1996. Urban drainage impacts on receiving waters. *Journal of Hydraulic Research* 34 (6): 771–83.

Fletcher, T., and A. Deletic, eds. 2007. *Data requirements for integrated urban water management, urban water series: UNESCO-IHP.* Boca Raton, FL: Taylor & Francis.

Miller, I., and J. E. Freund. 1965. *Probability and statistics for engineers.* Englewood Cliffs, NJ: Prentice Hall, Inc.

Osuna, M. B., E. D. van Hullenbusch, M. H. Zandvoort, J. Iza, and P. N. L. Lens. 2004. Heavy metals in the environment: Effect of cobalt sorption on metal fractionation in anaerobic granular sludge. *Journal of Environmental Quality* 33:1256–70.

Strecker, E. W., M. M. Quigley, B. R. Urbonas, J. E. Jones, and J. K. Clary. 2001. Determining urban storm water BMP effectiveness. *Journal of Water Resources Planning and Management* 127 (3): 144–49.

Tessier, A., P. G. C. Campbell, and M. Bisson. 1979. Sequential extraction procedure for the speciation of particulate trace metals. *Analytical Chemistry* 51: 844–51.

U.S. EPA. 1983. Final report of the nationwide urban runoff program (NURP). Washington, DC: U.S. Environmental Protection Agency, Water Planning Division.

Walpole, R. E., and R. H. Myers. 1993. *Probability and statistics for engineers and scientists,* 5th ed. Englewood Cliffs, NJ: Prentice Hall Inc.

Wiersma, G. B., ed. 2004. *Environmental monitoring.* Boca Raton, FL: CRC Press.

8 Urban Wet Weather Quality Management

The meaning of the word management is within the area of urban drainage not defined in clear terms but typically rather diffusely interpreted. It is often seen that methods of more conventional origin like treatment are considered management methods. In order not to mix up the understanding of management with more well-defined technologies dealt with in other chapters, it is important to discuss, in broad terms, what urban wet weather quality management is.

8.1 THE NATURE OF URBAN WET WEATHER QUALITY MANAGEMENT

Wet weather flows in urban areas may cause a number of problems for the society. It is basically a city planning task supported by specific input from hydraulic and environmental engineers to produce a management plan that observes the needs of the society and the adjacent environment. In this way quality management focus on what beneficial uses of the local environment are considered important for the public. At the same time a focal point is the pollution control and how it can be implemented locally in a cost-effective manner. Stormwater management at this level requires overview as well as detailed information, the latter by applying a number of methodologies and tools that in terms of quality are general focal points of this text.

The protection of the city against flooding and deterioration caused by extreme precipitation events is a major reason why the wet weather flows should be regulated. Management of these flows is crucial. In the context of urban wet weather quality, management methods that tend to protect the beneficial uses of water will be understood as

- Procedures and concepts that are undertaken to reduce adverse effects related to wet weather impacts, particularly in the case of pollutant discharges into the environment
- Methods for the beneficial uses of the runoff water
- Considerations that tend to improve the entire performance of an urban drainage system (e.g., related to the potential adverse impact of climate changes)

Although urban wet weather quality management cannot be expressed in terms of a short and clear definition, the broad understanding of it is useful. Management is particularly relevant as a term to use when dealing with rather complex mixtures of methods and procedures that, however, at the detailed level may require specific technologies. Such specific subjects are dealt with in other chapters of this text and not considered management methods. Methods for treatment of CSOs and SWR are subjects in Chapters 5 and 9, respectively, and regulations that more or less refer to rules and legislation issued by governmental, state, regional, or local authorities are subjects in Chapter 11. The specific procedures for design and construction of drainage systems and corresponding aspects of operation are not, in the context of this text, considered management methods. These procedures and methods are linked to their performance in terms of their scientific and engineering origin. However, the nonstructural types of best management practices (BMPs) are examples of methods that are basically management methods. Such nonstructural BMPs are therefore included in this chapter.

Although the basic idea of quality management concerns the use of "soft," qualitatively expressed procedures, concepts and considerations, the implementation of the ideas may typically require use of constructions in the drainage system and treatment technologies. Similarly, the use of quantitatively expressed tools like quality models can, in concrete situations, support the assessment of concepts and procedures originating from quality management.

Urban drainage quality management is important in a planning process of how to implement and to some extent also operate drainage systems in a sustainable way, taking into account integrated impacts and deteriorating effects in view of future needs and consumption of resources. Correctly applied, quality management has therefore something to do with good housekeeping. The following sections of this chapter will briefly deal with such methods.

Several management methods expressed in terms of sustainable urban drainage and rainwater harvesting—although given new names—concern well-known and well-accepted methods for good housekeeping of stormwater. However, by being integrated and accepted approaches to urban drainage they might bring new life to improve design of drainage systems and to management of the polluted wet weather flows.

8.2 RAINWATER HARVESTING AND WATER SENSITIVE URBAN DESIGN

Water is a resource and in several areas of the world it is a scarce resource. The use (reuse) of rainwater and stormwater runoff from urban areas and roads can therefore be beneficial. In areas where water is not a scarce resource the beneficial use of stormwater in terms of recreational purposes (the concept of "water in the city") or for upgrading urban surface water systems can be of interest. Main reasons for the use of rainwater and stormwater runoff, named as rainwater harvesting or stormwater harvesting, are

- Potable water uses
- Nonpotable water uses (e.g., for toilet flushing and washing)
- Industrial process water uses
- Uses for irrigation
- Recreational purposes in the city
- Upgrading of local surface waters (e.g., improved flow in local streams)

The concept of rainwater harvesting basically means that it is managed in an integrated way with other parts of the water cycle. This integrated use is often discussed within the framework of water sensitive urban design (WSUD) or integrated sustainable urban water management. The basic idea is that runoff water from urban areas including roofs and roads for several purposes has a potential value and should not necessarily be disposed of as fast as possible. In areas where water of high quality for in-door use is a scarce resource, rainwater harvesting can provide specific advantages.

The beneficial use of rainwater requires that a number of steps be considered. What steps are appropriate or needed depend on not just the intended use but also local conditions. The following are possible steps expressed in general terms (Mitchell et al. 2007):

- Collection
- Treatment
- Storage
- Distribution

These different steps can be integrated with other aspects of rainwater management (e.g., flow attenuation and flood control).

A number of methods for rainwater mitigation that also observes rainwater harvesting requirements in terms of treatment are dealt with in Chapter 9. In this chapter, a number of both traditional and advanced treatment methods are described. Different types of treatment and management systems are thereby available depending on what level of quality is needed for the intended use and depending on what system is locally the most appropriate to implement. In case rainwater is used for residential in-door purposes like potable water use or toilet flushing, further treatment will be needed (e.g., fine-mesh screen filtration, membrane filtration, activated-carbon adsorption, and disinfection). The microbial quality for harvested rainwater is typically—even for nonpotable in-door applications—not observed and disinfection may therefore be needed.

Several of the mitigation methods dealt with in Chapter 9 and referred to as BMPs or SUDS (sustainable urban drainage systems) are technologies that can be applied for WSUD. The understanding of this concept is that the use of so-called low technologies is appropriate for rainwater management and therefore also referred to as sustainable methodologies. In Chapter 9 such methods are characterized as "low-impact development approaches" that mimic a predevelopment hydrology of a site as much as possible. Furthermore, a great number of technical and management methods for urban stormwater control in terms of an adaptive decision support system are found in DayWater (2008).

8.3 INTERACTIONS BETWEEN SEWERS, WASTEWATER TREATMENT PLANTS, AND RECEIVING WATERS DURING RUNOFF PERIODS

Referring to Figure 8.1, it is readily seen that there are close relations between which parts of the wet weather flow from combined sewer networks that will end in the overflow and which part that will be diverted to a downstream wastewater treatment plant. These flow interactions must be considered in the design process of the

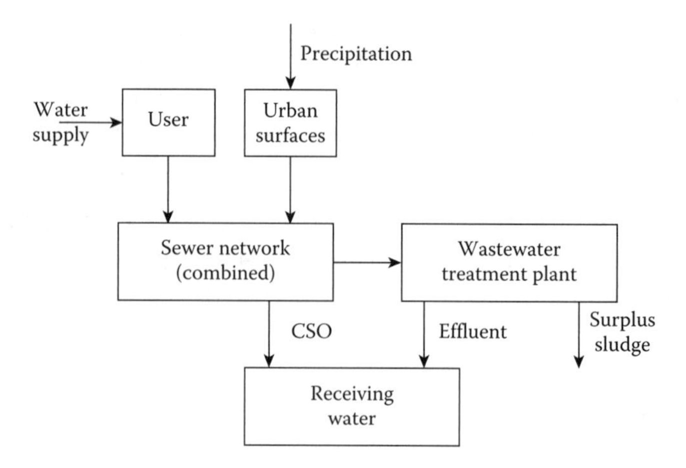

FIGURE 8.1 Overview of the main water and mass flows in a combined sewer system.

systems and in their operation. The entire performance of combined sewers in terms of CSO loads is a major subject dealt with in Chapter 5, and the potential effects of the CSOs are subjects of Chapter 6. In addition to these aspects, it is during dry and wet weather periods crucial to take a holistic view of those interactions that affect the entire performance of the sewer network, the corresponding wastewater treatment plant, and the impact on adjacent receiving waters.

A number of questions concerning the performance of the entire combined sewer network related to wastewater treatment and the receiving water quality should therefore be considered:

- What is an acceptable CSO load in terms of the potential adverse effects on the receiving water body?
- What is an acceptable load on the wastewater treatment plant taking into account the required treatment efficiency and operational performance? Should the treatment plant be extended or upgraded?
- Is extended detention of the runoff flow in equalization basins in the sewer network or in the network itself feasible?
- Is local treatment of the overflow an appropriate solution?

The questions raised are complex because several details of the possible solutions influence the performance of the different subsystems. The varying nature of the runoff makes it complex to find an optimal solution. In addition, an optimal solution to the total performance of the sewer network is influenced by a number of local constraints.

The potential impacts of wet weather flows on a downstream wastewater treatment plant are in the case of an activated sludge treatment plant with nutrient removal briefly:

- The flow during a runoff event—including the delayed flow caused by an extended detention of the runoff—may increase the hydraulic load on the secondary settlers, in particular the shallow clarifiers, to a level where they are not sufficiently effective. The result is a potential for increased concentration of sludge particles in the effluent (Harremoës et al. 1993).

- An increased flow in pipes and channels of the treatment plant may increase the shear stress on the sludge particles and cause reduced floc stability and increase the concentration of small particles in the effluent.
- The composition of the incoming wastewater during a runoff event is different compared with the dry weather flow. A reduced biodegradability of the substrate may affect sludge floc properties and reduce the potential for nutrient removal. Furthermore, inflow of both heavy metals and organic micropollutants may affect the sludge quality.
- Inflow of snowmelt will reduce the temperature and thereby affect the biological processes.

A number of attempts have been made to establish computer models for simulation of the integrated performance of wastewater collection and treatment systems including receiving water impacts (Rauch et al. 2002; Vollertsen et al. 2002). In principle, models for the subsystems that are linked and run in parallel should be established. In this way it is possible to provide an overview of the integrated phenomena. It is beyond the objective of this text to go into details on how integrated modeling, in terms of the linkage and feedback mechanisms between the subsystems, can be formulated. It is, however, clear that integration of models add another problem to those already known for modeling the subsystems. The state-of-the-art for integrated urban drainage modeling is at a rather basic level and it is therefore clear that when it comes to details, most phenomena must be formulated and solved in a narrow and less complex context. This statement does not change the fact that it is crucial to establish an insight in the performance of the entire integrated system.

It is common to design a wastewater treatment plant to treat twice the average flow. However, during a runoff event further capacity may be needed. Rather simple operating strategies for treatment plants are often used to manage the wet weather flows. The implementation of such methods depends on the actual treatment plant design and the permit limits of treatment. An example of a widely used strategy for wet weather operation of treatment plants is to store sludge from the clarifier temporarily in an (aerated) equalization tank to prevent excess sludge in the effluent. In other cases storage basins in the sewer system or additional capacity for the settlers will be needed.

8.4 NONSTRUCTURAL BMPs

The nonstructural BMPs are management methods for controlling adverse impacts of the wet weather flows. Nonstructural BMPs should be understood in a rather broad context. The group of nonstructural BMPs includes initiatives taken by various private and public institutions, rules and regulations issued by authorities, educational aspects, and a number of practices that tend to reduce pollution. The nonstructural BMPs also include our daily behavior in reducing improper littering. The list of different scenarios of nonstructural BMPs is long. The following only outlines some common control measures that mainly tend to reduce the negative impacts from stormwater runoff in urban areas and from roads although they are also relevant in the case of combined sewer networks:

- Public education
- Regulations
- Land use and urban development resource planning
- Landscaping
- Chemical use control (e.g., pesticide, herbicide, and fertilizer management)
- Litter and debris control
- Painting and use of metals in constructions
- Road and street sweeping
- Snow handling in cold climates

Source control measures have in several cases been efficient for reduction of the environmental deteriorating impacts from stormwater runoff. A classic successful example is the regulation of lead in gasoline. The result of this regulation has, in several countries, reduced the load of lead originating from stormwater runoff considerably, often to a level of 25–30% or even lower from the load in the 1980s. Other examples of potential source control measures are material replacement for copper roofs, banning the use of copper in brake linings of cars, the use of more environmentally friendly studs in tires, and changes in driving patterns reducing the PAH load from road wear (Ahlman et al. 2005).

As an example, road and street sweeping is in the list shown as a nonstructural BMP for reduction of pollutants in the runoff from urban and road surfaces. Street sweeping is, in general, considered appropriate for aesthetic reasons by removing debris and gross solids. However, a number of early studies have, because of a low efficiency to remove particles below 100 μm, questioned its effectiveness in controlling pollutants in urban runoff (U.S. EPA 1983). In regions with wet and dry seasons, sweeping prior to the wet season is considered appropriate. Road sweeping is probably also useful following a snowmelt period when excess solids still remain on the road surface. Kang and Stenstrom (2008), however, claim that datasets for evaluating the effectiveness of street sweeping typically had insufficient sample numbers for statistical analysis. Furthermore, because of a potential transport of pollutants to adjacent or distant locations, they consider the direct impact on the runoff water quality not fully including all the benefits of street sweeping.

In general, the nonstructural BMPs are important because they often result in more expedient procedures resulting in reduced impacts from the wet weather flows. However, it must be realized that the reduction of pollution in this way is often not sufficient to meet desired environmental quality targets. The additional uses of structural BMPs are therefore often needed.

8.5 CLIMATE CHANGES AND URBAN RUNOFF

Climate changes refer to a number of meteorological phenomena that, in a complex way, may change and affect the conditions of life and society. These meteorological phenomena include global changes in temperature, precipitation, and wind. Among other factors, the climate conditions of a specific region are crucial for the way society has developed over time and how it is performing. Climate changes have several faces depending on in what part of the world they appear. They might occur in terms of both excessive rainfalls or in a lack of rain and they affect urban as well as rural areas. In the context of urban and highway stormwater pollution, it is the rainfall pattern and its potential impacts that are in focus.

The important point is that climate conditions may seriously affect the functioning of the urban areas and their infrastructures. Climate changes mean that the performance of the drainage network (i.e., pipes, channels, overflow structures, manholes, basins, treatment systems, and different devises) may no longer be appropriate functioning according to its design. Depending on how the nature and extent of the climate changes appear at a specific site, the capacity of the drainage network and thereby its performance in terms of safe transport and environmentally acceptable impact of the runoff water may not be sufficiently observed. Malfunctions may result in surface flooding, erosion in channels and pipes, and excessive sediment transport and deposition. In extreme situations, the associated problems are physical damage of the surrounding land, human health risks, increased discharges of pollutants, and hydraulic deterioration and physical habitat changes of receiving waters downstream.

In the context of this text on urban wet weather quality aspects, a focal point is the relation between the characteristics of rainfall, the corresponding runoff, the performance of drainage networks, and the environmental impact. The entire text therefore includes concepts and tools that can also be applied when solving quality problems related to climate changes in terms of both assessment of effects and technologies for reduction of impacts. The following subsections therefore only serve the objective to give an overview and a starting point when approaching quality problems of urban and road runoff in view of climate changes.

8.5.1 Urban and Road Runoff Quality Problems Related to Climate Changes

Quality problems that are closely related to changes in the precipitation and runoff pattern—particularly in the case of heavy rainfall—refer in principle to the following two main phenomena:

- Transport and discharge of increased water volumes
- Increased flow rates during runoff events

It should be noted that neither the actual nor the potential future sources of pollutants being input to these flows have basically nothing to do with the quality related impacts caused by climate changes. From a pragmatic point of view, however, it might be appropriate to take into account to what degree such pollutant sources contribute to these hydraulic phenomena.

Each of the two hydraulic phenomena mentioned may cause environmental problems but they do so in an integrated way. Erosion of sediments and deposits occurring at both the urban surfaces and in pipes and channels are directly related to the two phenomena. Increased transport of pollutants and a risk of discharges into the receiving environment are the immediate impacts that might result in deterioration. As an example, physical habitat changes of receiving waters downstream might occur in terms of both erosion and deposition of particulate materials.

It is crucial to understand that solutions to problems caused by climate changes follow what otherwise will be relevant in similar cases, well known from different geographical locations. The solutions to the problems and the tools applied to assess these solutions are therefore already included in this text. What is new is the potential worldwide occurrence of climate changes and a corresponding widespread need for changes of the drainage systems.

A number of problems associated with climate changes (e.g., a potential increased sea level and a corresponding risk of more or less permanent flooding of urban lowlands), will not be dealt with in this context.

8.5.2 Climate Changes and Performance of Combined Sewers

Management of extreme flows and water volumes in combined sewers basically cause increased potential for erosion of the deposits in the network and a corresponding increased transport of both water and pollutants. The risks of increased discharges from the overflows into adjacent receiving waters and overloads of wastewater treatment plants downstream are evident. Furthermore, surcharges of the wastewater in the sewers and a corresponding impact caused by flooding of cellars, for example, may cause potential health risks and aesthetic deterioration.

The direct solution to these problems is to increase the capacity of the combined sewer network. A capacity increase can be done by increasing the dimensions of the pipes or by adding storage capacity like detention basins and tanks. Furthermore, the use of technologies based on real time control (RTC) and use of weather radars can improve the prediction of the runoff flow pattern and possible detention of the water in selected parts of the network. Another possible solution to impacts caused by climate changes is a complete renovation by constructing a separate sewer system where the sanitary wastewater is separated from the stormwater.

8.5.3 Climate Changes and Performance of Storm Sewers

In principle, a separate sewer network is better suited to cope with increased water volumes and flows than a combined sewer system. Provided that sufficient space is available, different mitigation methods, BMPs, as dealt with in Chapter 9, can in a relatively simple way improve the capacity for detention. An improved separate sewer system that, to a great extent, makes use of such mitigation methods by substituting underground pipes with channels and ponds will not only improve the robustness of the network but also add a recreational element to the city. The robustness is twofold in terms of providing detention capacity as well as being a potential for treatment of the runoff.

8.5.4 Climate Changes and Further Reasons for Upgrading Drainage Systems

As previously stressed, it is a general expressed objective for engineering of a drainage system to design the network according to the needs, irrespective of the origin of the problem that causes the need. The potential adverse effects of urban runoff caused by climate changes and the corresponding solutions needed for their solutions have therefore not created a new approach for urban drainage, just another reason for its upgrading. Basically, the same concepts, analytical tools, and technologies should be applied irrespective of these needs and are caused by an increased population (increased paved areas) or it is the climate that creates the needs. This basic understanding has been a fundamental and central criterion for the development of this text.

Looking at climate changes in a narrow context related to urban wet weather pollution, the reason why we should engineer the drainage system is new, not the technologies applied. The challenge, however, is to predict the extent of climate changes in view of the capability of the drainage system to adapt to future and basically unknown impacts. Compared with this situation it is easier to predict what changes of a network are needed if a catchment area is increased from x to y ha. The uncertainties related to the extent of climate changes, particularly in terms of the precipitation pattern, therefore call for robust technologies to be implemented. Such technologies include both the transport system and the associated treatment measures (e.g., wet ponds and infiltration systems). If well designed and constructed, the improved separate system is robust. The contents of Chapter 9 are, in this respect, very central.

REFERENCES

Ahlman, S., A. Malm, H. Kant, G. Svensson, and P. Karlsson. 2005. Modelling non-structural best management practices: Focus on reductions in stormwater pollution. *Water Science and Technology* 52 (5): 9–16.

DayWater. 2008. *DayWater: An adaptive decision support system for urban stormwater management*, ed. D. R. Thevenot. London: IWA (International Water Association) Publishing.

Harremoës, P., A. G. Capodaglio, B. G. Hellström, M. Henze. K. N. Jensen, A. Lynggaard-Jensen, R. Otterpohl, and H. Soeberg. 1993. Wastewater treatment plants under transient loading: Performance, modelling and control. *Water Science and Technology* 27 (12): 71–115.

Kang, J.-H., and M. K. Stenstrom. 2008. Evaluation of street sweeping effectiveness as a stormwater management practice using statistical power analysis. *Water Science and Technology* 57 (9): 1309–15.

Mitchell, V. G., A. Deletic, T. D. Fletcher, B. E. Hatt, and D. T. McCarthy. 2007. Achieving multiple benefits from stormwater harvesting. *Water Science and Technology* 55 (4): 135–44.

Rauch, W., J.-L. Bertrand-Krajewski, P. Krebs, O. Mark, W. Schilling, M. Schütze, and P. A. Vanrolleghem. 2002. Deterministic modeling of integrated urban drainage systems. *Water Science and Technology* 45 (3): 81–94.

U.S. EPA. 1983. Final report of the nationwide urban runoff program (NURP). Washington, DC: U.S. Environmental Protection Agency, Water Planning Division.

Vollertsen, J., T. Hvitved-Jacobsen, Z. Ujang, and S. A. Talib. 2002. Integrated design of sewers and wastewater treatment plants. *Water Science and Technology* 46 (9): 11–20.

9 Stormwater Pollution Control and Mitigation Methods

The impacts onto the environment caused by both combined sewer overflows (CSOs) and stormwater runoff (SWR) call for control and mitigation methods. In the preceding chapters, a number of construction and operational methods and management procedures have been dealt with for reduction of the environmental impacts. In continuation of that, this chapter aims at methods that are normally considered as BMPs (best management practices) and thereby focuses on procedures that are particularly directed to the management of SWR from urban surfaces and roads. Methods for management and treatment of CSOs are in this text dealt with in Chapter 5. The BMPs dealt with in this chapter can, to some extent, also be implemented in parking areas and along roads located in combined sewer catchments and thereby reduce the wet weather flows in the network.

The BMPs—in the United Kingdom also known as sustainable urban drainage systems (SUDS)—are both structural and nonstructural technologies that tend to reduce impacts caused by SWR from impervious areas (i.e., from urban developments, roads, and highways). A central characteristic of a BMP for stormwater management is its ability to mimic the predevelopment hydrology of a site. The small-scale structural BMPs are technologies that are often characterized as approaches for low-impact development (LID). In general, structural BMPs combine measures to reduce flooding and erosion of SWR with the removal of pollutants that are associated with the runoff. The nonstructural BMPs include a wide range of measures (e.g., regulations, management, and public education with the purpose of reducing the spread of pollutants into the environment). These nonstructural BMPs are briefly dealt with in Section 8.4.

In brief, a BMP for stormwater management is defined as follows (Strecker et al. 2001):

> A devise, practice, or method for removing, reducing, retarding, or preventing targeted stormwater runoff quantity, constituents, pollutants, and contaminants from reaching receiving waters.

It is not the purpose of this chapter to deal with all possible structural types of BMPs. The reference list for this chapter includes a number of selected references that give a more comprehensive description, Debo and Reese (2003) and Field et al. (2005). Furthermore, DayWater (2008) describes a great number of technical and management methods for urban stormwater pollution control in terms of an adaptive decision support system. In this text, it is considered important to describe design and performance characteristics of selected BMPs that are representative for a number of BMPs for stormwater management operating at a low technology basis.

From practice, it is well known that there are several obstacles for efficient performance of BMPs and no ideal treatment device for SWR exists. This text cannot deal with all possible and often locally relevant constraints but it is considered important to describe main characteristics of BMPs including central limitations in their use.

9.1 PHYSICAL, CHEMICAL, AND BIOLOGICAL CHARACTERISTICS OF POLLUTANTS RELATED TO TREATMENT

The basic characteristics of pollutants that are important in terms of effects, management, and treatment of SWR are dealt with in the preceding chapters. In particular, the basic wet weather characteristics will be found in Chapters 1 through 3. A number of overall principles are, in terms of BMP performance, important to focus on.

Firstly, a fundamental characteristic of urban and road runoff treatment is the fact that large volumes of runoff water can be generated over short periods of time. Secondly, low concentrations—but not sufficiently low in terms of effects onto the environment—must be reduced to even lower levels. In addition to these fundamentals, a large number of pollutants with different characteristics in terms of treatment must be dealt with. All these facts call for treatment methods that in principle must be performed differently compared with well-known treatment of the more or less continuous municipal and industrial wastewater flows. The fact that the generation of stormwater per definition occurs widespread is also a constraint that affects the practical implementation and costs of the treatment systems and their maintenance. Treatment technologies used for management of the wet weather flows should therefore be simple and robust.

For each specific case, it is important to know what pollutants are the most important to treat. The following is to consider what BMPs will best observe a given requirement for treatment. Pollutants that originate from urban and road runoff in relatively large quantities and that also result in adverse effects onto the environment are found among the nutrients, the heavy metals, and the organic micropollutants (cf. Chapter 6). Treatment must therefore be performed for pollutants with accumulative effects although it is realized that pollutants with toxic effects may also exert acute effects, particularly in the case of relatively high concentrations. As a first approach, it becomes important to secure efficient treatment over a period of time rather than for a single event. If a first flush phenomenon might occur, it is relevant to limit a treatment capacity of a BMP to capture the first and most polluted part of the runoff. In some countries (e.g., the United States), a "first flush volume" is typically defined as being about 13 mm (0.5 inch) of runoff.

In addition to these fundamental characteristics of a BMP for treatment of the normal runoff from rain events, it is important that discharges from spills and accidents in the catchment can be managed. It is expected that such events will occur rather infrequently but when they take place they may result in discharges of large quantities of pollutants. To omit corresponding fatal and often long-lasting impacts in adjacent receiving waters, it is important that the dual purpose of runoff-related treatment and management of discharges from spills be observed of a BMP. In principle a BMP should not be designed for spills and accidents but it is important that it is constructed robustly to protect the environment temporarily. The possibility to capture and detain a highly polluted volume of water or liquid is therefore crucial.

In continuation of the basic characteristics of a BMP, the following are major points to consider when selecting a management and treatment method:

- Methods having kinetic characteristics in terms of relatively high transformation rates for the governing treatment processes.
- Methods that are efficient at relatively low concentration levels.
- Methods having a rather high efficiency for treatment of pollutants with rather different physical, chemical, and biological characteristics.
- Methods having a detention capacity.

Main physical, chemical, and biological characteristics and processes for pollutants originating from urban and road runoff are briefly outlined in relation to the four points mentioned. The details, in this respect, are dealt with in Chapters 1 through 4:

- *Physical characteristics and processes*
  Sedimentation and filtration are central for treatment because particulate matter in urban and road runoff plays a dominating role. Furthermore, pollutants are often, to a rather high extent, associated with particles. Pollutants can thereby be removed from the water phase and accumulated at a bottom surface or retained in a filter. The removal from the water phase can be associated with degradation (e.g., in case of organic matter), but often "treatment" must be understood as "accumulation" and "removal," which takes place for phosphorus and heavy metals.
- *Chemical characteristics and processes*
  It is in practice often not possible to distinguish clearly between a chemical reaction and adsorption, a physicochemical process. For practical reasons (e.g., in case of modeling) it is often appropriate to consider adsorption as a chemical process that proceeds in a BMP. Adsorption of soluble, colloidal, and particulate species is often central and interaction between chemical and physical characteristics and processes are important. A number of such interactions are dealt with in Chapter 3.
- *Biological characteristics and processes*
  Compared with physical and chemical processes, biological processes often proceed with a rather low transformation rate. The runoff flow rate through a BMP is typically high and does therefore not support low-rate biological processes being efficient for control and treatment of pollutants. However, in case detention of the runoff volume is possible—and integrated with the BMP—biological treatment processes may turn out to be significant.

For several reasons (e.g., because of the number of different pollutants and the varying conditions under which they occur), it is important to make use of the physical, chemical, and biological processes in combination whenever possible for control of stormwater and road runoff.

9.2 OVERVIEW OF STRUCTURAL TYPES OF BMPs

The following list outlines main types of structural BMPs intended for management and treatment of urban and road runoff. It must be stressed that the number of types of BMPs is basically legion and that varying names and characteristics for the different types are in use. The list focuses on main characteristics of structural BMPs, to some extent following the definitions used by United States and Canadian agencies (cf. FHWA 1996). In Section 8.4, a number of nonstructural BMPs are listed. Details for design and performance of selected structural BMPs are found in Sections 9.3 through 9.8.

9.2.1 Extended Detention Basins

This type of BMP is also called a dry pond expressing that the *pond* (berm-encased area, excavated pond, or tank) does not necessarily have a permanent water pool between storm runoff events. The basin temporarily stores the SWR or a portion of it to attenuate peak runoff flows for protection of facilities or receiving waters located downstream. The extended detention basin mainly serves the purpose of a hydraulic control measure with a structure that restricts outlet discharges. The water in the pond is mainly discharged following the runoff event, however, may also to some extent undergo evaporation/transpiration and infiltration. At least to some extent, settling of suspended particulate materials takes place.

9.2.2 Wet Ponds

This type of pond is basically similar to an extended detention basin except that it is designed with a permanent pool of water and a temporary storage volume above this permanent pool. This type of BMP is also called a "wet detention pond" or a "retention pond." The design makes it like a small and shallow lake—also between storm events—with sufficient residence time for the water to allow for a number of pollutant removal processes to proceed. The wet pond thereby acts as a BMP for removal of particulate and to some extent soluble pollutants. In addition to the hydraulic control and the improved water quality of SWR, a wet pond can also have a recreational value (e.g., obtained by proper design and use of plants in the pond and integrated with the surrounding environment).

9.2.3 Constructed Wetlands

Constructed wetlands are characterized by being dense vegetated with low water depth areas of typically 0.1–0.3 m. The water depth and the entire structure and performance varies dependent on the rainfall pattern and the season of the year. The structure of a wetland is diverse with free water table, dense vegetated surface waters, and even small islands (i.e., characterized as a shallow wet ecosystem). Dependant on the detailed structure and its variability, the relative importance of physical, chemical, and biological processes in a wetland system may vary.

9.2.4 Infiltration Trenches

Traditionally, an infiltration trench is an excavation that is lined with a filter fabric and backfilled with stones to form an underground basin. In the more recent types, the underground basin is established by piling up plastic boxes with a high degree of cavity. From the infiltration trench, the runoff either exfiltrates into the surrounding soil or enters a perforated pipe from where the flow is routed to an outflow facility. In a complete trench, all runoff is exfiltrated into the soil whereas in a partial trench with a perforated pipe, only part of the runoff exfiltrates. In a water quality exfiltration system, only the first part (first flush volume) of the runoff is typically managed.

9.2.5 Infiltration Basins

Runoff water is temporarily stored in an open infiltration basin—also called infiltration pond—from where infiltration takes place into the underlying soil. Often an infiltration basin is designed to capture only a first flush volume.

9.2.6 Filters

Sand filters are designed to remove particulate matter (i.e., sediments and associated pollutants from the runoff). Removal of pollutants in the runoff is enhanced by a biofilm attached to the filter medium. A filter media can also be selected having specific adsorption characteristics (e.g., a limestone material for removal of phosphorus or an olivine filter—a silicon mineral where iron and manganese can be replaced by heavy metals originating from the runoff). A bioretention system (cf. the following brief description of filter strips and rain gardens) is a vegetated filter for enhanced removal of pollutants.

9.2.7 Water Quality Inlets

These types of BMPs cover a variety of devices that appear technological compared with the previously described units that more or less mimic the elements of nature. Often these water quality inlets are designed having a chamber structure. The outflow from these devices can be routed to other BMPs (i.e., the structure acts as a pretreatment unit). In some cases, these types of BMPs are also called hydrodynamic devices referring to oil and grit separators, sand traps, swirl separators, and underground sand filters. Also more advanced treatment units for stormwater belong to this group (e.g., the ballasted flocculation, cf. Example 5.2).

9.2.8 Swales

Swales are shallow vegetated channels used to convey stormwater. Pollutants that are transported with the runoff flow can be partly removed by settling or by infiltration into the soil. However, effective removal of pollutants requires that a swale has a rather low slope and is well drained.

9.2.9 Filter Strips, Bioretention Systems Biofiltration Systems, and Rain Gardens

The different names refer to vegetated infiltration and percolation systems for control of SWR. These technologies are often used on a small scale and can therefore be characterized as LID practices. Filter strips, also known as vegetated buffer strips, are to some extent like grassed swales except that they typically have flat and wide banks. A bioretention and biofiltration system or a rain garden is typically a small footprint BMP for attenuation and percolation of stormwater. Such designed landscape features are often dispersed located in parking areas, along swales, and even in dense populated areas. The use of deep-rooted perennial plants in such systems can enhance the infiltration. Several processes, sedimentation, adsorption, filtration, volatilization, biological uptake in plants, and decomposition proceed in such systems.

9.2.10 Porous Pavement

A porous pavement, also called permeable pavement, consists typically of asphalt or concrete materials through which stormwater is quickly transported into a layer of a high-void material (e.g., gravel). The runoff is stored in this layer until it either infiltrates into an underlying soil or is routed through a drain to a stormwater conveyance system or an infiltration trench.

In total, the structural types of BMPs dealt with in Sections 9.2.1 through 9.2.10 constitute different types of techniques. Some are treatment systems that remove the pollutants by degradation or accumulation depending on the nature of the pollutant. An example of this technique is the wet pond. Other methods (e.g., the porous pavement) basically tend to reduce the runoff volume by detention and following diversion of the runoff. In Sections 9.3 through 9.8, a number of these BMPs will be further dealt with.

9.3 POND SYSTEMS FOR STORMWATER TREATMENT

Basically two main types of detention ponds exist: the extended detention basin (dry pond) and the wet pond (wet detention pond). The dry pond is basically a sedimentation basin of the same type as described in Section 5.6.2. In the following, the wet pond will be focused on.

9.3.1 Wet Ponds

Wet ponds, or wet detention ponds, are designed to collect stormwater and to drain it slowly, often holding it for days. By doing so, treatment processes for the inflowing water similar to those that naturally take place in small, shallow lakes may occur (e.g., pollutant accumulation in the pond sediments, transformations of biodegradable substances, and uptake of pollutants in the vegetation). The main characteristics of a wet pond receiving urban SWR or road runoff:

- It has a permanent water volume.
- It holds the inflowing stormwater and reduces hydraulic effects like erosion of receiving water systems located downstream at the same time pollutant removal processes proceed.
- It reduces nonpoint source loadings of pollutants onto receiving water bodies located downstream.

A wet pond is typically equipped with a catch basin (forebay) located at the inlet where coarse materials (sand) can settle and not cause siltation of the pond.

The main types of pollutant removal processes in wet ponds are

- Sedimentation and a following accumulation of pollutants in the sediments
- Uptake of (soluble) pollutants in the vegetation followed by degradation or accumulation in the sediments when the vegetation dies and decays
- Adsorption of fine particulate (colloidal) materials on fixed surfaces (e.g., plants and bottom sediments)

To some extent, degradation of pollutants in the water phase may take place. However, compared with these three types of processes, it is normally less important because of a typical low biodegradability of the incoming organic matter.

In addition to observing hydraulic and pollutant removal purposes, a wet pond can be designed as a natural lake and thereby contribute to the recreational value in an urban area. A wet pond is therefore a potential element of "water in the city" observing dual purposes. A wet pond will over time typically be populated with plant and animal species. It is, however, important to stress that it is a treatment device that should remain as such and that authorities should not convert it to an ecological system and define corresponding quality criteria.

In principle, the water volume of a wet pond consists of two parts, a permanent water volume that defines a minimum dry weather water depth and an overlaying storage volume that corresponds to the water volume that was temporarily accumulated in the pond originating from the inflowing runoff water, Figure 9.1. This storage volume is also called treatment volume. In practice, a number of phenomena and processes result in modifications of this simple approach. As examples, evaporation can further reduce the permanent volume during a long dry weather period and during extreme events, the maximum water level can be further raised or overflow can take place.

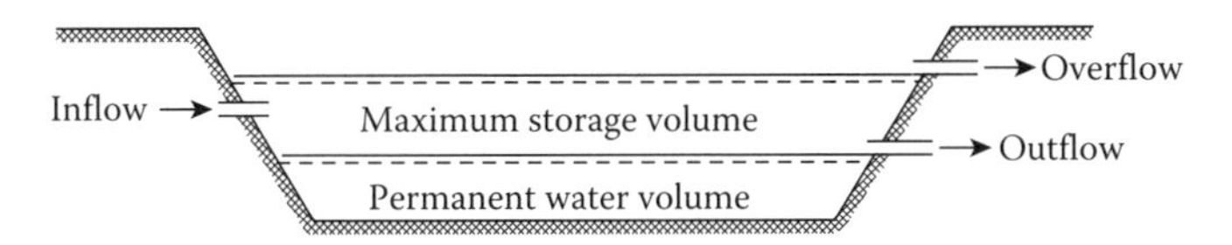

FIGURE 9.1 Principle of a wet pond showing the permanent water volume and the maximum storage volume for the runoff water.

FIGURE 9.2 A wet pond receiving highway runoff. The pond is located in a property development, Fornebu, in Oslo, Norway. The area was, until 1998, the airport of Oslo. The "permanent water volume" is in this case dominating compared with the "storage volume" (cf. Example 9.1).

Furthermore, the permanent water volume and the storage volume will be mixed. More correctly, it is the entire complex water balance of the pond and the flow pattern that will determine the distribution, mixing, and residence time of the pond water.

Continuous outflow from a pond will occur as long as the water level exceeds the level of the permanent water volume. Although mixing between the two water volumes takes place, it is important to understand the basic approach of division between two types of volumes in terms of performance and design (cf. Example 9.1).

Figure 9.2 shows an example of a wet detention pond receiving highway runoff.

9.3.1.1 Design Principles for Wet Ponds

There are several concepts for design of a wet pond that observes the objectives and characteristics mentioned in the start of Section 9.3.1. In Figure 9.3, two different principles for construction are depicted. Due to risk for clogging of the filter bottom, the type with the horizontal flow is most commonly implemented.

Further design details are depicted in Figure 9.4. This figure also shows a flood control volume used for peak discharges.

In the design phase it is, in addition to the treatment performance, important to consider those environmental problems that can occur in wet ponds and also those aspects that are related to the operation and maintenance. The following are major environmental problems that are frequently observed in wet detention ponds:

- Eutrophication
- Wet ponds as breeding grounds for mosquitoes and mosquito-borne deceases like the West Nile virus

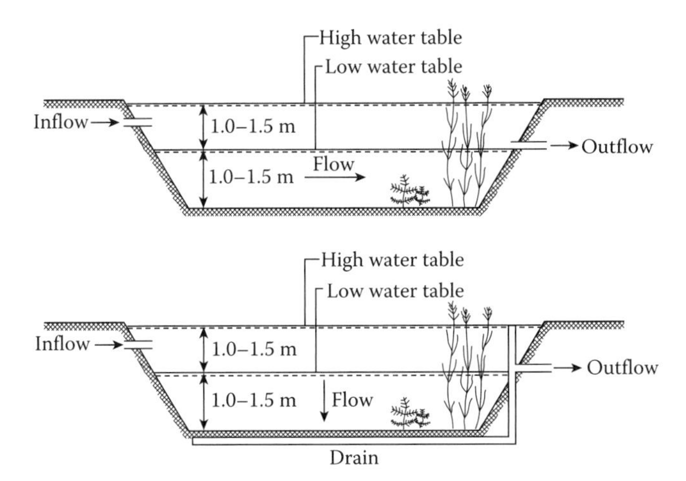

FIGURE 9.3 Principle of two wet ponds with horizontal net flow and vertical net flow through a sand or stone filter, respectively. Both submerged plants and marsh plants can be present. Values for high and low water tables are indicated.

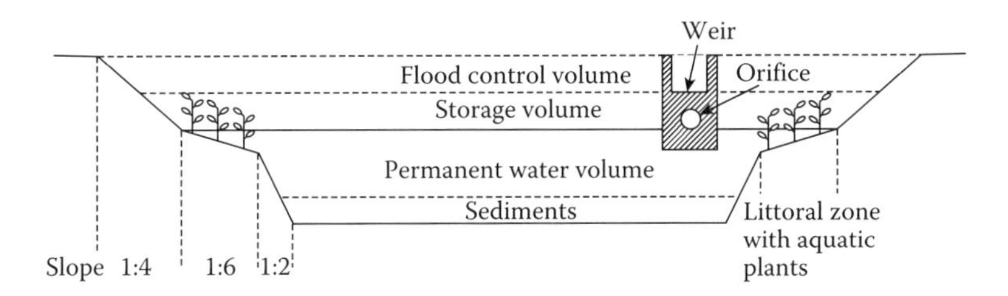

FIGURE 9.4 Example showing construction details of a wet pond (cf. Figure 9.1–9.3).

Furthermore, the following should be considered:

- Occurrence of excessive siltation
- Avoiding areas where more or less permanent stagnant waters exist
- Maintaining aerobic conditions in all parts of the pond
- Design the systems that draw in birds and predators like mosquito fish and dragonflies whereby the occurrence of harmful insects can be reduced
- Creating as much as possible a natural system

In principle, the first step in the design process of a wet pond is to select a site for its location where inflow and outflow can occur by natural flow and without any use of pumps. The next step is to determine the volume of the wet pond. A number of additional steps should be considered to observe hydraulic, process, and operational related criteria for the pond.

The following four principles illustrate what methods normally are applied for determination of the pond volume. In Sections 9.3.1.2 through 9.3.1.5, these principles will be further illustrated:

1. Design based on the specific pond area in terms of a recommended surface area of a pond per impervious unit area of the contributing catchment (e.g., determined in units of m^2 ha^{-1}, cf. Section 9.3.1.2).
2. Design based on empirical knowledge on the pollutant removal efficiencies in detention ponds versus a dimensionless variable describing the pond volume relative to the runoff volume from the local mean storm event (cf. Section 9.3.1.3).
3. Design based on a minimum duration of the interevent dry period between successive storm events and a corresponding return period for exceeding this criterion (cf. Section 9.3.1.4).
4. Design based on a rainfall-runoff model for the catchment and an empirically based expression for removal of selected pollutants in the detention pond (cf. Section 9.3.1.5).

Although these design principles are mentioned in order of increasing complexity, they are all based on empirical knowledge. A conceptual description of all the important processes for the performance of a wet pond does not exist at a level where it is realistic and true.

These four principles for determination of the wet pond volume are not exclusively given to illustrate the application of different tools for this specific purpose. The objective is in more general terms to illustrate that the use of different empirical approaches can lead to appropriate solutions for the same problem. It is important because it must be realized that pollution from urban catchments and roads is subject to considerable variability and that one of the implications is that quantification must rely on empirical and not conceptual formulated knowledge. Contrary to applying a conceptually based tool that relies on well-defined theoretical knowledge on the governing processes, an empirical approach as a means to solve a problem can have different starting points.

Last but not least it is crucial to stress that the volume of water in a wet pond varies considerably, in principle between a minimum value (a permanent pool of water volume) and a high value including both the permanent water volume and the overlaying storage volume. Each of the pond volumes determined by the four design methods must therefore be interpreted by taking this fact into account. Furthermore, changes in the water volume will be affected by the outflow rate from a wet pond. This outflow can be restricted in case a hydraulic impact onto downstream receiving waters is relevant. These aspects will be discussed in Section 9.3.1.6 and Example 9.1.

In addition to the determination of the volume of the wet detention pond, there are a number of other design criteria that must be considered. In Sections 9.3.1.7 and 9.3.1.8, these criteria will briefly be dealt with. To avoid siltation of the wet pond, it is important to add an inlet structure (catch basin) that is often designed as a settling tank for sand particles (grit chamber). For practical reasons it is often designed as an integral part of the wet pond (cf. Section 9.8).

9.3.1.2 Size of Wet Ponds Based on a Specific Surface Area, Method #1

A rather simple design principle for a wet detention pond is to relate its surface area to the area of the catchment (i.e. to determine a specific pond surface area in units of m^2 per impervious contributing catchment, ha). To some extent this design principle must reflect the hydraulic load on the pond and a recommended value for the pond/catchment area ratio will therefore vary with the rainfall pattern from one region to another. Indirectly, the principle is based in a fixed water depth of the pond, typically varying in the interval 1–1.5 m.

The design principle will be illustrated by an example. Based on measured long-term pollutant removal efficiencies, Pettersson, German, and Svensson (1999) have studied the importance of the surface area of a wet pond relative to the catchment area. They performed investigations in two pond systems, one that was high loaded and one with a relatively low surface load. Table 9.1 and 9.2 show main results from this study.

The two ponds have average depths of about 1.2 m. The volume of the high loaded pond and the low loaded pond corresponds to a capacity for detention of the precipitation of about 5 and 30 mm of rainfall, respectively.

Investigations that have been performed in other wet ponds under similar rainfall conditions and with a specific pond area larger than 250 m^2 ha^{-1} have not shown removal efficiencies exceeding what is reported in Table 9.2. From a practical point of view, this specific area of a wet pond, with a water depth of about 1–1.5 m, is therefore considered an upper limit for the pollutant removal capacity of a simple pond. The results that are shown in Figure 9.5 illustrate this fact.

TABLE 9.1
Inflow and Outflow Concentrations of Selected Pollutants and Corresponding Removal Efficiencies for a High Loaded Wet Pond with a Specific Surface Area of 40 m^2 ha^{-1} Impervious Area of the Catchment

Pollutant	Inflow SMC	Average Outflow Concentration	Removal Efficiency (%)
TSS (g m^{-3})	55	17	42
VSS (g m^{-3})	16	6	39
Zn (mg m^{-3})	120	83	24
Cu (mg m^{-3})	53	37	24
Pb (mg m^{-3})	13	7	30
Cd (mg m^{-3})	0.55	0.48	12
Tot N (g m^{-3})	2.0	1.9	8
Ortho P (g m^{-3})	0.07	0.04	27

Note: Due to the high load, an overflow structure at the inlet diverts runoff water directly to the receiving water during heavy storm events. The number of events included is 65.

TABLE 9.2
Inflow and Outflow Concentrations of Selected Pollutants and Corresponding Removal Efficiencies for a Low Loaded Wet Pond with a Specific Surface Area of 240 m^2 ha^{-1} Impervious Area of the Catchment

Pollutant	Inflow SMC	Average Outflow Concentration	Removal Efficiency (%)
TSS (g m^{-3})	153	25	84
VSS (g m^{-3})	27	6.5	76
Zn (mg m^{-3})	135	24	82
Cu (mg m^{-3})	34	9	75
Pb (mg m^{-3})	26	5	82
Cd (mg m^{-3})	0.41	0.20	50
Tot N (g m^{-3})	0.9	0.6	33
Ortho P (g m^{-3})	0.09	0.02	74

Note: The number of events included is 13.

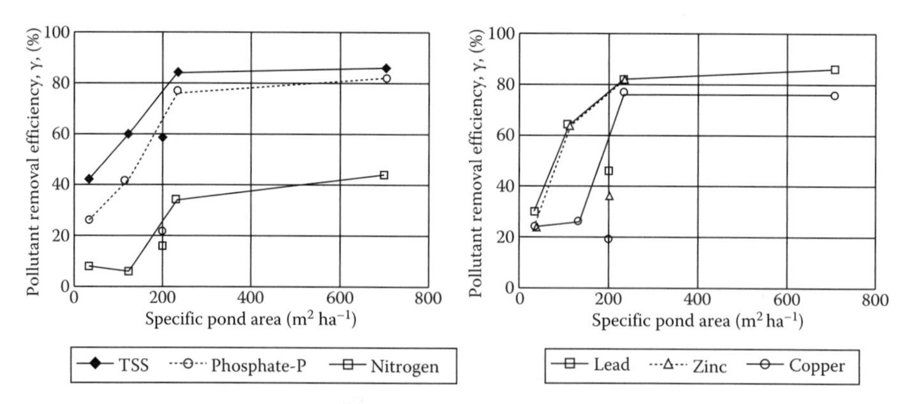

FIGURE 9.5 Measured and estimated pollutant removal efficiencies for selected pollutants versus the specific pond area. (Modified from Pettersson, T. J. R., German, J., and G. Svensson, Pollutant Removal Efficiency in Two Stormwater Ponds in Sweden, Proceedings of the 8th International Conference on Urban Storm Drainage, Sydney, Australia, August 30–September 3, 1999.)

9.3.1.3 Determination of Pond Volume Based on Pollutant Removal Efficiency and a Mean Storm Event, Method #2

This empirical design principle is based on rather simple information of local rainfall characteristics that is combined with general knowledge on the pollutant removal efficiency observed in wet ponds located at different sites. The following is the information originating from different wet ponds contributing to the design basis:

- The rainfall depth for a mean rainfall event at the site
- The volume of the wet pond
- The pollutant removal efficiency (%) observed for the pond

Compared with the design principle explained in Section 9.3.1.2 that is based on determination of a specific pond area, the estimation of the pond volume by this alternative method takes into account the local rainfall pattern in a simple way. The following formula defines, in this respect, a central parameter, n, by combining the volume of the wet pond with information on the local rainfall:

$$n = \frac{V}{v} \tag{9.1}$$

where

n = dimensionless water volume of the pond (–)
V = volume of a wet pond determined per unit impervious area of the catchment (m^3 ha^{-1})
v = volume of water for a mean storm event per unit impervious area of the catchment (m^3 ha^{-1})

As shown in Figure 9.1, the pond volume can be subdivided in a permanent water volume and a storage volume. In Section 9.3.1.6 the interpretation of V in Equation 9.1 will be further dealt with.

Based on information available from a number of wet ponds, an empirical relationship between the dimensionless pond water volume, n, and the pollutant removal efficiency, γ, can be established, Figure 9.6.

$$\gamma = f(n) \tag{9.2}$$

where

γ = pollutant removal (treatment) efficiency (%)

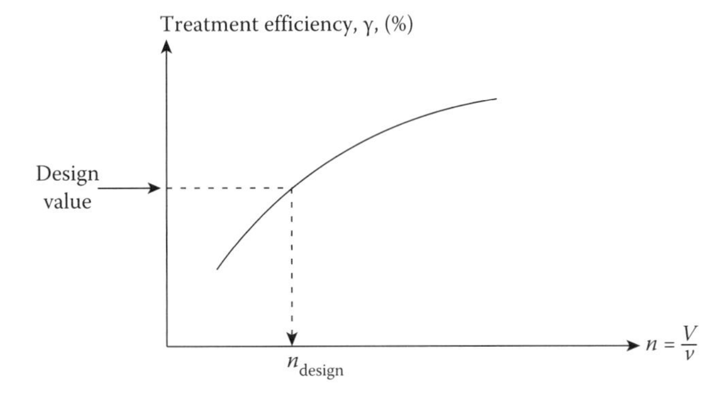

FIGURE 9.6 Principle for determination of a wet detention pond volume, V, based on the volume of water for a mean storm event, v, and required removal (treatment) efficiency, γ, for a pollutant.

The starting point for determining the pond water volume is the empirical relationship $\gamma = f(n)$ for a selected pollutant that is assessed to be central and for which a specific removal efficiency should be met (cf. Figure 9.6). Based on the selected value of γ, a corresponding dimensionless pond water volume, n, is found and following the pond water volume, V, according to Equation 9.1. In case of rainfall conditions for Northern Europe, it is typically observed that an optimal value of γ corresponds to a value of n in the interval 6–8.

The principle depicted in Figure 9.6 is exemplified for total suspended solids (TSS) and total phosphorus (tot. P), Figure 9.7. The design curves are based on information originating from U.S. EPA (1986), Hvitved-Jacobsen, Keiding, and Yousef (1987); Hvitved-Jacobsen, Johansen, and Yousef (1994); and Hvitved-Jacobsen (1990).

A mean rainfall depth can be defined in different ways. In case of the results shown in Figure 9.7, the mean (arithmetic mean) value is based on a rainfall series with rainfalls > 0.4 mm and a minimum interevent dry weather period of one to two hours. Due to the empirical nature of the design principle, other parameter values for a criterion can be selected as well.

As discussed in this Section 9.3.1 and illustrated in Figure 9.1, the water volume of a wet pond consists of two parts: a permanent water volume and a storage volume. As a consequence of this division, it becomes important to further discuss the explanation of the water volume, V, defined by Equation 9.1. In principle, the interpretation depends on the type of underlying data for the empirical Equation 9.2. In case of Figure 9.7, the wet ponds that were included in this design basis operated with a rather small storage volume. Results based on this figure therefore refer to similar conditions and the volume V should, as a first estimate, consequently be interpreted as a permanent water volume. However, the interpretation is more complex. Further—and to some extent contradictory—details will be discussed in Section 9.3.1.6.

From Figure 9.7 it is seen that a removal efficiency, $\gamma = 0.65$ for total P corresponds to a dimensionless pond water volume, $n = 7.5$. Under Northern European

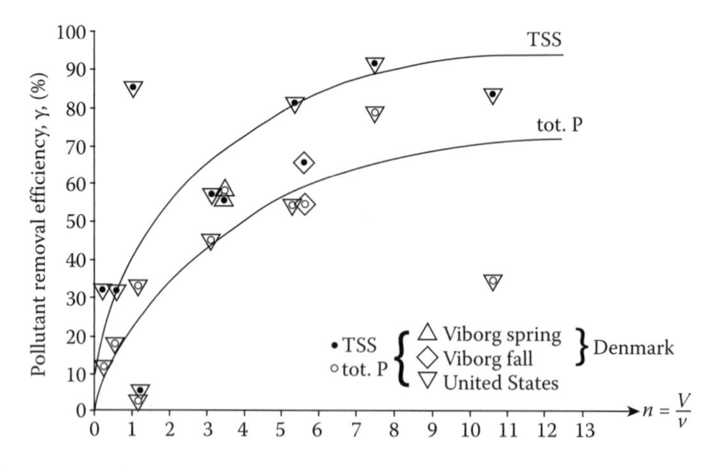

FIGURE 9.7 Empirical based design curves for determination of a wet pond volume, V, based on required pollutant removal efficiency, γ, and information on the mean rainfall volume, v, for the site. Design curves are shown for total suspended solids (TSS) and total phosphorus (tot. P).

conditions with a typical mean storm event, $v = 30\ m^3\ ha^{-1}$, the (permanent) water volume is therefore:

$$V = n\,v = 7.5 \times 30 = 225\ m^3\ ha^{-1}.$$

9.3.1.4 Wet Pond Design Based on Interevent Dry Period Characteristics, Method #3

The central idea behind this design method is the observation that high pollutant removal efficiency in a wet pond will occur if sufficient residence time is provided and if conditions with a minimum of mixing in the water phase can be established. The reason is that sedimentation and pollutant uptake in plants and other transformation processes require time to proceed and because optimal sedimentation is supported by rather quiescent conditions in the water phase. The combination of sufficient residence time, hydraulic residence time (HRT), and quiescent condition in the water phase can be expressed in terms of the length of the interevent dry period between two consecutive rainfall (runoff) events.

It has been observed that under temperate climate conditions, two to three days of interevent dry period is typically required to fulfill what can be considered optimal pollutant removal efficiency. Such optimal conditions means that not just settling of solids will occur but also to some extent biological uptake of soluble species (e.g., soluble nutrients), having low reaction rates. Observations show that more than two or three days of interevent dry period normally, only to a minor degree, improves the removal of a pollutant, probably because some kind of complex equilibrium is established with both uptake and release of accumulated substances.

Applying this design principle, the concept exemplified in Figure 2.2 can be used for determination of the pond volume (Hvitved-Jacobsen and Yousef 1988). This figure shows the relation between rainfall depth (rainfall volume), interevent dry period and frequency of the event at low return periods. The curves in Figure 2.2 are based on statistical analysis of a 33 year rainfall series from the city of Odense, Denmark, and must, for specific design purposes, be replaced by similar local rainfall statistics.

As an example, the following design criteria for determination of the volume of a wet pond volume can be selected:

- A frequency of overflow of untreated runoff water from the pond: $3\ yr^{-1}$ (i.e., a return period of four months)
- An interevent dry period between two consecutive rainfall events equal to 72 hours (i.e., a period where conditions with a minimum of mixing in the water phase can be expected)

The example in Figure 2.2 shows that these two criteria result in a design rainfall equal to 24 mm. If the rainfall volume and the runoff volume from an impervious area are almost equal for such a storm, the volume of the wet pond per unit area of the catchment is

$$V = 24 \times 10^{-3} \times 10^4 = 240\ m^3\ ha^{-1}$$

As it was the case for Method #2, this pond volume shall be interpreted in terms of the permanent water volume and/or the storage volume. For this Method #3, one of the central criteria for the design concerns the frequency of overflow of untreated runoff water (i.e., outflow of water that has not observed the criterion of sufficient residence time). It is therefore clear that the volume V_r includes the storage volume but also, depending on the hydraulic design of the pond, a portion of the permanent water volume. As will be further discussed in Section 9.3.1.6 including Example 9.1, good hydraulic performance of a wet pond will ensure that a portion of a treated permanent water volume will be discharged during a runoff event and replaced by the incoming runoff water.

The result in terms of the wet pond volume is by applying Method #3 highly dependent on the actual rainfall pattern and the criteria selected (i.e., the frequency of overflow and the interevent dry period). The criteria that are selected in the example resulting in a wet pond volume of 240 $m^3\ ha^{-1}$ therefore relate to the climate and process conditions in Denmark that is situated in the northern European temperate zone. If, and as an example, these criteria are applied for Florida in the United States with a subtropical climate, the wet pond volume turns out to be about 4 times larger (Hvitved-Jacobsen, Yousef, and Wanielista 1989). It is probably not realistic—and not a correct choice of criteria—because these should be chosen to observe optimal pollutant removal under the conditions that prevail in the actual case. Specific values for such climate depending criteria must therefore be evaluated and carefully selected for each region.

9.3.1.5 Wet Pond Design Based on Model Simulation for Pollutant Removal, Method #4

The physical, chemical, and biological processes that proceed in a wet pond are as previously mentioned the same as those of a shallow lake. For modeling purposes of wet ponds, the governing pollutant removal processes are not sufficiently known in details, particularly not in case the design concerns new systems. It is expected to be so because of the complex nature of the pollutant removal processes. As a substitute, a simple 1′ order understanding of pollutant removal in a wet pond can be applied (Hvitved-Jacobsen, Johansen, and Yousef 1994; FHWA 1996):

$$C = C_0 e^{-kt} \tag{9.3}$$

where

C = pollutant concentration at time t (g m^{-3})
C_0 = pollutant concentration of the incoming stormwater (g m^{-3})
t = residence time in the pond (d)
k = first-order removal rate of the pollutant (d^{-1})

The kinetics shown in Equation 9.3 can be included in a model for pollutant removal in a wet pond.

Based on results from investigations on pollutant removal in wet ponds, the following values for k were estimated (Hvitved-Jacobsen, Johansen, and Yousef 1994):

dissolved phosphorus: $k = 0.1\ d^{-1}$
particulate phosphorus: $k = 0.35\ d^{-1}$
TSS: $k = 0.5\ d^{-1}$

As an example, Equation 9.3 was used in a model for prediction of pollutant removal in a wet pond receiving runoff from a highway near Oslo, Norway (Vollertsen et al. 2007a). Based on continuous measurements of inflow and outflow from the pond during a year, pollutant removal rate constants were estimated. In this case the yearly average removal rate constant for TSS ($k = 2.0\ d^{-1}$) and for tot. P($k = 0.14\ d^{-1}$) was determined. Because of the variability that exists for the runoff phenomena and treatment processes, results based on Equation 9.3 are not relevant at a time scale corresponding to a single runoff event but must be interpreted as an average value for a number of events. For pollutants with accumulative effects, it is not an important obstacle for the design.

Design Method #4 differs from the three previously described methods (#1, #2, and #3) by the fact that it is based on model simulation. It is therefore a methodology appropriate for analysis of different scenarios (cf. Section 9.3.1.6). Furthermore, the input to the model for both water and pollutants can be described in terms of a runoff model for the catchment. Such model formulations are typically based on historical rainfall series. As an example, the residence time distribution of the inflowing runoff water to the pond is a possible and useful model result (Vollertsen et al. 2007a).

Method #4 requires a number of input data (e.g., an initial value of the pond volume determined by one of the other methods).

Because Method #4 is a model simulation procedure that takes into account hydraulic phenomena, it is readily possible to subdivide the pond volume in two parts, the permanent water volume and the storage volume.

9.3.1.6 Pollutant Removal and Hydraulic Performance of Wet Ponds

Wet ponds designed for pollutant removal have their origin in ponds designed for hydraulic purposes (i.e., for detention of the runoff from mainly impervious urban areas). The dual purpose of runoff detention and at the same time observing sufficient residence time for pollutant removal processes to proceed was obvious. The word "wet detention pond" signals a hydraulic as well as a water quality related purpose. Although a hydraulic performance set requirements for a storage volume and not necessarily a permanent water volume, it is an important issue that procedures for design of wet ponds must observe and thereby integrate both hydraulic and quality related objectives.

When implementing and assessing the results obtained with the three design Methods #1 through #3 for the pond volume, it is therefore, as already discussed in Sections 9.3.1.3 and 9.3.1.4, important to combine the potential of pollutant removal with the hydraulic performance of the wet pond. Method #4 dealt with in Section 9.3.1.5 should be differently interpreted because it includes a simulation procedure and thereby directly integrates pollutant removal with an analysis of the hydraulic performance.

After having used one of the design methods, the starting point of process design of a wet pond is a pond volume (i.e., a volume that ensures both the hydraulic objective and the water quality objective. An important design step concerns the determination of how this volume should be interpreted in terms of both pond construction and treatment performance. In this respect the important question is if the pond volume determined is a *storage* volume, a *permanent* water volume, the total of these volumes or some other kind of *effective* volume. In the following, this question will be further focused on and discussed in terms of the hydraulic performance of the pond.

The pond volume determined by the design Methods #1 through #3 is in principle based on what water volume can become active for the removal of pollutants. These three methods are empirically based and the underlying data originate from real systems where different phenomena (e.g., specific hydraulic characteristics) prevail and are of importance. When implementing the result of the design in practice, the actual hydraulics of the wet pond must therefore be considered, in particular the flow regime in the pond and the outflow relative to the inflow. It is therefore important to observe the following three integrated objectives or requirements in the design procedure:

- Hydraulic performance in general.
- Sufficient residence time of the captured runoff volume in a wet pond must be ensured to support the treatment processes.
- Required protection of downstream receiving waters.

In case the flow out of a wet pond is not restricted (e.g., if hydraulic protection of downstream facilities or receiving waters is not needed), the incoming runoff water will, to some extent, almost simultaneously replace a portion of the treated water from the permanent water volume. In this scenario it is crucial that the geometry of the pond facilitates plug flow so that incoming water is not diverted directly to the outlet. The storage volume in such a pond is small and basically only needed to ensure a proper hydraulic function of the outlet. In this case the pond volume determined by the design Methods #1 through #3 is the permanent wet volume of the pond.

In case the flow is strongly restricted (e.g., if hydraulic protection of downstream facilities or receiving waters is needed), the incoming water will swiftly become completely mixed with the treated water from the permanent water volume. In this scenario, the pond geometry is of less importance compared to a pond without outlet flow restrictions. The more restricted the outlet of the pond is, the more storage volume is needed to detain the incoming stormwater. At the same time, the residence time of the storage volume of the pond is sufficient for treatment processes to proceed. The extent of the treatment processes occurring in the storage volume therefore approach the treatment processes occurring in the permanent water volume of a pond without restricted outlet. In this case, the pond volume determined by the design Methods #1 through #3 must consequently be interpreted as the sum of the permanent volume and a fraction of the storage volume. How large a fraction of the storage volume should be included depends on the flow restriction. For a given set of local climate conditions, this fraction can be calculated applying a numerical modeling approach similar to design Method #4.

In Example 9.1, these two different scenarios for interpretation of the volume for design of wet ponds will be briefly illustrated.

Example 9.1: Pollutant Removal and Hydraulic Performance of Two Wet Ponds

Two cases are outlined in Figure 9.8. Case a illustrates a situation where a receiving water body sensitive to erosion is located downstream of the outlet from the wet pond and the outlet flow is restricted to 0.6 L s^{-1} ha^{-1}. In case b the receiving

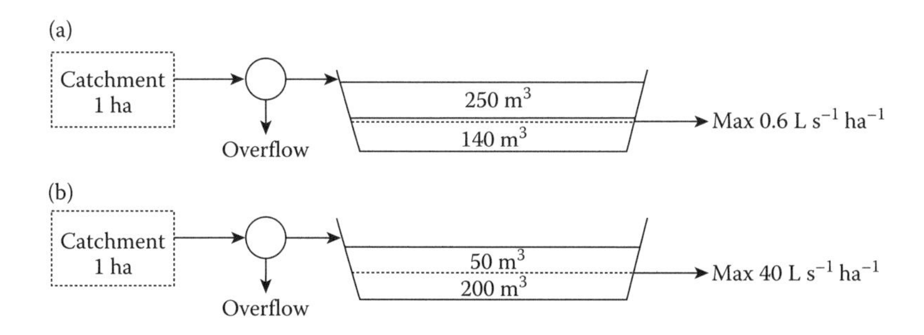

FIGURE 9.8 Illustration of two different pond design principles.

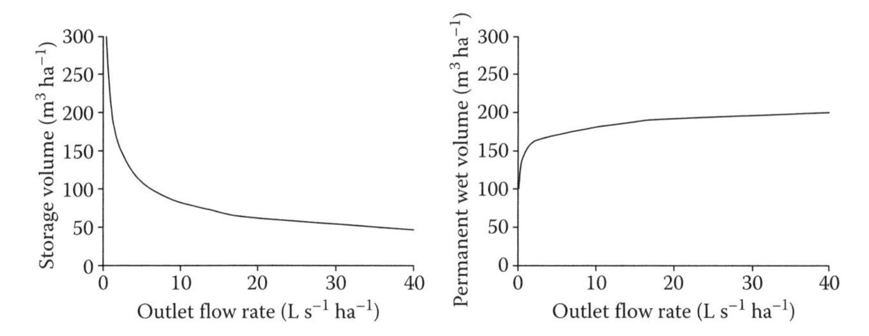

FIGURE 9.9 Pond volumes for varying outlet flow rates and constant pollutant removal rates.

water body is not sensitive to erosion, but for practical reasons the outlet flow has been restricted to 40 L s^{-1} ha^{-1}.

In both cases the contributing catchment is 1 ha and design Method #3 (Section 9.3.1.4) is used. Based on results from Figure 2.2 with criteria of an interevent dry period equal to 48 hours and three months return period, the design volume of the wet pond is 200 m^3 for each of the two ponds.

For the region where the ponds are located, numerical simulations of the treatment performance versus the pond volume have been made for long, historical rain series. Figure 9.9 shows one set of such design graphs in terms of storage volumes and permanent wet volumes needed to ensure constant pollutant removal rates at varying values of retardation of the outlet flow.

According to Figure 9.9, the requirement for an overlying storage volume is in case a, 250 m^3. From this figure it is readily seen that even small changes in the outlet flow (0.6 L s^{-1} ha^{-1}) will considerably change the requirement for storage volume. In case b, where the outlet flow rate requirement is 40 L s^{-1} ha^{-1}, the storage volume is 50 m^3. The wet pond volume of 200 m^3 determined by design Method #3 corresponds to a pond with little retardation of the outlet flow (i.e., to the wet pond volume of case b). When restricting the outlet flow to 0.6 L s^{-1} ha^{-1}, the wet pond volume can be reduced by 60 m^3, as some treatment will occur in the storage volume, Figure 9.8.

An important point of Example 9.1 is that the result in terms of the pond volumes determined from the design Methods #1 through #3 should be further assessed by addressing both the hydraulic performance of the pond itself and that of the downstream receiving systems. Although the pond design concept as illustrated in Example 9.1 is based on the same pond volume originating from design Method #3, it is readily seen that the total pond volume necessary to obtain a certain pollutant removal will not turn out to be equal. By applying the simulation Method #4 outlined in Section 9.3.1.5, further details of the two scenarios can be analyzed, taking into account inflow according to a local historical rainfall series, hydraulic characteristics of the outlet structures, and mass balances for both water and selected pollutants.

9.3.1.7 Specific Hydraulic Characteristics of Wet Ponds

As dealt with in Section 9.3.1.6 and illustrated in Example 9.1, the hydraulic performance of a wet pond is important for two main reasons, namely for the efficiency of pollutant removal and because of a potential hydraulic impact on receiving waters located downstream. Substances that are associated with particulate matter must be subject to conditions where they can settle and not be eroded. Furthermore, sufficient residence time for the pollutant removal processes to proceed must be observed and the flow out of the pond must be controlled. If possible, an appropriate hydraulic function of a wet pond should be observed without the use of pumps and other types of unit operations that require extensive maintenance, energy consumption, and use of manpower.

In brief, the following generally expressed design rules may result in an appropriate hydraulic function of a wet detention pond:

- The inlet structure, the pond itself, and the outlet must be designed to manage the design flow from the catchment. An overflow structure for extreme runoff events must be an integral part of the entire system.
- The inlet structure and the flow regime in the pond itself must observe conditions for sedimentation and protection against erosion by applying energy dissipators.
- Plug flow through a pond is normally recommended to provide sufficient and uniform residence time to support the pollutant removal processes. Ideally, the pond water should, during a runoff event, be substituted with the incoming water. To provide such conditions, a length to width ratio of the pond between 3:1 and 4:1 is typically recommended. However, investigations have shown that even under such conditions, the flow regime in the water phase may gradually result in conditions corresponding to a more or less completely mixed reactor (Vollertsen et al. 2008). Other investigations have also shown that the pond geometry has no significant effect on the accumulation rate of pollutants in the sediments (Starzec et al. 2005). However, it is important that the placements of inlet and outlet of a pond do not result in a short circuit of the flows.
- Flow control out of the pond must observe requirements to protect downstream channels, constructions, and receiving waters against a hydraulic load. For sensitive receiving waters, a typical recommended flow out of the pond is 0.5–2 l s^{-1} ha^{-1} of the total contributing urban and rural catchment.

- A *natural* flow of water through the pond with no use of pumps requires that an appropriate location of the pond between the contributing catchment and the receiving water can be found.

Wind induced currents and waves can, in a wet pond under both dry and wet weather conditions, affect the shear stress at the bottom and thereby cause resuspension, or prevent efficient settling, of particles. The extent and impact of the wind-induced forces on the treatment performance depend on the location and a number of construction details of a pond. For a pond located in the temperate zone of prevailing west winds, Bentzen, Larsen, and Rasmussen (2009) recommend a dry weather water depth in a pond not lower than 0.6–0.8 m. Referring to Section 9.3.1.8, an optimal water depth of a wet pond is thereby determined by a balance between protection against resuspension of bottom materials and sufficient reaeration to maintain aerobic conditions in both the bulk water phase and at the near bed region.

9.3.1.8 Specific Design Characteristics for Wet Detention Ponds

In addition to the volume of the wet pond and the hydraulics that are dealt with in Section 9.3.1.7, a number of design characteristics must be observed.

A main requirement is related to maintaining aerobic conditions with a DO concentration typically > 4 g m^{-3} and not lower than 2 g m^{-3}. Under such conditions, the redox potential at the sediment-water interface is typically sufficiently high for retaining phosphorus and metals. Furthermore, good biological quality can be observed and an obnoxious smell caused by degradation under anaerobic conditions is avoided. Although a wet pond is a technical installation, it is also a potential recreational element and habitat for several animals including fish. The DO mass balance is therefore also crucial for such reasons. However, ecological water quality standards should not be used for a treatment device.

Main processes in the DO balance are the reaeration (the air–water transfer of oxygen), the photosynthesis of the plant community, and the total respiration. A simple DO mass balance for a wet pond is shown in Equation 9.4 (Madsen, Vollertsen, and Hvitved-Jacobsen 2007):

$$\frac{dC}{dt} = K_L a(C_S - C) + P(t) - R - \frac{SOD}{h}, \quad (9.4)$$

where

C = DO concentration (g m^{-3})

t = time (d)

$K_L a$ = overall volumetric mass transfer coefficient between water and air—reaeration rate coefficient (d^{-1})

C_S = DO saturation concentration (g m^{-3})

$P(t)$ = photosynthesis (g m^{-3} d^{-1})

R = respiration in the water phase mainly caused by plants (g m^{-3} d^{-1})

SOD = sediment oxygen demand (g m^{-2} d^{-1})

h = pond water depth (m).

In addition to oxygen produced by plants, Equation 9.4 shows that the reaeration plays a central role for input of oxygen. Madsen, Vollertsen, and Hvitved-Jacobsen (2007) showed that the reaeration rate in wet ponds was particularly affected by the wind speed at the pond water surface and that K_La at low and critical values of the wind speed was a linear function. Consequently, an ice cover of a pond that prevents reaeration will have a negative effect on the performance (German et al. 2003). Fortunately, the rate of the DO consuming processes is, during a winter period, rather low. An acceptable DO concentration in a shallow pond (i.e., a pond with a maximum 1–1.5 m of water depth during a dry weather period) is typically observed under the following conditions: occurrence of a balanced vegetation of rooted plants, a relatively low concentration of phytoplankton, and insignificant occurrence of areas with stagnant water. The critical period of low DO concentrations will, under temperate climate conditions, typically occur during a relatively long and dry summer period.

Another design criterion is related to the water balance in the case of a long and dry period. A membrane—artificial or produced in clay—is therefore important.

Designed as a shallow lake, a wet detention pond has potential as a recreational element valuable for the local community. The shape of the pond, its vegetation, the ecological quality, and the plantation of the surroundings are, in this respect, important elements. Important for safe performance is sloping banks and a possibility to establish reed vegetation in parts of the pond (cf. Figure 9.4).

To avoid siltation of a wet pond, a sediment trap (catch basin) is normally placed at the inlet to the pond and often as an integral part of the design (cf. Section 9.8). An efficient removal of sand particles in an inlet structure will considerably reduce the need for sediment removal in the wet pond itself in 20–30 year intervals.

9.3.1.9 Pollutant Removal and Sediment Management of Wet Ponds

Pollutant removal efficiencies in wet ponds receiving SWR from urban catchments and roads have been discussed as an integral part of the design methods and particularly exemplified in Section 9.3.1.2. Further comments will be given in relation to Example 9.2.

Pollutant removal in wet ponds is, except for what is degraded, closely related to a corresponding accumulation in the deposits at the bottom. Furthermore, accumulated sediments will reduce the effective water volume of a wet pond. Management of the polluted sediments is therefore an integral aspect of pond operation and maintenance.

Example 9.2: Pollutant Removal In a Wet Pond

A wet detention pond located in Oslo, Norway that receives highway runoff from a 2.2 ha impervious area is selected as an example (Vollertsen et al. 2007a). The pond volume is designed according to the principle described in Section 9.3.1.4 with a criterion of 72 hours of interevent dry period and a frequency of 3–4 overflow events per year from the pond. It operates with a rather small storage volume and the design characteristics dealt with in Section 9.3.1.7 and 9.3.1.8 have been followed. The wet pond is therefore designed according to rather pragmatic criteria

TABLE 9.3
The Yearly Average Pollutant Removal Efficiency and Flow Weighted Yearly Average Pollutant Concentrations in the Inlet to a Wet Detention Pond and the Outlet from the Pond

Pollutant (Unit)	Average Inlet Concentration (SMC)	Average Effluent Concentration	Removal Efficiency (%)
TSS (g m^{-3})	246	43	82.5
Total N (g m^{-3})	1.49	1.05	29.5
Total P (g m^{-3})	0.674	0.262	61.1
Bioavailable P (g m^{-3})	0.388	0.146	62.4
Oil and fat (g m^{-3})	5.0	0.9	82.0
Total PAH (mg m^{-3})	1.77	0.26	85.3
Pb (mg m^{-3})	17.1	4.1	76.1
Zn (mg m^{-3})	272	78	71.3
Cu (mg m^{-3})	86	36	58.1
Cd (mg m^{-3})	0.21	0.08	61.9
pH (–)	7.4	7.6	–
Conductivity (mS m^{-1})	39	42	–

following what might be considered possible in practice, still observing a potential for optimal removal of pollutants.

The example is selected because the design follows well-defined criteria and because continuous monitoring at the inlet and the outlet of the pond have been performed during a one-year period. Totally, 87% of the incoming stormwater during this year has thereby been monitored. Main results from the monitoring program is shown in Table 9.3.

In general, the results shown in Table 9.3 correspond in terms of removal efficiencies to what is often observed for well-functioning wet detention ponds in Northern and Central Europe and in the United States (Shueler, Kumble, and Heraty 1992). The results therefore exemplify what must be considered possible with this type of BMP.

Because the data used cover a whole year of continuous monitoring, the example illustrates what treatment level of a number of pollutants in terms of outlet concentrations is possible to achieve with a well-functioning wet detention pond.

The treatment efficiency of a BMP is traditionally expressed as a percent reduction of a pollutant concentration relative to the level in the incoming SWR (cf. Section 3.3). However, in a treatment device—including the BMPs dealt with in this chapter—the outcome of the treatment processes in terms of an outlet concentration from a well-designed device is more or less constant irrespective of the inlet concentration. The efficiency given as percentage of pollutants removed is therefore not the best measure for assessment of the treatment performance of a BMP because a high inflow concentration favors a high value of removal efficiency. A mean or median effluent concentration from a BMP receiving SWR is the most relevant measure for what is possible with a given technology (Strecker et al. 2001). Unfortunately, this fact has not been emphasized in most literature on BMPs for stormwater management (see Table 9.7).

The sediment removed in the forebay (catch basin) of a wet pond consists mainly of sand and coarse materials. If well operated, only a limited amount of fine particles and thereby associated organic matter and toxic pollutants will be found in such sediments.

However, and as intended, the sediments in a wet pond are contaminated with micropollutants and potentially pose ecotoxicological risks when disposed off. Basically, the frequency of sediment removal from a wet pond can be determined by its potential toxic impact onto the ecosystem but is mostly determined by loss of detention volume (Heal, Hepburn, and Lunn 2006). In terms of a potential pollution of the local environment, it is important that there is a high affinity of heavy metals (and organic micropollutants) to the bottom sediments of a wet pond, which is typically the situation when aerobic conditions in the pond prevail. Transport of metals to the underlying soil and groundwater is generally very slow (Yousef et al. 1994b).

The sediment accumulation rate in wet ponds is typically 0.5–2 cm yr^{-1} and it is feasible to remove the accumulated sediments for periodic maintenance of a pond when the effective volume is reduced by 10–15% (Yousef et al. 1990; Yousef et al. 1994a; Heal, Hepburn, and Lunn 2006). As an example, a net sediment accumulation rate of 1 cm yr^{-1} will, during a period of 20 years, produce 20 cm of sediments. In a pond with a dry weather water depth of 1.5 m it corresponds to about 13% reduction of the volume. The sediment removed is potentially hazardous waste and should therefore be assessed against local soil and sediment quality guidelines (cf. also Table 9.4).

To some extent, the distribution of pollutants in sediments from wet ponds will depend on the particle size (cf. Table 9.4). Compared with coarse sediments, the table shows an increase in the concentration of organic matter and heavy metals in fine

TABLE 9.4
Examples of Size Distribution, Organic Matter and Heavy Metal Contents in Sediments from Two Ponds in France Receiving Stormwater Runoff

Origin of Wet Pond Sediments	Size Distribution, d_{10}, d_{50}, d_{90} (μm)	Fraction < 63 μm (%)	Organic Matter (%)	Cupper (μg g^{-1} DW)	Lead (μg g^{-1} DW)	Zinc (μg g^{-1} DW)
Paris sediments						
Fine	3, 25, 210	71	12.0	139	244	631
Coarse	24, 800, 2930	13	5.2	72	69	268
Nancy	55, 1110, 4110	10	4.1	111	59	315
Dutch target value			–	36	85	140

Source: Data from Pétavy, F., Ruban, V., Conil, P., and Viau, J. Y., Reduction of sediment micro-pollution by means of a pilot plant. Proceedings of the 5th International Conference on Sewer Processes and Networks, Delft, the Netherlands, August 29–31, 297–306, 2007; Spierenburg A., and Demanze, C, *Environnement et Technique* 146, 79–81, 1995.

Note: The pollutant contents are mean values and compared with Dutch standards for polluted soils.

sediments. Indirectly, the data therefore also indicate the importance of removing only sand and coarse materials in forebays to allow a use or disposal of sediments from it without environmental risk.

As expected, Table 9.4 shows that both size distribution and pollutant content in contaminated sediments from wet ponds may vary considerably. Corresponding size distributions and pollutant contents for street sweeping sediments are shown in Table 4.5. Not surprisingly, there are, in terms of pollutant contents, similarities between street dust and the more coarse sediments from wet ponds.

In an investigation by Marsalek, Watt, and Anderson (2006), the concentration levels of heavy metals in sediments from a number of stormwater BMPs are reported. They found that according to local guidelines between 80 and 100% of the samples were marginally to intermediately polluted with several heavy metals and also noted severe levels at several facilities (e.g., 100–200 $\mu g\ g^{-1}$ TS for Cr and Cu).

9.3.2 Extended Wet Ponds for Advanced Treatment

Although soluble and colloidal materials to some extent can be removed in wet ponds (e.g., by plant uptake and adsorption), the major process for treatment is sedimentation and following accumulation of particulate bound pollutants in the sediments. Wet ponds may, however, be extended with facilities for advanced treatment, in this context, defined as unit processes to enhance removal of dissolved and colloidal pollutants remaining in the water phase after traditional treatment.

Typically, advanced treatment of stormwater is of relevance in case of discharges into sensitive receiving waters or if the stormwater is used as a source for potable water, water for irrigation, or toilet flushing. In principle, advanced treatment is possible by combining BMP units that in terms of treatment performance are complementary (cf. Section 9.1). In particular, it is important that a storage volume is an integral part of the BMP unit. It is possible to apply well-known technologies from municipal wastewater treatment (e.g., accomplished by addition of chemicals or by filtration; Tchobanoglous, Burton, and Stensel 2003).

The selection of chemicals, types of filter materials, and layout of systems that can be applied for advanced treatment of stormwater is legion. Advanced treatment of SWR that has been subject to pretreatment in wet ponds can be performed by addition of chemicals or by filtration. In Sections 9.3.2.1 and 9.3.2.2, respectively, such possibilities will be exemplified.

9.3.2.1 Chemicals Added to Enhance the Removal of Dissolved and Colloidal Pollutants

Chemicals are widely used for both drinking water clarification and wastewater treatment and can be used for lake restoration purposes too. Aluminum salts are particularly common for drinking water clarification and salts of aluminum, calcium, and iron are used for wastewater treatment. Precipitation, coagulation, and flocculation are the central unit processes for the purification (cf. Sections 3.2.4.1, 3.5.2, and 3.5.3, respectively).

In the case of wet ponds, the effect of Al, Ca, and Fe salts for the removal of phosphorus, metals, and organic micropollutants—including particles like algae—are particularly important. In the following, the mechanisms of these salts will be briefly outlined (Stumm and Morgan 1996; Cooke et al. 2005).

9.3.2.1.1 Aluminum

Aluminum can be added to the runoff water at the pond inlet as aluminum sulfate (alum) $Al_2(SO_4)_3$, 14 H_2O, where the aluminum ion is immediately hydrated, typically within a few minutes:

$$Al^{3+} + 6\,H_2O \rightarrow Al(H_2O)_6^{3+}. \tag{9.5}$$

Depending on the pH, different hydrated forms of aluminum hydroxide will be formed by liberation of a hydrogen ion:

$$Al(H_2O)_6^{3+} + n\,H_2O \rightarrow Al(H_2O)_{6-n}(OH)_n^{(3-n)+} + n\,H^+ \tag{9.6}$$

where
n = 1, 2, or 3 (–)

The different hydrated forms of aluminum hydroxide appear in the form of amorphous flocs with characteristics in terms of precipitation, coagulation, adsorption, or flocculation of a number of different pollutants and particles. These flocs will settle and thereby transfer pollutants to the sediments in the pond.

At pH values between 5 and 8, the insoluble $Al(OH)_3$ will dominate (cf. Figure 9.11). Above pH 8.5, as might occur during intense photosynthesis, there is a risk of producing the soluble—and toxic—aluminate ion ($Al(OH)_4^-$) and thereby also a corresponding release of adsorbed pollutants. In low alkalinity pond waters (i.e., in not well-buffered waters), it is not recommended to apply aluminum for pollutant control. The $Al(H_2O)_6^{3+}$ ion is also toxic but pH values below 4.5, where this ion is a dominating species, should not occur in wet ponds. The occurrence of short periods with high pH values that might exist during daytime in eutrophic surface waters seems less problematic than more permanent occurrence of low pH values in waters affected by acid rain.

The sorption characteristics of $Al(OH)_3$ is inert to changes in the redox potential.

9.3.2.1.2 Calcium

Calcium can be added at the pond inlet in the form of calcium hydroxide (lime). However, most often Ca is not applied this way but as $CaCO_3$ in a filter (cf. Section 9.3.2.2). Efficient removal of phosphorus by precipitation of hydroxyapatite ($Ca_{10}(PO_4)_6(OH)_2$) requires a relatively high pH value (high alkalinity). The use of lime added for removal of pollutants in wet ponds is therefore, contrary to what is the case for certain types of wastewater treatment plants, not appropriate.

9.3.2.1.3 Iron

As can be seen from Figures 3.6 and 3.7, $Fe(OH)_3$ is at a relatively high redox potential (i.e., under aerobic conditions), a stable compound in an aqueous system having a low solubility between pH 7 and 10. Although the solubility and stability of $Fe(OH)_3$

are more complex under the conditions that exist in a wet pond receiving SWR, these two figures depict the real conditions rather well. In general, a major part of iron therefore accumulates in the sediment as $Fe(OH)_3$ together with other low solubility iron species like FeO(OH) and $FePO_4$. A high redox potential in the upper part of the sediment is crucial for not risking dissolution of Fe(III) as ferrous iron (Fe^{2+}). Furthermore, FeS can be produced if anaerobic conditions in the sediments result in sulfate reduction.

The pollutant removal efficiency of iron salts applied for treatment in wet ponds relies on the chemical characteristics of iron as mentioned above. In general, $Fe(OH)_3$ can provide sorption sites for a number of pollutants and Fe(III) salts can form precipitates or other low solubility complexes with a number of pollutants and thereby accumulate these in the sediments. Aerobic conditions are crucial for two reasons: by not risking dissolution of Fe(II) and by not inactivating iron as FeS.

The entire sequence of chemical and physicochemical reactions with Fe in an aqueous system are very complex and because of that, in this context, not realistic to quantify in details. The amount of iron required to remove pollutants from the water phase in a wet pond should therefore be determined empirically. In the case of phosphorus, investigations have shown that the equilibrium sorption capacity of lake sediments follows a Langmuir adsorption isotherm. A relation between the equilibrium P concentration in the water phase and that accumulated in the sediment is thereby established (cf. Section 3.2.4.1 and Jacobsen 1977). Empirically it is known that these sorption isotherms depend on the sediment composition in terms of its iron content. Figure 9.10 is an example that originates from analysis of aerobic sediments from a number of shallow lakes (Jacobsen 1977). As shown in this figure, an increase in the iron content does not cause a proportional increase in the phosphate sorption capacity, probably because the specific and active area of iron hydroxide will decrease.

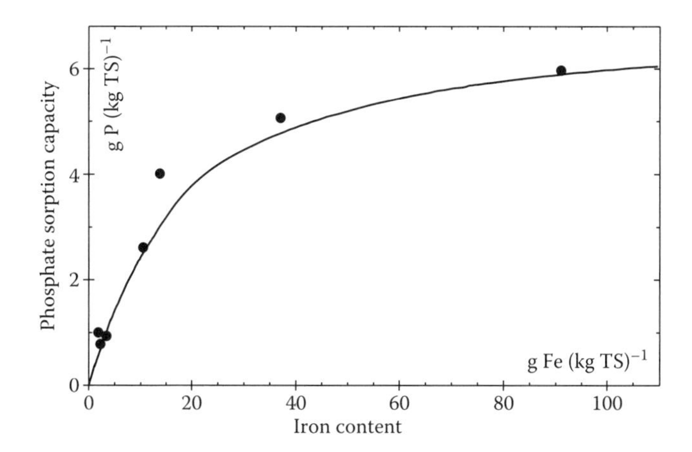

FIGURE 9.10 Phosphate sorption capacity versus the iron content of aerobic lake sediments. The results originate from investigation of the upper 5 cm of sediments from eight shallow Danish lakes. (Modified from Jacobsen, O. S., *Vatten*, 3, 290–98, 1977.) Total Fe is determined by extraction with oxalate and thereby represents a measure for reactive iron in terms of sorption characteristics.

It is expected that the similarities between shallow lakes and wet detention ponds makes such relation valid for sediments in wet ponds too.

Figure 9.10 shows that the iron/phosphate ratio for sediments is a central parameter that controls the capacity for sorption of phosphate and thereby its removal from the water phase. As an example, the figure shows that a reactive iron content (e.g., as $Fe(OH)_3$ and FeO(OH)) in the sediments of 100 g Fe (kg TS)$^{-1}$ results in semioptimal conditions for adsorption of phosphorus. In this case the Fe/P ratio is approximately 17 g Fe (g P)$^{-1}$. A ratio in the order of 15–20 g Fe (g P)$^{-1}$ in the sediments is therefore considered an appropriate design value when adding iron to the sediments in a wet pond for improved treatment (cf. Example 9.3 and Jensen et al. 1992). This design criterion only concerns treatment for phosphorus. However, it is likely that oxides and hydroxides of iron also provide sorption sites for a number of other pollutants (e.g., heavy metals).

As already mentioned, $Fe(OH)_3$ has its lowest solubility between pH 7 and 10. Sorption of phosphorus onto the surfaces of iron hydroxide, however, is optimal at about pH 5–7. In addition, other low-solubility phosphates (e.g., $FePO_4$ and $AlPO_4$), are formed and precipitate. It illustrates that complex chemical interactions determines the behavior of the entire system.

9.3.2.1.4 *Solubility of Chemicals Added*

As already dealt with in this section, the pH value plays a central role for the removal of pollutants by the formation of insoluble precipitates or by formation of adsorption sites. Furthermore, the removal pattern in terms of precipitation or adsorption followed by the settling of flocs is complex and not possible to quantify in details. As an example, phosphate may—at high redox potentials—form insoluble precipitates with iron (e.g., as $FePO_4$), but also be removed by adsorption to $Fe(OH)_3$.

By only focusing on hydroxide and inorganic carbon species (carbonates), Figure 9.11 illustrates the influence of pH on the solubility of Al, Ca, and Fe in stormwater. The solubility is calculated at 25°C and the total inorganic carbon concentration is C_T = 1 mM (12 g C m^{-3}). Furthermore, the curves are based on solubility equilibria

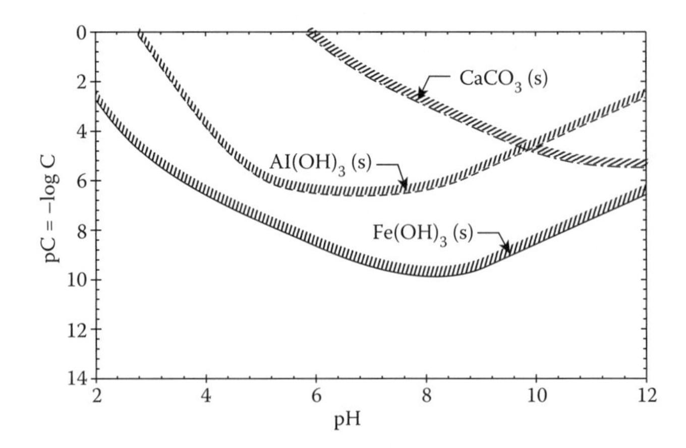

FIGURE 9.11 Solubility diagram for Al, Ca, and Fe at 25°C assuming a concentration of total inorganic carbon C_T = 1 mM (12 gC m–3).

for relevant species as illustrated for $Fe(OH)_3$ in Figure 3.6. The carbonate system plays an essential role in terms of formation of carbonate but also as a buffer (cf. Section 3.2.4.2). Although Figure 9.11 is an incomplete and simplified description of what happens when applying chemicals to a wet pond, it shows what pH intervals are generally needed to observe the minimum of solubility to enhance pollutant removal and to avoid toxic soluble species of aluminum.

Example 9.3: Amount of Iron Added to the Sediments In a Wet Pond to Improve P Removal

The example concerns a methodology for advanced phosphorus treatment in a wet pond receiving SWR. In addition to the settling of particulate bound phosphorus as a major removal process in a wet pond, a substantial part of the soluble and colloidal phosphorus should therefore also be removed. In this example, an iron salt is added to the pond sediments to improve the sorption capacity for phosphorus.

The sediment surface area of the pond is 240 m^2 per impervious ha of the contributing catchment with a yearly runoff volume corresponding to 470 mm of rainfall. The iron salt will be dosed during a dry weather period as a concentrated suspension that will settle. After gentle mixing of the sediments, iron is expected to accumulate in the upper 5 cm of the sediments. It is the intention that the amount of iron salt dosed should be equivalent to two years of phosphorus accumulation in the sediments. The removal efficiency of phosphorus in the wet pond is, before iron addition, 65%, but is expected to increase to 90% when improving the sorption capacity of the sediments. The yearly mean value of tot. P concentration in the inflowing stormwater is 0.35 g P m^{-3}.

Furthermore, it is assumed that an iron/phosphorus ratio in the sediments of 20 g Fe (g P)$^{-1}$ is needed to secure removal efficiency as intended. When calculating the amount of iron salt that will be dosed, the initial amount of iron and phosphorus in the pond sediments should be taken into account. It is expected that sulfate respiration (H_2S-production) in the sediments will not occur, and that iron therefore can be added as iron sulfate ($FeSO_4$, 7 H_2O).

Analysis of the initial iron content in the sediments is performed by extraction with oxalate. It is therefore, in the following calculations, assumed that this fraction of iron is reactive and has sorption characteristics as shown in Figure 9.10. Furthermore, it is assumed that the inflow of iron from the SWR has no further capacity available for sorption and is therefore not taken into account.

It is a point of discussion if the amount of iron dosed should correspond to 90% or (90 − 65) = 25% of the incoming phosphorus, in the latter case assuming that pond treatment processes proceed unchanged when adding iron. It is likely that the correct value is between these two extremes. However in this example, it is decided to add iron corresponding to 90% of the inflowing amount of phosphorus.

Analysis of the upper 5 cm of the sediments in the wet pond before iron addition:

Total P in dry sediment matter: 0.24 g P (kg TS)$^{-1}$
Iron in sediment dry matter: 6 g Fe (kg TS)$^{-1}$
Dry matter: 0.58 g TS (g sediment)$^{-1}$
Sediment density: 1.7 g sediment cm^{-3}

The initial iron/phosphorus ratio in the sediment is therefore $6/0.24 = 25$ g Fe (g P)$^{-1}$ (i.e., > 20 g Fe (g P)$^{-1}$. Compared with the design value of the iron/phosphorus ratio, there are free sites expected to be available for sorption. The amount of reactive iron (i.e. what is assumed to be available for sorption of phosphorus) can be calculated based on the total amount of both Fe and P in the upper 5 cm of the sediments.

Initial amount of Fe in the upper 5 cm of the sediments per 240 m^2 (i.e., sediment surface area receiving runoff from 1 ha of impervious contributing catchment):

$$0.05 \times 240 \text{ (m}^3\text{ of sediments)} \times 1.7 \times 10^3 \text{ (kg m}^{-3}\text{ of sediment)} \times 0.58 \text{ (kg TS (kg sediment)}^{-1}) \times 6 \times 10^{-3} \text{ (kg Fe (kg TS)}^{-1}) = 71 \text{ kg Fe.}$$

Initial amount of P in the upper 5 cm of the sediments:

$$0.05 \times 240 \text{ (m}^3\text{ of sediments)} \times 1.7 \times 10^3 \text{ (kg m}^{-3}\text{ of sediment)} \times 0.58 \text{ (kg TS (kg sediment)}^{-1}) \times 0.24 \times 10^{-3} \text{ (kg P (kg TS)}^{-1}) = 2.84 \text{ kg P.}$$

Amount of iron that is assumed to be available for sorption of phosphorus per 240 m^2 of sediment surface area:

$$71 \text{ (kg Fe)} - 2.84 \times 20 \text{ (kg Fe)} = 14 \text{ kg Fe.}$$

The total inflow of phosphorus during two years from an impervious catchment area of 1 ha:

$$0.35 \text{ (g P m}^{-3}) \times 0.470 \text{ (m yr}^{-1}) \times 10^4 \text{ (m}^2\text{ ha}^{-1}) \times 2 \text{ (yr)} = 3.29 \times 10^3 \text{ g P ha}^{-1}$$

If 90% of the P-load is transferred to the sediments, the accumulated amount is:

$$0.9 \times 3.29 \times 10^3 = 2.96 \times 10^3 \text{ g P ha}^{-1}.$$

The corresponding required amount of iron per unit area of catchment is:

$$20 \text{ g Fe (g P)}^{-1} \times 2.96 \times 10^3 \text{ g P ha}^{-1} = 59 \text{ kg Fe ha}^{-1}.$$

The initial amount of iron available for sorption (14 kg) will reduce the amount of iron added to about 45 kg Fe per ha of the catchment. This amount corresponds to about 225 kg of iron sulfate ($FeSO_4$, 7 H_2O) per ha of the catchment area or a dose of about 1 kg ($FeSO_4$, 7 H_2O) m^{-2} of the sediment surface in the pond. It is thereby assumed that sufficient sorption capacity is available for improving phosphorus removal during a period of two years.

As already mentioned, there are several reasons why the calculation in this example should be considered a tentative estimate. The fact that other pollutants in the stormwater inflow can be adsorbed to or react with iron hydroxide might decrease the capacity available for phosphorus sorption. On the other hand, the calculation of the amount of iron added to the sediments was done based on removal of 90% of the phosphorus inflow, not taking into account that 65% would be removed without a dose of iron.

9.3.2.1.5 Dose of Chemicals

Estimation of an appropriate dose of chemicals for pollutant control in wet detention ponds is from a theoretical point of view complex. The reason is that the dose basically depends on what pollutants are central, the level of concentration and—as already mentioned in this section—a number of external conditions (e.g., determined by parameters like the alkalinity and the redox potential). Furthermore, the mode of the control (i.e., event-based pollutant removal by direct addition of chemicals to the inflowing stormwater or dose for inactivation of the pollutants from a number of events) will play a role. Determination of an appropriate dose therefore relies on experimental procedures (c.f. Cooke et al. 2005) with particular reference to phosphorus removal in lakes.

Table 9.5 gives an impression on what doses have been used for P removal in lakes and refers to what has been typically applied (Cooke et al. 2005). However, the data does not refer to a well-defined period for which the added amount is assumed to be effective.

It is important to notice that the data shown in Table 9.5 refer to lakes. For comparison with wet ponds, it is important to take into account the flow and residence time pattern of the inflowing stormwater. In general, it might be appropriate to add the chemicals directly and flow proportional to the stormwater during the runoff events. Because of that, concentration levels for efficient pollutant control are lower than expected from levels shown in Table 9.5.

The coagulating effect of the conventional metal salts can be improved by modification. An example is poly-aluminum chloride (PAC), a partially prehydrolyzed Al(III) coagulant containing polymeric species and produced by titrating $AlCl_3$ with a base (Trejo-Gaytan, Bachand, and Darby 2006).

9.3.2.2 Filter Media with Binding Efficiencies and Capacities for Pollutant Removal

In addition to the hydraulic conductivity, important process characteristics for filter materials are the kinetics and the capacity (cf. Section 3.2.4.1). A high sorption rate of a pollutant will reduce the need for residence time in the filter medium and thereby increase the area specific flow through it. Furthermore, a high capacity filter will reduce

TABLE 9.5
Amount of Chemicals That Have Been Typically Used for P Removal in Lakes

Chemical	Dose	Comments
Al	5–20 g Al m^{-3}	Added to the water phase
Al	50–300 g Al m^{-2}	Added to the sediments
Ca	10–100 g Ca m^{-3}	Added to the water phase
Ca	–	–
Fe	2–5 g Fe m^{-3}	Added to the water phase
Fe	100 g Fe m^{-2}	Added to the sediments

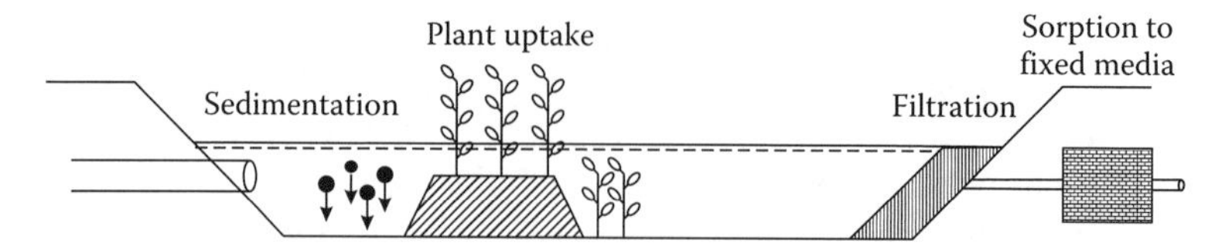

FIGURE 9.12 Sketch of a wet pond equipped with filters for advanced treatment of stormwater.

the need for volume of the filter material and prolong its effectiveness and lifetime. Additional aspects and characteristics of filter materials are subjects of Section 9.4.

A main problem in applying filter media for sorption of soluble and colloidal pollutants is the risk for clogging the filter due to the contents of particles associated with stormwater. To reduce the risk for clogging that otherwise will reduce the flow and ultimately stop it, a prefilter with coarse sand is generally recommended. The design of such sand prefilters will not be dealt with in this respect. General information on filters can be found in Tchobanoglous, Burton, and Stensel (2003) and Shilton (2005). Specific and more detailed knowledge on sand filters related to treatment of SWR is available in FHWA (1996).

Figure 9.12 shows a sketch of a wet pond integrated with a prefilter of sand and a filter for sorption of soluble and colloidal pollutants. To reduce the risk for an inconvenient headloss through the filter, equipment for backwash can be installed. For practical purposes, the filter systems will typically be designed to manage only a portion of the inflowing stormwater (e.g., a first flush volume), and that overflow of water from the pond that has not been subjected to advanced treatment is accepted in case of extreme events.

Several materials have been tested for their potential binding characteristics for soluble and colloidal pollutants (Färm 2002; Liu, Sansalone, and Cartledge 2004, 2005a, 2005b; Westholm 2006). Both natural filter substrates, modified materials, and artificially produced filter media can be used. The following list includes a few selected examples of filter materials:

- Materials where the active component is $CaCO_3$ and/or $MgCO_3$ (i.e., materials like marble, limestone, dolomite, and different types of shells effective for P removal).
- Silicon containing rock materials (e.g., zeolite adsorbing metal ions).
- Iron oxide-coated sand; adsorption of particularly metal ions.
- Manganese oxide-coated polymeric media.
- A number of natural and synthetic organic fiber materials.

Different media show different adsorption kinetics and capacity to the different pollutants. As examples, calcium carbonate filters may exert good sorption characteristics for phosphates, whereas manganese oxide-coated polymeric media particularly adsorb metals. In general, the capacity of a filter material is highly dependent on the pH value and the cation exchange capacity (CEC). A filter material that consists of mixed media can support improved pollutant removal characteristics and hydraulic properties compared with the single components.

For practical purposes, the pollutant removal capacity of a filter material relative to the residence time in the filter is crucial. The variability of these properties is highly affected by both filter media and stormwater characteristics. It is therefore needed that sorption characteristics in terms of the kinetic properties be determined experimentally in each specific case.

It is crucial that a filter material observes efficient sorption characteristics at the very low pollutant concentrations that exist for stormwater exposed to treatment. Such sorption characteristics are important for a wide range of pollutants. Filter media with a sorption capacity for a pollutant about 200–1000 g m^{-3} of filter material (or 0.1–0.5 mg g^{-1} of filter material) must—depending on what pollutant it concerns—be considered acceptable. Furthermore, a rate corresponding to 20–30 minutes of contact time for the stormwater in the filter to obtain this efficiency is, as a first estimate, considered appropriate for practical use (Barbosa and Hvitved-Jacobsen 1999; Färm 2002).

The hydraulic surface load—reflecting the pollutant load—of a filter is a simple design parameter. A more advanced, and also more relevant, design criterion is based on the hydraulic conductivity of the clogging (colmation) layer (Vollertsen et al. 2007b). It is the silt-like particles that accumulate in the clogging layer and particularly those particles with a diameter < 5 μm that might limit the transport of water through a filter.

9.4 INFILTRATION OF STORMWATER

An infiltration system for stormwater is a BMP where the runoff water is temporarily detained until it is recharged into the ground at the site where it is generated. It is thereby a method that mimics nature's way of handling rainwater. Two main types of infiltration systems exist: infiltration ponds (infiltration basins) where the water is typically stored in an excavation from where the infiltration takes place, and infiltration trenches where the water is stored in voids between gravel or synthetic produced materials from where it percolates into the surrounding soil. In addition to these two main types of infiltration systems, infiltration may also take place in swales and at porous pavements. Parallel with the infiltration and transport of the water through the soil, a number of pollutant transformations and removal processes take place. The basic transport processes that proceed in porous media like soils are dealt with in Section 3.4.3.

9.4.1 Main Characteristics of Treatment by Infiltration

In general, filtration (physical retention), biological transformation (degradation), and adsorption to soil particles are main processes that reduce the pollutants in stormwater subject to infiltration. Figure 9.13 shows the principle of the performance of an infiltration pond. Although the principle in terms of pollutant removal for infiltration ponds and infiltration trenches are the same, an infiltration trench should basically be characterized as exfiltration of the runoff into the surrounding soil.

The basic theory for flow of water is for porous media and under saturated conditions described by Darcy's law, Equation 3.56. For an infiltration pond like the one shown in

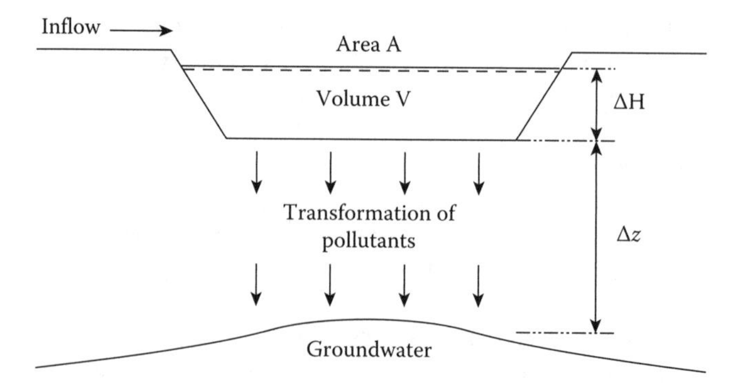

FIGURE 9.13 Principle of stormwater infiltration from an infiltration pond.

Figure 9.13, a first approximation of Darcy's law is formulated according to Equation 3.57. Equation 9.7 is approximately valid for saturated soils and for vertical flow:

$$u = -K\frac{\Delta H + \Delta z}{\Delta z} \tag{9.7}$$

where

u = one-dimensional flow velocity (m s^{-1})
K = hydraulic conductivity (m s^{-1})
ΔH = water depth in the pond, the water pressure (m)
Δz = length of travel (m)
$\Delta H + \Delta z$ = pressure head or hydraulic head (m)

In addition to the hydraulic conductivity, the central system parameters determining transport of water appear from Equation 9.7. The levels of the hydraulic conductivity shown in Table 9.6 are valid for pure water but are relevant for stormwater with a low content of particulate matter.

The efficiency of an infiltration pond can be reduced considerably by clogging caused by both settling of particulate materials from the runoff water and accumulation of small particles in the pore volume of the soil. If the hydraulic resistance mainly occurs in a thin clogged layer—a colmation layer—at the bottom, Darcy's law, Equation 9.7, can be reformulated as follows:

$$u = -K_c\frac{\Delta H}{\Delta l} = -L_l\Delta H \tag{9.8}$$

where

K_c = hydraulic conductivity of the colmation layer (m s^{-1})
Δl = thickness of the colmation layer (m)
$L_l = K_c/\Delta l$ = leakage factor for the colmation layer (s^{-1})

The leakage factor—or its inverse value: the hydraulic resistance, R—is a central parameter that describes the hydraulic property of a colmation layer.

TABLE 9.6
Typical Levels of Hydraulic Conductivity, K, and Porosity, ε, for Selected Soil Types

Soil Type	Hydraulic Conductivity, K (m s^{-1})	Porosity, ε (%)
Gravel	10^{-1}–10^{-3}	25–30
Coarse sand	10^{-3}–10^{-4}	25–35
Fine sand	10^{-4}–10^{-5}	40–50
Silt	10^{-5}–10^{-8}	50–60
Clay	10^{-8}–10^{-11}	50–60

A number of factors affect infiltration of stormwater. First of all, the permeability of the soil, also expressed as the hydraulic conductivity, is central. From Equation 9.8 it is evident that risk of clogging can cause major problems. As an example, a colmation layer of just 1 cm with a hydraulic conductivity of 10^{-8} m s^{-1} will, in a pond with 1 m water depth, theoretically result in a vertical flow velocity of only 3.6 mm hr^{-1}. Occurrence of particulate materials in stormwater may therefore require pretreatment in an inlet structure (cf. Section 9.8). However, such pretreatment may not sufficiently remove the small particles that can cause clogging.

A successful infiltration also requires a groundwater level located minimum 1–3 m below the bottom of an infiltration basin. Vegetation at the infiltration site might have a positive effect in terms of evaporation, pollutant uptake, and formation of a root structure that enhances the vertical flow of water.

One of the environmental advantages of infiltration of SWR is that recharge of groundwater at the site is possible and that it also can reduce the risk of settlement of the soil in the case of groundwater depletion. Furthermore, the pollutant level and flow peaks affecting sensitive downstream receiving waters are reduced. In addition, infiltration will also reduce the need for capacity of the storm sewer network. A disadvantage is, however, a potential risk of soil and groundwater contamination.

In principle, any type of BMP needs detention capacity to obtain time for the pollutant removal processes to proceed. In a wet detention pond, settling and adsorption of the small particles are limiting processes determining the pond volume and thereby the residence time before the treated water is discharged through an outlet. In an infiltration pond, the infiltration rate will determine when sufficient capacity exists to store the incoming volume of water from a runoff event. Example 9.4 will briefly exemplify this situation.

Example 9.4: Time to Empty an Infiltration Pond

An infiltration pond for SWR has a surface area $A = 510$ m^2 and the maximum water depth is $\Delta H = 1$ m (cf. Figure 9.13). The soil under the pond consists of coarse sand that is saturated during infiltration. The hydraulic conductivity is $K = 0.3 \times 10^{-4}$ m s^{-1} and the occurrence of a colmation layer is neglected. The distance to the groundwater level from the soil surface is $\Delta z = 2.5$ m.

According to Equation 9.7, the volumetric flow rate out of the full pond is:

$$Q = AK\frac{\Delta H + \Delta z}{\Delta z} = 510 \times 0.3 \times 10^{-4}\frac{3.5}{2.5} = 0.0214\,m^3\,s^{-1}.$$

Because of a reduced water depth during infiltration, the volumetric flow rate, Q, is gradually reduced. If the average water depth in the pond is 0.5 m, estimated values of Q and the time to empty the pond, *t*, are:

$$Q = AK\frac{\Delta H + \Delta z}{\Delta z} = 510 \times 0.3 \times 10^{-4}\frac{3.0}{2.5} = 0.0184\,m^3\,s^{-1},$$

$$t = \frac{V}{Q} = \frac{510}{0.0184}\frac{1}{3600} = 7.70\,hr.$$

Details of the layout for infiltration systems are legion with infiltration trenches and infiltration ponds being the two main types (cf. Section 9.2.1 and Figures 9.14 and 9.15). From an infiltration trench, the stormwater exfiltrates into the surrounding soils where, from the open infiltration pond, it infiltrates into the underlying soil.

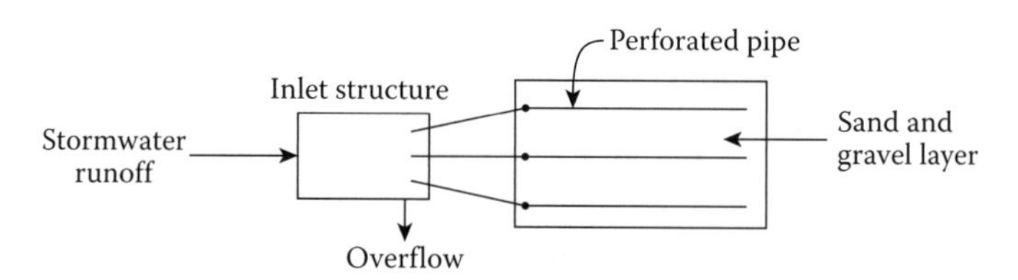

FIGURE 9.14 Outline of an infiltration trench (cf. Section 9.2.1).

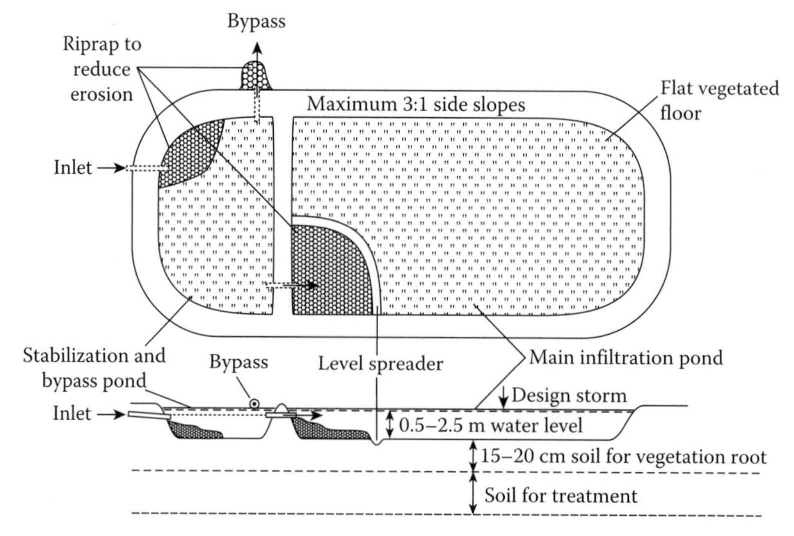

FIGURE 9.15 Example of an infiltration pond layout with a bypass located in the first compartment.

9.4.2 Design of Infiltration Ponds

When designing infiltration systems for stormwater treatment, a number of possible procedures can be applied, very simple and more complex. The following four steps include criteria that are considered central for design of a well-functioning infiltration pond.

1. Detention capacity of the pond (cf. Section 9.4.2.1)
2. Infiltration in soils and clogging (cf. Section 9.4.2.2 and Equation 9.8)
3. Capacity for accumulation (sorption) of pollutants in the soil under the pond (cf. Section 9.4.2.3)
4. Sorption kinetics of selected pollutants (cf. Section 9.4.2.4)

These four basic design steps can be implemented in different ways. In the following, they will be further described and discussed.

9.4.2.1 Detention Capacity of an Infiltration Pond

The detention capacity of an infiltration pond determines what portion of the runoff water should be treated (infiltrated) and thereby to what extent overflow should be diverted to an adjacent surface water body. The following three methods, a, b, and c, are commonly used in this respect. The three methods for determination of the retained runoff volume represent different levels of complexity. The first and simple method is based on a situation where the pond is totally empty before a runoff event takes place. Method b takes into account local rainfall statistics. In addition to that, Method c focuses on the water mass balance of the pond during a selected period of time.

a. The pond volume determines, in principle, what portion of each runoff event can be infiltrated (e.g., determined as a first flush volume where a relative essential part of the pollutants is expected to be present). This volume depends on local conditions but is often chosen corresponding to 10–15 mm of runoff from the catchment. Based on this simple approach, it is clear that the criterion selected will not always be observed (e.g., in case the infiltration pond is not sufficiently empty to capture the runoff volume designed for).
b. The size of the pond is based on a selected return period for overflow by applying a runoff model with a local historical rainfall series. Taking the model results from Figure 2.2 as an example, it is seen that for events defined by 24 hours of minimum interevent dry period, a pond with a detention capacity of about 18 mm of rainfall will in average result in three to four overflow events per year. A detention capacity of 18 mm of rainfall corresponds to a pond volume of 180 m^3 ha^{-1} of contributing catchment.
c. The pond volume is based on a water mass balance taking into account all (important) inputs and outputs like inflow, infiltration, evaporation, and overflow. A simulation model with input form a historical rainfall-runoff series and based on a first estimate of a pond volume from Method a or b is possible. The criterion for the volumetric size of the pond can be determined in terms of the relative amount of overflow from the pond during a selected period of time.

In addition to the pond volume, the maximum water depth of the pond must be determined. Typically, a shallow pond with a depth between 0.5 and 1.5 m will be selected.

9.4.2.2 Infiltration Rate of the Soil

The infiltration rate of the runoff water in an infiltration pond with saturated underlying soil is theoretically determined by the hydraulic conductivity of the soil and the pressure head by applying Darcy's law (cf. Table 9.6, Equation 9.7, and Example 9.4). In practice, the infiltration rate is, however, often governed by clogging in a colmation layer at the bottom (cf. Equation 9.8).

No specific recommended value exists for the magnitude of the infiltration rate. However, an infiltration rate that results in a situation where the pond is almost dry one to three days after a runoff event corresponding to the capacity of the pond is a criterion that might be considered appropriate.

Fine sand has, according to Table 9.6, an initial hydraulic conductivity as low as 10^{-5} m s^{-1} (36 mm h^{-1}). A soil type with less than 25 to 30% of silt and clay and less than 10 to 15% of clay is sometimes considered a first estimate for successful infiltration. However, it should, in the design process, be taken into account that clogging might occur (e.g., corresponding to a 50–90% decrease of the initial infiltration rate). It is therefore recommended that pretreatment (settling of particles) of the runoff water in swales, vegetated filter strips, or catch basins can take place, although fine particles are not efficiently removed.

9.4.2.3 Capacity for Sorption of Pollutants in Soils

When treating the runoff flow in infiltration systems, it must be ensured that accumulation of relevant pollutants is possible in the upper part of the soil. The thickness of the active soil layer is determined by

- The sorption capacity of the soil for the relevant pollutants
- The area specific loading rate of pollutants
- The period of operation for efficient removal of the pollutants

If the capacity for pollutant accumulation of the soil at the site is not judged sufficient, another soil or filter type can be applied.

Each soil type has its specific characteristics that determine to what extent pollutants can be adsorbed. The assessment of the sorption process is complex because different pollutants are present in SWR and also compete for sorption sites. In practice, a simple empirical approach based on experimental results is therefore important for determination of the sorption capacity.

However, basic knowledge exists for selection of an appropriate soil type. In general, the fine soil types have better sorption characteristics for both metals and nutrients than coarse soils. Unfortunately, the fine soils have low hydraulic conductivity. Furthermore, soils with an organic content have also good sorption characteristics.

In addition to the physical characteristics of a soil in terms of hydraulic conductivity and porosity that affect the water transport, physicochemical characteristics are important for sorption. In particular, the pH and the CEC are important parameters.

The pH of the soil influences the electrical charge of the soil particles and thereby the potential for sorption of ions and molecules with dipoles. On most surfaces, adsorption of heavy metals will increase with increasing pH. The CEC reflects the fact that cations like Na^+ and K^+ and to some extent Ca^{2+} are associated with the soil particles and can be substituted by bivalent metals like Cu, Pb, and Zn. The following example gives an impression of the sorption capacity: In a well-designed stormwater infiltration pond, 1 m of soil layer with an organic carbon content of 0.5–1% and with minimum CEC = 5 meq per 100 g of soil will provide sufficient heavy metal sorption capacity for several years. In general, the most mobile heavy metal is Zn. The Zn ion is therefore an appropriate tracer for the risk of heavy metal transport through a soil column.

The sorption capacity for soils vary considerably with both soil characteristics and constituent adsorbed. As an example, Westholm (2006) refers to investigations on the sorption capacity of phosphorus for soils and filter media. Based on these results, typical sorption capacities for soils subject to stormwater infiltration are in the order of 0.2–2 mg P $(gTS)^{-1}$. For heavy metals, the sorption capacities are in general a factor of 5–10 lower (Barbosa and Hvitved-Jacobsen 1999). For specific filter materials, sorption capacities can be higher.

Theoretically, the sorption capacity of a soil or a filter media is, under equilibrium conditions, expressed in terms of an isotherm (e.g., a Langmuir or a Freundlich isotherm, cf. Equations 3.5–3.8). Figure 9.16 exemplifies, by Langmuir, fits results from laboratory experiments of phosphorus sorption for different soil media under equilibrium conditions (Hsieh, Davis, and Needelman 2007).

Although equilibrium does not necessarily exist in the soil during infiltration, sorption isotherms are helpful for estimation of the potential sorption capacity under dynamic conditions (cf. Figure 9.17).

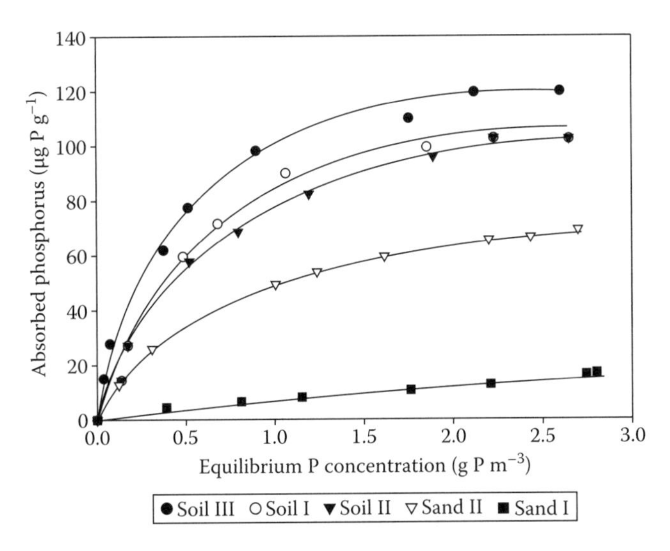

FIGURE 9.16 Phosphorus sorption and corresponding Langmuir isotherms for different soil and sand media at pH = 7 and initial P concentration 2.85 g m^{-3}.

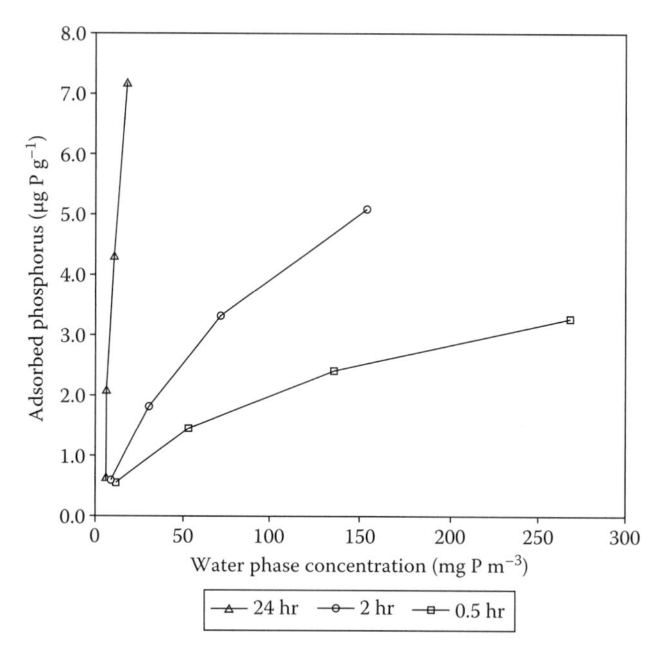

FIGURE 9.17 Isotherm and dynamics for sorption of phosphorus on shell sand. (Data from Vollertsen, J., Lange, K. H., Nielsen, A. H., Nielsen, N. H., and Hvitved-Jacobsen, T., Treatment of Urban and Highway Stormwater Runoff for Dissolved and Colloidal Pollutants, Proceedings of the 6th International Conference on Sustainable Techniques and Strategies in Urban Water Management (NOVATECH), Lyon, France, June 25–28, 2007b.) The sorption capacity is shown after different time length of contact.

9.4.2.4 Pollutant Sorption Rate of a Soil

As exemplified in Figure 9.17, the sorption kinetics of pollutants from infiltrating stormwater is important for the design of treatment systems. Particularly, it is important for the efficiency of pollutant removal at high infiltration rates and low soil depths.

From a theoretical point of view, the kinetics of sorption can be formulated in terms of a second-order reaction that takes place between pollutants in the water phase and active sorption sites at the surface of soil particles (cf. Table 3.5; Liu, Sansalone, and Cartledge 2004, 2005a):

$$\frac{dC_t}{dt} = -k(S_e - S_t)(C_t - C_e) \tag{9.9}$$

where

C_t = pollutant concentration at time t (g m^{-3})
t = time (s)
k = second-order reaction constant (g s^{-1})
S_e = number of active sites occupied at equilibrium per unit amount of adsorptive material (g^{-1})

S_t = number of active sites occupied at time *t* per unit amount of adsorptive material (g^{-1})
C_e = pollutant concentration at equilibrium ($g\ m^{-3}$)

Although potentially important, the kinetics for sorption of stormwater is typically not included in design procedures for infiltration ponds. In practice, a residence time for stormwater in the soil between a few hours and one day is considered acceptable and sufficient for approaching an equilibrium state.

Example 9.5: Simple Design of an Infiltration Pond

An infiltration pond for SWR is designed for accumulation of phosphorus in the soil under the pond. Focus is on the determination of pond volume and the active soil depth for sorption of phosphorus. The basic conditions for the design are below.

SWR from a 3 ha impervious urban catchment area is diverted to an infiltration pond. The maximum water depth of the pond is 1 m and when empty, the required storage capacity of the pond is 17 mm of runoff. Calculations have shown that the yearly average amount of runoff is 620 mm and that an amount corresponding to 450 mm of runoff yr^{-1} is temporarily stored. The difference (i.e., 170 mm of runoff yr^{-1}) at the inflow to the pond is discharged into an adjacent surface water system. The total P concentration (SMC) of the stormwater at the inlet is 0.3 $g\ m^{-3}$ with one third being in a particulate form that will settle and therefore not require capacity for soil sorption. The concentrations of the different fractions of phosphorus are considered equally distributed in the runoff volume. The sorption capacity of the soil is 0.25 mg P $(gTS)^{-1}$ and the soil bulk density is 1.25 g TS cm^{-3}. The infiltration system must be designed for operation during a period of 40 years.

The pond volume, V, and the soil surface area, A, are

$$V = 17 \times 10^{-3} \times 3 \times 10^{4} = 510\ m^{3}$$

$$A = 510\ m^{2}$$

The yearly inflow of phosphorus that require sorption capacity of the soil is two thirds of the tot. P inflow to the pond

$$M_{year} = 3 \times 10^{4} \times 450 \times 10^{-3} \times 0.3 \times \frac{2}{3} = 2700\ g\ yr^{-1}$$

The corresponding area loading is

$$M_{area} = \frac{2700}{510} = 5.29\ g\ m^{-2}\ yr^{-1}$$

The sorption capacity per unit volume of the soil

$$c_{soil} = 0.25 \times 10^{-3} \times 1.25 \times 10^{6} = 313\ gP\ m^{-3}$$

An average yearly soil depth needed for accumulation of phosphorus

$$h = \frac{5.29}{313} = 0.017 \text{ m yr}^{-1}$$

The soil depth needed for operation during a 40 year period

$$z = 0.017 \times 40 = 0.68 \text{ m}$$

9.4.2.5 Model Simulation

Although the four steps for design of infiltration ponds (Sections 9.4.2.1–9.4.2.4) to some extent are individually dealt with, they are in terms of an integrated and optimal performance of a design closely related. Model simulation by use of a historical rainfall series can assist in a more detailed assessment of such relations and overall performance affected by interacting runoff events. An infiltration pond model can be expressed in different terms and at different levels of detail. What is important is that mass balances for water and central pollutants be observed.

9.4.3 Pollutant Removal in Infiltration Ponds

A large number of parameters defined by the type of infiltration system, the specific design of it, and the local external conditions like rainfall pattern will determine to what extent pollutants can be retained. It is therefore not possible to give "characteristic" values for pollutant removal. However, Table 9.7 as an example that originates from FHWA (1996) shows what are estimated levels depending on the storage capacity. As an example, the infiltration system can thereby be designed to capture the first flush volume of the runoff.

TABLE 9.7
Pollutant Removal Efficiencies (%) for Infiltration Ponds with Different Detention Capacities

Pollutant	Storage Capacity: 13 mm	Storage Capacity: 26 mm	Storage Capacity: 52 mm
Sediments	75	90	99
Total phosphorus	50–55	60–70	65–75
Total nitrogen	45–55	55–60	60–70
Trace metals	75–80	85–90	95–99
BOD	70	80	90

Source: Data from FHWA. Evaluation and management of highway runoff water quality. U.S. Department of Transportation, Federal Highway Administration, Publication No. FHWA-PD-96-032, 1996; Shueler, T. R., Controlling urban runoff: A practical manual for planning and designing urban BMPs. Washington, DC: Department of Environmental Programs, Metropolitan Washington Council of Governments, 1987.

It is important to understand the meaning of "removal efficiencies" in Table 9.7 correctly. Removal means that a pollutant is not discharged to the adjacent environment but accumulated in the sediments of the pond, adsorbed on the soil particles, or transported downward to the groundwater. Concerning the mobility of a pollutant (i.e., its potential for groundwater contamination), there is contradicting information in the literature. The risk of groundwater contamination is, however, not only related to the soluble pollutant fractions that potentially might be adsorbed to the fixed soil particles, but can be due to adsorption on the mobile colloidal phase found in the pore water (Durin et al. 2007).

9.5 FILTERS AND BIORETENTION SYSTEMS

Because of their porous structure and potential for adsorption, filters perform to some extent like infiltration systems (cf. Sections 9.3.2.2 and 9.4). However, the effluent from a filter is typically diverted to surface water, Figure 9.18. A filter media is often selected with specific adsorption characteristics for efficient removal of a pollutant. To account for possible clogging, a filter can be designed as a technical system with a possibility for backwash. Methods and details for design of such filters are well known (e.g., Tchobanoglous, Burton, and Stensel 2003).

A bioretention system is a vegetated filter where the filter material typically consists of a mixture of sand and an organic media. The pollutant removal processes include filtration, adsorption, ion exchange, and biological uptake. A bioretention system is typically designed at approximately 5% of the contributing (paved) drainage area. In a built environment, a bioretention system can add to the landscape diversity value.

9.6 CONSTRUCTED WETLANDS

A constructed wetland performs like a complex ecosystem with open water, marsh, and areas typically dry (cf. Section 9.2.3). In general the demand for area is relatively high and for proper pollutant removal typically in the order of 5% of the contributing catchment.

Because of a complex structure, the design and corresponding treatment efficiency of constructed wetlands cannot be well defined. For nutrient removal Shueler (1987) recommends area loads for phosphorus < 5 g P m^{-2} yr^{-1} and for nitrogen

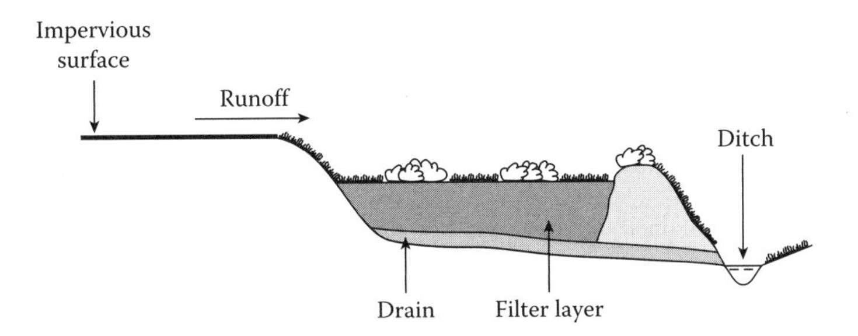

FIGURE 9.18 Outline of an open filter with a layer of sand. In a typical configuration the filter layer is 0.5–1 m.

< 26 g N m^{-2} yr^{-1}. Another example of a pragmatic design criterion refers to a settling velocity of 0.2 m d^{-1} for removal of particulate matter (about 50% removal of particulate P) corresponding to a minimum HRT of about one day in the wetland system.

9.7 SWALES AND FILTER STRIPS

Swales and filter strips observe the dual purpose of a transport and detention system for the runoff and to some extent also a treatment system for pollutants, Figures 9.19 and 9.20. Particles may settle and accumulate at the surface and infiltration transporting soluble substances to the soil system can take place. However, swales are typically less effective for pollutant removal compared with wet ponds and constructed wetlands. Although the pollutants can be trapped in a grass-covered swale, they are generally not permanently bound to the vegetation or the soil particles and erosion of bottom materials is a possible process (Bäckström, Viklander, and Malmqvist 2006).

9.8 INLET STRUCTURES

For almost all types of BMPs, the following are crucial to observe prior to a following and final treatment of the runoff water:

- *Detention capacity*
  Under design conditions, the treatment rate of a BMP is in general considerably lower than the maximum acceptable flow rate at the inlet. A detention basin, if it is not an integral part of the design of the BMP, is therefore needed.

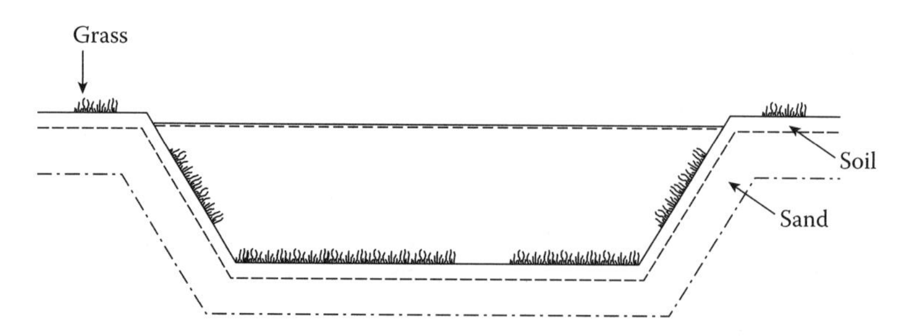

FIGURE 9.19 Cross section of a vegetated swale during a runoff event.

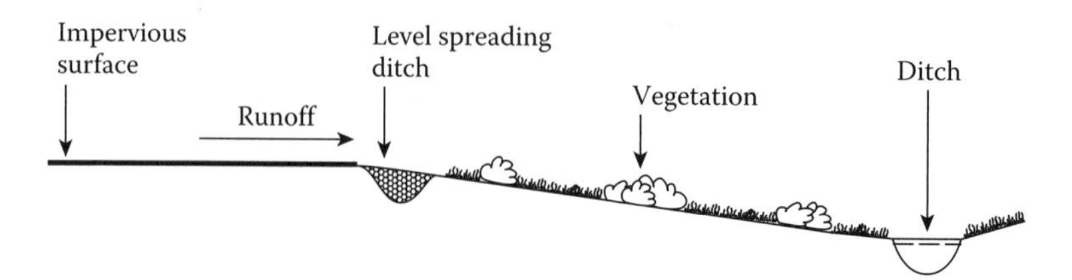

FIGURE 9.20 Principle of a filter strip.

- *Particle removal*
  For a proper overall performance, several BMPs require that particles—at least to some extent—have been removed. For wet detention ponds, it is appropriate to include a sediment forebay (catch basin, sand trap, sediment trap, or gully pot) where a major part of the sand particles can settle. A sand trap will reduce the risk for siltation of the main pond and thereby reduce the need for sediment removal in it. For infiltration systems and filters, the preremoval of particles is even more important to avoid inappropriate clogging. For such systems it is important to remove particles bigger than 10–20 μm. However, it is in practice difficult to observe such high efficiency.

Depending on the particle contents of the inflowing stormwater, a sand trap requires periodic cleaning for removal of sediments (e.g., one to two times per year). If not well maintained, a sand trap may become a source of sediments and not a sink in case of extreme runoff flows.

Sand traps can be designed using the limiting settling velocity, w, for the particles that should be removed. A simple design method (the Camp–Hazen equation) is based on the determination of the required residence time for the stormwater in the catch basin to remove particles with a settling velocity > w:

$$T = \frac{h}{e\,w} \tag{9.10}$$

where
T = hydraulic residence time of the catch basin (s)
h = water depth in the catch basin (m)
e = efficiency constant for particle settling (–)
w = settling velocity of the particles (m s^{-1})

Depending on the design inflow rate to the catch basin, its surface area is

$$A = \frac{Q}{e\,w} = Q\frac{T}{h} \tag{9.11}$$

where
A = surface area of the catch basin (m^2)
Q = design flow rate of the inflowing stormwater to the basin (m^3 s^{-1})

Example 9.6: Simple Design of a Catch Basin

A catch basin should be designed for removal of sand particles with a diameter > 100 μm. The design inflow rate to the basin is 0.3 m^3 s^{-1} and the water depth of the basin is 0.5 m. These conditions should be used to determine the surface area of the basin.

According to Figure 3.12, the settling velocity of 100 μm spherical particles with a density of 2.65 g cm^{-3} is about 10^{-2} m s^{-1}. The efficiency constant, e, for the settling of the particles is typically in the range from 0.3 to 0.6 depending on the turbulence level and the design of the basin. The value of e is, in this example, selected equal to 0.4. The HRT of the basin is therefore (cf. Equation 9.10)

$$T = \frac{0.5}{0.4 \cdot 10^{-2}} = 125 \text{ s}$$

Based on Equation 9.11, the area of the catch basin is

$$A = 0.3\frac{125}{0.5} = 75 \text{ m}^2$$

9.9 COMPARISON OF POLLUTANT REMOVAL PERFORMANCES OF BMPs

The treatment performance of BMPs has, in the preceding sections of this chapter, been a focal point exemplified in numerous cases and also discussed for specific types of BMPs. Assessment and comparison of the potential treatment performance of one BMP type over another would be essential for a selection procedure in a specific case. However, it must be stressed that the importance of partly unknown site conditions and the variability in governing design and operational parameters make a general and solid comparison absurd and irrational. In spite of that, there is —as also discussed in the preceding sections—characteristics for the different BMPs that, given specific constraints and requirements, will support the selection of an appropriate type. In a study performed by Revitt, Scholes, and Ellis (2008), the effectiveness of 15 structural BMP types has been assessed and compared.

Keeping in mind that a solid criterion to exclude a BMP does not exist, it is possible to analyze statistically available data and assess on this basis. The BMP database (www.bmpdatabase.org) that includes measurement data for more than 300 different BMPs provides such possibilities (Clary et al. 2007). Table 9.8 is just a single example based on results from such statistical analysis of these data. As previously discussed

TABLE 9.8
Median Value, 5% Lower Confidence Level (LCL) and 95% Upper Confidence Level (UCL) of Characteristic Effluent TSS Concentrations for Stormwater by BMP Type

BMP Types	Number of BMPs (–)	Median Value (g m^{-3})	LCL (g m^{-3})	UCL (g m^{-3})
Extended detention basins	11	33.0	26.9	40.5
Wet ponds	24	12.0	10.5	13.8
Wetland basins	9	7.6	5.9	9.6
Filter strips, etc.	40	24.0	21.3	27.0
Media filters	19	15.0	12.2	18.3
Hydrodynamic devices	14	36.0	27.6	47.0

in Section 9.3.1.9, the effluent concentration and not percentage removal efficiency should be selected for the assessment.

In spite of the difficulties to extract general concluding results from such analysis, there are indications that the area demanding BMP types (wet ponds and wetlands) result in rather low effluent concentrations for TSS. Also the media filters have, in general, low effluent values.

REFERENCES

Bäckström, M., M. Viklander, and P.-A. Malmqvist. 2006. Transport of stormwater pollutants through a roadside grassed swale. *Urban Water Journal* 3 (2): 55–67.

Barbosa, A. E., and T. Hvitved-Jacobsen. 1999. Highway runoff and potential for removal of heavy metals in an infiltration pond in Portugal. *The Science of the Total Environment* 235: 151–59.

Bentzen, T. R., T. Larsen, and M. R. Rasmussen. 2009. Predictions of resuspension of highway detention pond deposits in interrain event periods due to wind-induced currents and waves. *J. Environmental Engineering*, 135(12): 1286–1293.

Clary, J., M. Quigley, J. Jones, and E. Strecker. 2007. International stormwater BMP database enhancements and updated findings. In Proceedings WEFTEC 80th Annual Water Environment Federation Technical Exhibition and Conference, San Diego, CA, October 13–17.

Cooke, G. D., E. B. Welch, S. A. Peterson, and S. A. Nichols. 2005. *Restoration and management of lakes and reservoirs*. Boca Raton, FL: CRC Press.

DayWater. 2008. *DayWater: An adaptive decision support system for urban stormwater management*, ed. D. R. Thevenot. London: IWA (International Water Association) Publishing.

Debo, T. N., and A. J. Reese. 2003. *Municipal stormwater management.* Boca Raton, FL: Lewis Publishers.

Durin, B., B. Béchet, M. Legret, and P. Le. Cloirec. 2007. Role of colloids in heavy metal transfer through a retention-infiltration pond. *Water Science and Technology* 56 (11): 91–99.

Färm, C. 2002. Metal sorption to natural filter substrates for storm water treatment: Column studies. *The Science of the Total Environment* 298: 17–24.

Field, R., A. N. Tafuri, S. Muthukrishnan, B. A. Acquisto, and A. Selvakumar. 2005. *The use of best management practices (BMPs) in urban watersheds*. Lancaster, PA: DEStech Publications, Inc.

FHWA. 1996. Evaluation and management of highway runoff water quality. U.S. Department of Transportation, Federal Highway Administration, Publication No. FHWA-PD-96-032.

German, J., G. Svensson, L.-G. Gustafsson, and M. Vikstrom. 2003. Modelling of temperature effects on removal efficiency and dissolved oxygen concentrations in stormwater ponds. *Water Science and Technology* 48 (9): 145–54.

Heal, K. V., D. A. Hepburn, and R. J. Lunn. 2006. Sediment management in sustainable urban drainage system ponds. *Water Science and Technology* 53 (10): 219–27.

Hsieh, C.-H., A. P. Davis, and B. A. Needelman. 2007. Bioretention column studies of phosphorus removal from urban stormwater runoff. *Water Environment Research* 79 (2): 177–84.

Hvitved-Jacobsen, T. 1990. Design criteria for detention pond quality. In *Urban stormwater quality enhancement: Source control, retrofitting and combined sewer technology*, ed. H. C. Torno, 111–30. Reston, VA: ASCE (American Society of Civil Engineers).

Hvitved-Jacobsen, T., N. B. Johansen, and Y. A. Yousef. 1994. Treatment systems for urban and highway run-off in Denmark. *The Science of the Total Environment* 146/147: 499–506.

Hvitved-Jacobsen, T., K. Keiding, and Y. A. Yousef. 1987. Urban runoff pollutant removal in wet detention ponds. In *Urban storm water quality, planning and management*, eds. W. Gujer and V. Krejci, 137–42. Proceedings of the 4th International Conference on Urban Storm Drainage, Lausanne, Switzerland, August 31–September 4.

Hvitved-Jacobsen, T., and Y. A. Yousef. 1988. Analysis of rainfall series in the design of urban drainage control systems. *Water Research* 22 (4): 491–96.

Hvitved-Jacobsen, T., Y. A. Yousef, and M. P. Wanielista. 1989. Rainfall analysis for efficient detention ponds. In *Design of urban runoff quality controls*, eds. L. A. Roesner, B. Urbonas, and M. B. Sonnen, 214–22. Reston, VA: ASCE (American Society of Civil Engineers).

Jacobsen, O. S. 1977. Sorption of phosphate by Danish lake sediments. *Vatten* 3: 290–98.

Jensen, H. S., P. Kristensen, E. Jeppesen, and A. Skytthe. 1992. Iron:phosphorus ratio in surface sediment as an indicator of phosphate release from aerobic sediments in shallow lakes. *Hydrobiologia* 235/236: 731–43.

Liu, D., J. J. Sansalone, and F. K. Cartledge. 2004. Adsorption characteristics of oxide coated buoyant media ($\rho_s < 1.0$) for storm water treatment. II: Equilibria and kinetic models. *Journal of Environmental Engineering* 130 (4): 383–90.

Liu, D., J. J. Sansalone, and F. K. Cartledge. 2005a. Adsorption kinetics for urban rainfall-runoff metals by composite oxide-coated polymeric media. *Journal of Environmental Engineering* 131 (8): 1168–77.

Liu, D., J. J. Sansalone, and F. K. Cartledge. 2005b. Comparison of sorptive filter media for treatment of metals in runoff. *Journal of Environmental Engineering* 131 (8): 1178–86.

Madsen, H. I., J. Vollertsen, and T. Hvitved-Jacobsen. 2007. Modeling the oxygen mass balance of wet detention ponds receiving highway runoff. In *Highway and urban environment*, eds. G. M. Morrison and S. Rauch, 487–97. The Netherlands: Springer.

Marsalek, J., W. E. Watt, and B. C. Anderson. 2006. Trace metal levels in sediments deposited in urban stormwater management facilities. *Water Science and Technology* 53 (2): 175–83.

Pétavy, F., V. Ruban, P. Conil, and J. Y. Viau. 2007. Reduction of sediment micro-pollution by means of a pilot plant. Proceedings of the 5th International Conference on Sewer Processes and Networks, Delft, the Netherlands, August 29–31, 297–306.

Pettersson, T. J. R., J. German, and G. Svensson. 1999. Pollutant removal efficiency in two stormwater ponds in Sweden. In *Proceedings of the 8th International Conference on Urban Storm Drainage*, I. B. Joliffe and J. E. Ball, Sydney, Australia, August 30–September 3, 866–73.

Revitt, D. M., L. Scholes, and J. B. Ellis. 2008. A pollutant removal prediction tool for stormwater derived diffuse pollution. *Water Science and Technology* 57 (8): 1257–64.

Shilton, A. 2005. *Pond treatment technology*. London: IWA Publishing.

Shueler, T. R. 1987. Controlling urban runoff: A practical manual for planning and designing urban BMPs. Washington, DC: Department of Environmental Programs, Metropolitan Washington Council of Governments.

Shueler, T. R., P. Kumble, and M. Heraty. 1992. *A current assessment of urban best management practices: Techniques for reducing nonpoint source pollution in the coastal zone*. Washington, DC: Anacostia Research Team, Metropolitan Washington Council of Governments.

Spierenburg A., and C. Demanze. 1995. Pollution des sols: Comparaison et application de la norme hollandaise [Soil pollution: Comparison and application of the Dutch standards]. *Environnement et Technique* 146: 79–81.

Starzec, P., B. B. Lind, A. Lanngren, A. Lindgren, and T. Svenson. 2005. Technical and environmental functioning of detention ponds for the treatment of highway and road runoff. *Water, Air and Soil Pollution* 163: 153–67.

Strecker, E. W., M. M. Quigley, B. R. Urbonas, J. E. Jones, and J. K. Clary. 2001. Determining urban storm water BMP effectiveness. *Journal of Water Resources Planning and Management* 127 (3): 144–49.

Stumm, W., and J. J. Morgan. 1996. *Aquatic chemistry: Chemical equilibria and rates in natural waters,* 3rd ed. New York: John Wiley & Sons, Inc.

Tchobanoglous, G., F. L. Burton, and H. D. Stensel. 2003. *Wastewater engineering: Treatment and reuse*, 4th ed. New York: McGraw-Hill.

Trejo-Gaytan, J., P. Bachand, and J. Darby. 2006. Treatment of urban runoff at Lake Tahoe: Low-intensity chemical dosing. *Water Environment Research* 78 (13): 2487–2500.

U.S. EPA. 1986. Methodology for analysis of detention basins for control of urban runoff quality. U.S. Environmental Protection Agency, Report No. USEPA 440/5-87-001.

Vollertsen, J., S. O. Aasteboel, J. E. Coward, T. Fageraas, H. I. Madsen, A. H. Nielsen, and T. Hvitved-Jacobsen. 2007a. Monitoring and modeling the performance of a wet detention pond: Treatment of highway runoff. Proceedings from the International Urban and Highway Symposium, Cyprus, June 12–14.

Vollertsen, J., K. H. Lange, A. H. Nielsen, N. H. Nielsen, and T. Hvitved-Jacobsen. 2007b. Treatment of urban and highway stormwater runoff for dissolved and colloidal pollutants. Proceedings of the 6th International Conference on Sustainable Techniques and Strategies in Urban Water Management (NOVATECH), Lyon, France, June 25–28, 877–884.

Vollertsen, J., K. H. Lange, J. Pedersen, P. Hallager, A. Buus, A. Laustsen, V. W. Bundesen, H. Brix, A. H. Nielsen, N. H. Nielsen, T. Wium-Andersen, and T. Hvitved-Jacobsen. 2008. Removal of soluble and colloidal pollutants from stormwater in full-scale detention ponds. Proceedings of the 11th International Conference on Urban Drainage, Edinburgh, Scotland, UK, August 31–September 5, 1–10.

Westholm, L. J. 2006. Substrates for phosphorus removal: Potential benefits for on-site wastewater treatment? *Water Research* 40: 23–36.

Yousef, Y. A., T. Hvitved-Jacobsen, H. H. Harper, and L'Yu Lin. 1990. Heavy metal accumulation and transport through detention ponds receiving highway runoff. *The Science of the Total Environment* 93: 433–40.

Yousef, Y. A., T. Hvitved-Jacobsen, J. Sloat, and W. Lindeman. 1994a. Sediment accumulation in detention or retention ponds. *The Science of the Total Environment* 146/147: 451–56.

Yousef, Y. A., L'Yu Lin, W. Lindeman, and T. Hvitved-Jacobsen. 1994b. Transport of heavy metals through accumulated sediments in wet ponds. *The Science of the Total Environment* 146/147: 485–91.

WEB SITES

www.bmpdatabase.org A database with information to be applied for stormwater management projects, in terms of calculation and evaluation of pollutant loads and design of control measures (BMPs).

10 Modeling of Wet Weather Water Quality

Water quality aspects of urban drainage phenomena are closely related to the transport of pollutants from their sources to the locations where they actually occur, temporarily or permanently. In terms of modeling, the description of the quality must therefore be linked to the transport of water and its constituents. The formulations in modeling terms of transformations and other quality related phenomena are therefore integrated with the hydrologic and hydraulic descriptions.

In general, a model is an approximation of reality that basically includes a cause-effect relationship. An important purpose of modeling is to analyze this relationship and simulate the performance of drainage systems (i.e., design the systems to cope with the wet weather flows). In this context, a model is defined as a formulation in mathematical terms of physical, chemical, or biological processes and phenomena. Typically, a model includes descriptions of processes and associated parameters that characterize the relevant system. In specific cases, a relation that is formulated in terms of a table or a graph can also be considered a model. Within a number of defined constraints, a model can thereby be used to simulate given characteristics or the performance of a system (e.g., a drainage system or part of it), the processes that proceed in a system or the impacts of CSOs or SWR onto the surroundings. Specific objectives are often to simulate how a technical system can be operated or managed to observe specific objectives or by giving information on how the system should be designed. When dealing with pollution from urban drainage systems, an important objective is often to simulate the pollutant loads from the wet weather discharges or the effects in the receiving environment. In this way, modeling of the receiving environment like streams and lakes should basically result in information on how the performance of a drainage system affects the surroundings and thereby what specific technologies or construction details might be relevant to reduce the adverse effects to an acceptable level.

Computers have considerably changed design, management, and control of urban drainage systems. To a great extent, simple rules and procedures have been replaced by corresponding models and formulations solved by computers. The computers have thereby increased the speed of calculation and, from a modeling point of view, there is almost no limitation in the extent of both system descriptions and process details that can be included. There is no doubt that hydraulic computer models have improved the overall possibility for correct and detailed simulation of transport phenomena related to both runoff and overflows. However, when it comes to the corresponding quality aspects in terms of chemical and biological processes and phenomena, a similar success has not been reached. It is important to realize that such processes and phenomena typically are far more complex than the physically based details. A detailed deterministic description—particularly in case of biological phenomena—therefore includes numerous interacting processes and a corresponding large number of model parameters. Such parameters are often mutually depending and thereby not possible to determine separately by calibration.

Quality modeling can improve the extent of our possibilities for estimation and prediction and also the corresponding speed with which results can be produced. However, it must be realized that the truth of a statement may not always be improved compared with the results obtained from rather simple and empirically based models and formulations. The limitation in our possibility for estimation and prediction of the quality related aspects is in general still determined by an empirically based knowledge in terms of data that characterize the systems, the processes, and the phenomena. The following conclusion drawn by Harremoës and Madsen (1999) summarizes the fundamental characteristics of a good model: "Model complexity is not a virtue by itself. The best model is a model that provides a suitable simulation of reality with the least complexity and an appropriate set of data for calibration. Model application must include estimation of uncertainty of prediction."

The quantification of quality related processes and phenomena is central for this text on urban wet weather pollution. The preceding chapters therefore include such formulations (i.e., models) related to the corresponding subjects. Chapter 10 will therefore not deal with such aspects. The objective is to provide the reader with general model characteristics that are important when dealing with simulation of urban wet weather pollution.

Numerous computer models for design and analysis of drainage systems in terms of their hydrologic and hydraulic performance can be downloaded from the internet or acquired from private engineering software companies. Furthermore, texts are available on the fundamental aspects of modeling of the management and assessment of water resource problems, flood forecasting, and design of urban drainage systems like pipes and channels. Beven (2001) and UNESCO-IHE (2003) are examples of such texts that focus on the basic aspects for modeling the physical phenomena and processes of urban drainage systems. In addition, numerous texts on modeling the performance of environmental systems can be found (e.g., Schnoor 1996). As previously mentioned, it is crucial that the physical models form a solid basis for the quality models by describing transport phenomena of water and pollutants.

10.1 HOW TO MODEL AND WHY DO IT?

Let us try to answer the second question raised in this heading: WHY? Basically, modeling is performed because it is the engineers or planners way to answer questions in a quantitative way. A model is therefore their ultimate tool. The result of a modeling process is central when designing and managing a system that must observe specific requirements. The beneficial use of a model is its ability to solve a design problem or alternatively act as a tool for controlling a system or a process.

It is evident that the reliability of a model as a prediction and simulation tool is crucial. The first step is, in this respect, to consider the validity of the model:

- *Validation*
  Validation of a model means a process to ensure that sound scientific information, logically and sufficiently, has been included in the model, typically as mathematical expressions. Basically, it is expected that the developer of the model has taken into account aspects that are the most central for the prediction. The user of the model must consider if the validity of the model expression is observed in the actual case. Validation is not just to ensure that sound deterministic information is included but also that a stochastic description follows the nature of the phenomena and processes.

The first question raised follows the requirement concerning model validity. Throughout the entire text it has been a constant problem when dealing with the wet weather pollution from urban areas and roads that the variability in time and place is extremely high. Our prediction tools—the models—must therefore, even in the ideal state, be equipped with this basic characteristic. It is only our empirical knowledge based on monitoring and experimental observations that provide concrete and sufficient information on the parameters of the models. In order to observe high reliability of the model we apply, it is crucial that the data used for estimation of the parameters reflect a corresponding degree of variability. It is a clear requirement that a model should never be used outside its area of definition (i.e., in a situation where no data have ever proved the statement of the model).

Because of the variability, extensive data sets that correspond with the following use of the quality model are typically needed. Such data are normally divided into two groups and applied for the following two procedures, respectively:

- *Calibration*
  The objective of this procedure is to determine specific values of the model parameters based on empirically determined data that characterize the system and processes in question. The procedure represents a goodness of fit and the model parameters are adjusted until a certain (optimal) level of accuracy of the model output is obtained. The procedure results in values of model parameters that are defined as calibrated.
- *Verification*
  The model with the calibrated parameters is tested on the part of the data set that was not used for calibration. The level of accuracy should thereby be assessed under conditions where the model will be applied.

The extent of the data set used for calibration and validation will depend on the actual situation. It is clear that complex models with several parameters require more information than simple models to comply with the possible conditions under which the model should operate. As an example, and in the case of quality modeling of urban drainage systems, Mourad, Bertrand-Krajewski, and Chebbo (2005) recommend that data originating from at least 20 events be available and that the major part of these data be used for calibration.

There are different approaches for how calibration should be understood and performed. The ideal approach is that a parameter should express a specific system characteristic (e.g., that a growth rate parameter exclusively expresses the true growth characteristics of a given biological community). However, the real situation is that different system characteristics via empirically determined data influence the value of a calibrated parameter. Calibration becomes thereby complex. A model may require a large amount of data for its calibration and more than one set of parameters might observe a given criterion in terms of model performance. Further details of overparameterized models will be discussed in Section 10.2.3. The stochastic approach dealt with in Sections 10.2.4 and 10.2.5 is a possible answer to this problem.

It is outside the objective of this text to discuss how calibration and validation can be performed in practice. Several basic texts on mathematical modeling and statistics deal with this aspect. A number of papers (e.g., Mourad, Bertrand-Krajewski, and Chebbo 2005, and Bertrand-Krajewski 2007), exemplify and discuss specific aspects relevant for quality modeling within the area of urban wet weather pollution.

10.2 TYPES OF MODELS FOR TRANSPORT AND TRANSFORMATION OF CONSTITUENTS

Quality models that are applied within the area of urban wet weather pollution often include prediction of a pollutant mass load, the rate of formation or disappearance of constituents and prediction of where a pollutant actually occurs. In this way the following become core elements of a quality model:

- Stoichiometric characteristics of pollutant transformations
- Kinetic characteristics of pollutant transformations
- Transport characteristics of pollutants

Although formulated in broad terms, these characteristics constitute the base from which a model ideally should develop, typically with a mass balance as a central engineering approach (cf. Section 3.3). The description of the transport phenomena of both water and associated constituents must reflect the dynamic nature of the runoff. Unsteady flow conditions should therefore basically be considered, however, with steady-state flow as a first approximation (cf. Section 3.4).

The simple alternative to a model based on this approach is the "black box" model. In principle this type of model just formulates an empirically based relation (e.g., between input to and output from a system), and not governing processes and phenomena occurring within the system.

Models can be subdivided in several ways. In the previous chapters numerous models (i.e., mathematical expressions for calculation and prediction), have been presented. In most cases the goal was to quantify a certain phenomenon and thereby provide the user with a tool for calculation and prediction. Incidentally selected examples of such models are first-order decay models, dissolved oxygen stream models, models for prediction of wet weather pollutant loads, and eutrophication models for lakes and ponds. In the context of this chapter, a different approach in terms of grouping according to the general model performance will be dealt with:

- Empirical models
- Statistically formulated models
- Deterministic models
- Stochastic models
- Grey-box models

All five types of models are represented among those applied within urban wet weather pollution.

In the following sections, characteristics for these types will be outlined and commented. Although it is useful to subdivide models in such five groups, it is important to notice that a specific model can include characteristics from more than one of these groups.

10.2.1 Empirical Models

In principle, the formulation of an empirical model is based on results originating from experiments or systematic monitoring programs from where the data are organized according to the objective. No, or rather crude, knowledge of a phenomenon to be modeled is included in such a model. An empirical model can thereby predict a course-effect phenomenon that has previously been observed in similar systems and under similar conditions.

The degree of theoretical knowledge on a phenomenon that is modeled may vary considerably from one type of model to another. The very simple empirical models are black-box models (e.g., input–output models), which predict a phenomenon without knowledge about the governing processes. Other model types include some basic theoretical knowledge about the phenomena (e.g., the grey-box models, cf. Section 10.2.5).

As an example, Equation 4.5 ($M = A\ \alpha(y/A)^{\beta}$) is an empirical model for the prediction of the pollutant mass load, M, from a catchment during a runoff event based on information on the contributing catchment area, A, and an average runoff flow rate during an event, y. The (empirical) coefficients, α and β, are site-specific constants.

Empirical models are of particular importance when rather crude knowledge on the performance of a system exists. Such models are also useful when limited information in terms of data is available. What is, however, very important to note is that empirical models can only be used within their area of definition (i.e., the system or phenomenon modeled or predicted must be within the range at which the model was originally developed). Typically, the constants and parameters of an empirical model do not reflect any general understanding of underlying processes and phenomena. Such parameters must consequently be calibrated on-site and the calibration validated to obtain reliable results.

10.2.2 Statistically Formulated Models

Statistical models are based on a number of monitored data that are considered sufficient in terms of both quality and quantity to undergo a statistical analysis. The result of such a model is thereby expressed as a mean value and percentiles. Statistical models are by nature restricted to the conditions under which the data have been produced (e.g., limited by climate or catchment characteristics).

An example of a statistical model is the results of an analysis of historical rainfall series resulting in tables or curves relating values of intensity, duration, and frequency of the rainfall. The example in Figures 2.2 shows how an analysis of historical rainfall series can result in rainfall/interevent dry period/frequency curves. Another example in terms of a statistically formulated dissolved oxygen water quality criterion for CSO discharges into rivers is shown in Figures 6.7 and 6.8.

A well-known example of a statistically formulated model is the EMC lognormally distribution that, for practical reasons, often is expressed as a median value and the associated coefficient of variation. Numerous examples in this respect are found in Sections 2.4, 2.5, and 2.6 and in Table 2.2.

Statistical models are useful when a phenomenon is known to occur but a course-effect relationship cannot be directly determined. Neural network models that are developed by being taught from historical data are also examples of statistical models.

10.2.3 Deterministic Models

A deterministic model expresses the performance of a system in mathematical terms via a number of physical, chemical, or biological characteristics and processes. The mathematical formulation of the processes in a system and their interactions are often core elements of the deterministic description. A deterministic model is thereby based on theoretically well-accepted and relevant scientific knowledge that is applied for simulation of the performance of a system in process terms.

It is a main characteristic of a deterministic model that it is time invariant (i.e., when the input and initial conditions of a system is stated), the result of the model in terms of the output is always constant and unchanged.

A number of deterministic models within urban wet weather pollution have been developed. Particularly successful are the physically based hydrodynamic models (e.g., those describing nonstationary flow), but also stream and lake models including deterministic descriptions of major physicochemical, chemical, and biological characteristics and processes are known.

In general, deterministic models, and particularly those for simulation of quality aspects, include several parameters expressing stoichiometric and kinetic characteristics of governing processes. Deterministic models are strong tools but they require careful calibration and validation, often based on an extensive number of measured data. If such site-specific data are not available, the use of deterministic models cannot be recommended and the use of default values for model parameters may give misleading results. It is therefore crucial to understand that a complex deterministic model requires corresponding extensive data. If it is not the case, more simple model tools are strongly recommended.

Urban drainage phenomena are by nature complex and they are typically showing high variability—not necessarily high uncertainty. The fact that corresponding deterministic models therefore also turn out to be rather complex, often results in a structure that is overparameterized. A large number of parameters for expressing a given phenomenon is likely to be required to maintain a high complexity within a not too complex mathematical framework. It is therefore always a point to obtain an appropriate balance between the complexity of a model structure and the corresponding number of model parameters. At a pragmatic level, it is often needed to accept an overparameterized model in order to obtain an output that reflects the reality in terms of a correct response under varying external conditions. However, a large number of parameters may result in a difficult calibration and an extensive amount of data is often needed for the calibration process. Furthermore, one must be aware that good model fits to validation data are obtained, not just good fits to calibration data.

By nature, deterministic models lack the ability to handle the variability associated with the wet weather processes and parameters. In this respect, the stochastic models are stronger tools (cf. Section 10.2.4). However, the deterministic models are very useful for understanding the importance and impacts of specific processes and phenomena. Furthermore, deterministic models are part of the foundation of the grey-box models (cf. Section 10.2.5).

10.2.4 Stochastic Models

Stochastic models represent an extension of the deterministic models by including model parameters with a known or suspected statistic distribution. By using Monte Carlo simulation techniques, each single model sequence is repeated a large number of times—often several thousands. For each repetition of the Monte Carlo procedure, a random number generator is used for drawing a parameter value from the statistic distribution. The outcome of the modeling is analyzed statistically with the distribution of the output parameters being the model result. This result expresses the variability that is caused by the variability of the input parameters. As a consequence, stochastic models are suited for sensibility analysis.

As an example, a deterministic dissolved oxygen stream model simulating the effect of CSO discharges can be applied as a stochastic model if the input parameters are expressed in terms of a statistic distribution. A prerequisite is that the variability of the model parameters (e.g., the rate of oxygen transfer across the air–water interface) is known. Ideally, this variability should be known from relevant field investigations and determined a representative number of times.

The detailed requirements for using stochastic models within the area of urban wet weather pollution hardly exist and basically just as an approximation. Some model parameters may vary with a time resolution that in practice is not possible to determine. As an example, the quality of CSOs in terms of pollutant parameters varies with a time resolution of less than a minute. A pragmatic approach is to apply a simple and expected parameter distribution based on a limited number of investigations. It is often acceptable to apply such an approach, however, any prerequisite used must be taken into account when evaluating the model output. This interpretation is difficult because the parameter variability is affected by both the uncertainty in selecting the real statistic distribution and by the nature (e.g., variability over time) of the parameter in question.

The so-called regression models can be considered as stochastic models. These model types are also called black-box models because the parameters determined typically cannot be interpreted in a system or process related way.

10.2.5 Grey-Box Models

Grey-box models combine characteristics originating from both deterministic models and stochastic models. In this way, deterministic information in terms of physical, chemical, and biological knowledge of a system can be combined with stochastic information in terms of data by applying statistical methods for parameter estimation. The stochastic terms thereby include uncertainties in both model formulations and measurement data.

A grey-box model typically includes stochastic formulated differential equations with measurement noise terms (Bohlin and Graebe 1995). Unknown parameters of the model are estimated by applying a set of data describing the performance of the system. Statistical methods (e.g., a maximum likelihood method) can thereby be used for the estimation.

One of the benefits of using gray-box models is that typically few and physically meaningful parameters are included in the model. Furthermore, and compared with deterministic models, the use of measurement noise terms in a grey-box model means that process and measurement noise is less included in the model parameters. In general, more reproducible model results and less bias are therefore possible with the grey-box model concept.

10.3 TYPES OF URBAN DRAINAGE MODELS WITH SPECIFIC CHARACTERISTICS

The model types dealt with in Section 10.2 are used for a number of different purposes. The following grouping takes into account such specific applications (e.g., from a system point of view or based on the degree of complexity):

- *Models for the performance of technical systems*
  Models for batch and flow reactors are examples that can form a basis for technical systems (cf. Example 3.2). The design principle for a wet detention pond as depicted in Section 9.3.1.5 is another example.
- *Models for receiving water systems affected by CSOs or SWR*
  The dissolved oxygen stream model and nutrient model for lake eutrophication shown in Sections 6.3.3 and 6.4.4, respectively, are examples of such model types.
- *Models simulating the performance of subsystems*
  Models for pollutant transport in soils and models for load of pollutants from roads are examples of models that belong to this group. These model types can be subdivided and selected according to their degree of complexity.

As an example, Burton and Pitt (2002) subdivide stormwater models in the following three groups according to their degree of complexity:

- *Unit area loading models*
  Such models are basically load estimates corresponding to what is shown in Table 4.10.
- *Simple models*
  Examples of simple models in terms of pollutant buildup and wash-off at urban and road surfaces are dealt with in Section 4.5.
- *Complex models*
  In principle, models can become complex in two ways: they are formulated with either a complex hydrological or hydraulically description or they include a conceptual description of governing quality related processes. A pollutant buildup model as depicted in Equation 4.8 approaches this type of model.

The number of potential models within urban wet weather pollution is legion and the grouping selected is therefore not fully representative and well defined. However, for practical reasons it is often useful because a number of similarities exist within each group in terms of their performance, objective, or operation.

10.4 COMMERCIAL MODELS

Although it is not an objective of this text to deal with commercially available models that can be used for urban and road drainage purposes, it is considered useful to give a brief introduction and overview. The number of worldwide leading software developers within this area is rather limited. However, the number of models and corresponding versions they produce for specific purposes is rather large. These models, including their numerous versions, are subject to a continuous development.

Commercial models for analysis of wet weather hydraulic and hydrologic phenomena and processes have been available for at least the last three to four decades. The potential reliability of these models for predicting purposes is today rather high. A number of commercial hydrodynamic models, e.g., SWMM, MOdel for Urban SEwers (MOUSE), and HYDROWORKS have been developed. Typically such models are based on the use of local historical rainfall series and are calibrated and verified on flow data. Hydrodynamic drainage models may simulate runoff routing at the catchment surfaces and the flow in the drainage network, typically in terms of unsteady-state flow. Hydrodynamic models can be equipped with "add-on modules" or "blocks" that simulate selected aspects of drainage (e.g., sewer solids transport, advection–dispersion of constituents in the water phase, real-time control schemes, storage, and transformations of the constituents.

In commercial models, the term "water quality module" typically concerns a module that is designed to compute the generation and transport of pollutants in the sewer network, the pollutant load from a separate sewer catchment, or the discharge from a CSO structure. It is therefore important to note that a commercial water quality drainage model is typically a model for predicting pollutant loads in terms of computations of hydrographs and pollutographs that are generated for the discharges from a catchment or a sewer network. Models that are designed for prediction of effects in the receiving system do not belong to the group of drainage or sewer models. Exceptions are hydraulic models that simulate erosion in streams and channels. The pollutant load models are therefore typically restricted to simulate those water quality processes that concern inputs of pollutants to the wet weather flows (e.g., pollutant buildup and wash-off at urban surfaces, deposition and erosion in combined sewers, and the associated transport of the pollutants.

Examples of commercial available drainage models are

- SWMM (Storm Water Management Model) developed by the U.S. EPA (Environmental Protection Agency)
- InfoWorks CS developed by Wallingford Software in the United Kingdom
- Mike Urban CS developed by DHI (Danish Hydraulic Institute) in Denmark
- FLUPOL developed in France by AESN (Agence de l'Eau Seine-Normandie), SEDIF (Syndicat des Eaux d'Ile-de-France), and CGE (Compagnie Générale des Eaux)

These drainage models include hydrologic, hydraulic, and water quality processes. The models simulate transport processes through a system of urban surfaces, pipes, channels, and storage/treatment devices and they create hydrographs and pollutographs for discharges into a receiving system.

When using drainage models with their water quality add-ons, it is important to note that a complex model is not necessarily a better predicting tool than a simple model (Ahyerre et al. 2005). It is evident that complex models require extensive information on both the actual system and the relevant processes in terms of corresponding parameters. The variability and the dynamic behavior of urban wet weather phenomena must be accepted as central characteristics. These phenomena are fundamental obstacles for a deterministic description in model terms. The stochastic model concept is, in some cases, an appropriate solution in this respect.

REFERENCES

Ahyerre, M., F. O. Henry, F. Gogien, M. Chabanel, M. Zug, and D. Renaudet. 2005. Test of the efficiency of three storm water quality models with a rich set of data. *Water Science and Technology* 51 (2): 171–77.

Bertrand-Krajewski, J.-L. 2007. Stormwater pollutant loads modeling: Epistemological aspects and case studies on the influence of field data sets on calibration and verification. *Water Science and Technology* 55 (4): 1–17.

Beven, K. J. 2001. *Rainfall-runoff modeling: The primer*. Chichester: John Wiley & Sons, Ltd.

Bohlin, T., and S. F. Graebe. 1995. Issues in nonlinear stochastic grey-box identification. *International Journal of Adaptive Control and Signal Processing* 9: 465–90.

Burton, G. A., and R. E. Pitt. 2002. *Stormwater effects handbook: A toolbox for watershed managers, scientists, and engineers*. Boca Raton, FL: Lewis Publishers.

Harremoës, P., and H. Madsen. 1999. Fiction and reality in the modeling world: Balance between simplicity and complexity, calibration and identifiability, verification and falsification. *Water Science and Technology* 39 (9): 1–8.

Mourad, M., J.-L. Bertrand-Krajewski, and G. Chebbo. 2005. Calibration and validation of multiple regression models for stormwater quality prediction: Data partitioning, effect of dataset size and characteristics: *Water Science and Technology* 52 (3): 45–52.

Schnoor, J. L. 1996. *Environmental modeling: Fate and transport of pollutants in water, air and soil*. New York: John Wiley & Sons, Inc.

UNESCO-IHE. 2003. *Deterministic methods in systems hydrology: IHE Delft lecture note series*. UNESCO-IHE Institute for Water Education. Boca Raton, FL: CRC Press.

11 Legislation and Regulation

Pollution from urban stormwater runoff, combined sewer overflows, and highway runoff is by nature complex and site-specific. The stochastic nature of the rainfall, the variability of pollutant sources and loads, the influence of the infrastructure, the different types of systems that manage the runoff, and the effect of the local cultural behavior are just examples of what contribute to the complexity. These characteristics are, in several ways, different compared with corresponding aspects of the dry weather pollution from urban areas. A consequence is that legislation and regulation must be specifically formulated to comply with these wet weather characteristics.

The variability in the discharges in both time and place and the big number of discharge points that typically exist within a catchment are in this respect central and must be included when formulating discharge regulations. Performance and assessment procedures based on regular sampling and a data processing that is known for systems handling the more or less continuous dry weather point discharges (e.g., wastewater treatment plant outlets) are basically out of the question for the wet weather discharges. Therefore, a preassessed efficiency of a number of mitigation methods (e.g., the BMPs dealt with in Chapter 9), becomes central when selecting systems for regulation and control of pollutant discharges from the wet weather flows. In other words, wet weather runoff is basically regulated by a management strategy contrary to the dry weather discharges that are directly based on site-specific permits (e.g., as an amount or concentration of pollutant that is accepted being discharged). The formulation of specific water quality standards is central for the dry weather discharges; it is not the case for the wet weather flows.

An important issue in terms of regulation is the question of who is responsible for the pollutant discharges from the urban and road surfaces. For the combined sewers it is typically the water utilities or municipalities who operate the networks. The ownership of the separate system is often less well defined although it is typically a city or a department of transportation.

This book has—irrespective of the site-specific nature of the pollution—its focus on the basic and generally expressed quality characteristics, engineering methodologies, and tools for management of the urban wet weather discharges. Although it is crucial for the authors to produce a text that rests on such foundations, it is clear that requirements from the authorities in terms of legislation and regulations define important boundaries for the extent and need for engineering solutions. It is not the goal to focus on details of legislation and regulations that might exist in the different countries across the world. What is, however, important is to exemplify how the authorities formulate the details of legislation and regulations so they comply with the characteristics of pollution from the wet weather urban sources.

Public education on matters related to nonpoint source pollution cannot be overlooked and relevant information directed to the public is therefore crucial. Furthermore, it is clear that if efficient and up-to-date control of urban wet weather discharges should be successfully carried out, knowledge on both pollutant characteristics and control methods should be available for urban and road planners and engineers. Information via electronic media at a wide range of levels is therefore a central activity for governmental, state, and local authorities to support. Several agencies and institutions across the world are aware of the importance of Web sites and they inform the public and support professionals through the Internet.

Flood and erosion control have traditionally been undertaken in the development of urban areas, at least in the developed part of the world. Flood and erosion control have also widely been subject to regulation. Although the process is not yet completed, the runoff water quality from urban areas and roads has, during the past 30–40 years, been recognized as an environmental concern and included in legislation and regulation. In particular, countries in North America, Europe, Australia, and Japan have contributed to this development. In the following, examples of laws and regulations from the United States and Europe are focused on.

11.1 U.S. REGULATIONS

The U.S. federal and state laws and regulations related to urban and road runoff were developed early and enforced. Furthermore, massive investigations of the wet weather pollutant loads and their environmental effects have formed a rather solid basis for how to control and regulate the runoff. Due to that, the development in the United States has been a vehicle for what happened worldwide.

11.1.1 Clean Water Act

The Federal Water Pollution Control Act commonly known as the Clean Water Act (CWA) dates back to 1972. It established the basic structure for regulating discharges of pollutants into surface waters. It gave the U.S. Environmental Protection Agency (EPA) the authority to implement pollution control programs such as setting water quality standards for all contaminants in surface waters. It is thereby unlawful to discharge any pollutant from a point source without a permit.

Although the CWA originally focused on the chemical aspects of water quality, it is today more clearly devoted to restoring and maintaining chemical, physical, and biological quality aspects. Over the years, pollution from runoff has also been increasingly addressed. It is, furthermore, realized that stakeholder involvement and education of the public play a central role for both dissemination of information and for a successful implementation of means to reduce pollution.

11.1.2 Nationwide Urban Runoff Program

The U.S. EPA Nationwide Urban Runoff Program (NURP) was established in 1978 and a final report was produced in 1983. The program included extensive data collection related to urban drainage. To a great extent, the program has provided information that has affected not just the U.S. federal and state regulations mitigating and controlling pollution runoff in urban stormwater but also legislation and practice in other countries. Although the measurement program was performed 25–30 years ago, central parts of the results still play a significant role.

11.1.3 NPDES Permits and Measures to Manage Urban Runoff

The Clean Water Act requires that all point sources discharging pollutants into receiving waters must obtain a National Pollutant Discharge Elimination System (NPDES) permit. Most stormwater discharges and CSOs are considered point sources and therefore require coverage by an NPDES permit. As an example, the primary method to control stormwater discharges is in accordance with the program through the use of BMPs. The increased volume and rate of runoff from impervious surfaces as well as the discharge of pollutants are in this respect being addressed. Management programs that are developed at the regional level (i.e., by the different states) regulate discharges originating from the urban nonpoint sources.

From a regulation point of view, a number of measures to manage urban runoff are focused on

- *Plans for new developments*
  These plans are directed to protect sensitive ecological areas, minimize land disturbance, and retain natural drainage and vegetation. Structural controls and pollution prevention strategies should be implemented to achieve a pre-development level.
- *Plans for existing developments*
  Such plans include stepwise clean-up strategies starting with the reduction of priority pollutants and ending with an ecological restoration.
- *Plans for on-site disposal systems*
  Systems for disposal of different forms of waste should be properly designed and maintained and not be situated close to ecological sensitive areas and surface waters.
- *Public education*
  Public education programs can be directed to prevent pollution and keep the environment clean. Such programs are implemented in both schools and at a community level.

11.1.3.1 CSO Program

Under the NPDES permitting program, the CSO policy contains a number of fundamental principles that have developed over time. In addition to clear levels of control to meet health and environmental objectives, an important aspect is the flexibility to consider the site-specific nature of CSOs and find the most cost-effective way to control them.

The first milestone was a national CSO control strategy issued in 1989 and expressed in the following six minimum controls:

- Proper operation and regular maintenance of overflow structures
- Maximum use of the collection system for storage
- Review and modification of pretreatment programs

- Maximum flow to a treatment plant
- Prohibition of dry weather overflows
- Control of solids and floatable material in CSO discharges

The second milestone was the policy issued in 1994 for minimum technology-based controls required implemented in 1997. In principle, this policy became law with the passage of the Wet Weather Water Quality Act of 2000. The purpose was an elaboration on the control strategy issued in 1989 but also an expeditious compliance with the CWA. The corresponding modified controls are considered measures that reduce the prevalence and impacts of CSOs. However, they were not expected to require significant investigations or major constructions. As a result, the objectives for the local communities were to implement and submit documentation for the following nine minimum controls:

- Proper operation and regular maintenance programs for the sewer system and CSO structures
- Maximum use of the collection system for storage
- Review and modification of pretreatment programs to ensure that CSO impacts are minimized
- Maximization of flow to wastewater treatment plant, in general by primary treatment of the CSOs
- Prohibition of CSO discharges during dry weather
- Control of solids and floatable materials in CSO discharges
- Pollution prevention programs that focus on reduction activities
- Public notification to ensure information on CSO occurrences and CSO impacts
- Monitoring CSO controls to effectively characterize their impacts and efficiency

Following this phase with minimum controls, a community with combined sewer systems must develop a long-term control plan (LTCP) for CSO discharges that will comply with the CWA. According to this LTCP, the impact of CSOs are required to be monitored as a basis for assessment of appropriate control measures. It is still a cornerstone that such a plan is flexible according to the site-specific nature (e.g., by relocation of CSOs in sensitive areas). It is therefore important that different options of structural and nonstructural measures for CSO control exist (cf. Chapter 5). One of two possible approaches can be followed to assess if the LTCP goal is met. The first possibility is a performance criterion defined as either maximum four to six overflows per year or 85% capture by volume. The second possibility is to assess if required water quality standards are met.

Further information on the U.S. EPA program for CSO control can be found at the Web site cfpub.epa.gov/npdes/cso/cpolicy.cfm.

11.1.3.2 Stormwater Program

Following the CWA and the NURP, the U.S. EPA developed a Phase I program and a Phase II program of the NPDES Stormwater Program in 1990 and 1999, respectively. The due dates of these two programs were 1992 and 2003, respectively.

The Phase I program addressed sources of stormwater runoff that had the greatest potential to negatively impact water quality. Under this program, U.S. EPA required NPDES permit coverage for stormwater discharges in medium and large municipal separate storm sewer systems located in places with populations of 100,000 or more. Furthermore, certain industrial facilities of specific polluting potential were included. The Phase I regulations do not apply to discharges of parking lot runoff from nonindustrial areas or roof runoff.

Under Phase II, small and midsized communities were required to design and implement programs to curb stormwater from municipal, industrial, and construction sources as already required for larger communities under the Phase I rules.

The Phase II rule defines a stormwater management program to be implemented for Small Municipal Separate Storm Sewer Systems (referred to as Small M4Ss). The program comprises six elements that, when implemented, are expected to result in significant reductions of pollutants discharged into receiving water bodies. These six elements are termed *minimum control measures*:

- Public education (i.e., information to the citizens about the impacts of polluted stormwater runoff)
- Public participation (e.g., participation in program development and implementation)
- An illicit discharge detection and elimination (i.e., developing and implementation of a plan to detect and eliminate the illicit discharges to the storm sewer system)
- Construction site stormwater runoff control, particularly for erosion and sediment control. The use of a BMP is an example of a control measure
- Postconstruction runoff control. This element aims at controlling stormwater runoff from new development and redevelopment areas. Applicable controls include the use of structural BMPs. In doing so, municipalities were given a variety of BMPs to choose from
- Pollution prevention (i.e., good housekeeping). This sixth element includes a number of control measures with the goal of preventing or reducing pollutant runoff from municipal operations (e.g., reduction in the use of pesticides and cleaning of catch basins). The program thereby includes municipal staff training on pollution prevention measures and techniques

As can be seen, these six points show that public education, staff training, and the use of both structural and nonstructural BMPs are considered central elements in the control of pollution from stormwater runoff. Although it is required that the effectiveness of the chosen BMPs be evaluated, it is important to notice that monitoring is not required under the rule with the exception that the relevant authority has the discretion to require monitoring if deemed necessary. Regular monitoring for assessment of a control measure, as known from wastewater treatment facilities, is because of the number of stormwater outlets not feasible for a municipality. To reduce the discharge of pollutants to the maximum extent practicable, the general knowledge on the effectiveness of BMPs therefore becomes extremely important (cf. Chapter 9). The advice from the authorities concerning proper selection of BMPs and the availability of a corresponding database is therefore crucial (FHWA 1996; WEF 1998; ASCE 1998; Strecker et al. 2001, 2004).

Further details of the U.S. EPA program on stormwater control can be found via the following Web site: www.epa.gov/npdes/stormwater.

11.1.3.3 Proposals for Stormwater Regulation

Because urban stormwater in terms of its polluting impacts onto the environment has a rather short history, it is evident that there must be a standing discussion on the expediency on the current structure of regulation. As an example, the National Research Council (NRC) under the U.S. National Academies has proposed changes in the structure of stormwater regulation (NRC 2008). A central aspect of the proposal is that the current permitting structure be changed to a watershed-based permitting structure. It is clamed that this type of approach will more directly include all types of stormwater pollutant sources. Stormwater regulation is thereby placed at the municipal level and also more directly connected to land use and a number of potential control actions, such as conserving natural areas and reducing impervious urban areas.

The trend is that stormwater management systems switch from underground concrete pipes to ecological systems in terms of the different BMPs dealt with in Chapter 9. In this way, the concept is a return to some kind of predevelopment state of the urban hydrological cycle. The Phase II program focusing on the small and midsized communities should support this change.

11.2 EUROPEAN UNION LEGISLATION

European Union (EU) laws can be divided into different types. Among these types, the regulations, often initiated by the European Commission (EC), come into force throughout the EU from the moment they are passed. A different and important type of EU laws is the directive. Contrary to regulations, EU directives—also proposed by the Commission—are not directly implemented but they bind all Member States to implement national laws that observe the overall objective of a directive. However, the EU does not dictate how it should be done and the national governments themselves decide the means. In general, a directive is addressed to all Member States and specifies a date by which the Member States must have put the directive into effect. The European Court of Justice has provided guidelines for Member State judges on how to deal with cases where directives have not been transposed into national laws or have been transposed incorrectly.

The EC have issued several laws directed to protect humans and the environment. A great number of these are directly or indirectly important in case of pollution from urban and road runoff. Particularly important are the directives and, among them is the Water Framework Directive, are expected to play a central role in the future.

11.2.1 Water Framework Directive

In 2000, the EC issued its Water Framework Directive (WFD; EC 2000). The ambition was to establish a framework for protection of inland surface waters, transitional waters, coastal waters, and groundwater. In this respect it is the ultimate objective to achieve concentrations in the marine environment near background levels for naturally occurring substances and close to zero for manmade synthetic substances. As formulated in the WFD, "good chemical and ecological status" in all water bodies is what must be reached.

To achieve these goals, the WFD will work on catchment basis and thereby include all types of pollutant inputs to the water cycle (i.e., also inputs from sources for the urban and road runoff pollutants). The intention is that the WFD thereby establishes a strategic framework for managing the entire water environment holistically. The original plan was that by the end of 2015, all surface waters in Europe should fulfill the criteria set. However, it is realized that this deadline cannot be observed for all situations and an extension with two to six years until 2027 is therefore possible. The details of how to implement the WFD are not yet fully known and will demand extensive scientific, technical, and administrative skills, not to mention the resources that will be required.

The WFD does not include detailed plans for how to manage stormwater runoff and associated pollutants. However, pollutants originating from urban and road surfaces and their potential impacts on the water quality must, because of the requirements set by WFD, gain more attention. The increased concern within EU Member States for toxic chemicals in the environment is a step in the right direction.

Handling of the runoff from rain events by designing systems that are based on the management and control by natural occurring physical, chemical, and biological processes in a sustainable way is one of the objectives of this text. The CSO control measures and the different BMPs dealt with in Chapters 5 and 9, respectively, may in this respect play a central role. The WFD can, via its requirements, enhance a process of development that is not just directed to fulfill environmental and health demands but also esthetic and recreational purposes. From a pragmatic point of view it is important that these goals be observed in a cost-effective way.

The broad goals included in the WFD will probably encompass many of the individual quality standards set through specific legislation such as the Nitrates Directive and the directive concerning Urban Wastewater Treatment. In this way the WFD may contribute to integrated and more rational measures for water quality improvements.

11.3 WATER QUALITY CRITERIA FOR WET WEATHER DISCHARGES

Water quality is a characterization of water by use of qualitative or quantitative descriptors relevant to the state or use of water. The descriptors are most often called parameters and represent a selected and relevant set of pollutant descriptors (e.g., in terms of concentrations), together with other characteristics (e.g., kinetic and stoichiometric parameters for the pollutants) necessary for the description of the beneficial use of water.

A water quality criterion ideally represents a constituent concentration or level associated with a degree of environmental effect upon which judgment can be based. For practical purposes a criterion means a designated concentration of a constituent that, when not exceeded, will protect an organism, an organism community, or a prescribed water use or quality with an adequate degree of safety.

In general, water quality criteria are therefore relevant for surface waters and groundwater and not related to the mode of discharge. If this standpoint is taken for granted, it is not relevant to distinguish between the sources of the discharges. However, when it comes to the pragmatic solutions, the wet weather impacts are quite different compared with the impacts from the continuous point sources. Criteria that relate to the wet weather urban discharges—and different from those expressed for the dry weather discharges—can therefore be relevant. In this respect, the specific nature of the effects from wet weather sources were dealt with in Section 1.3 and Chapter 6. Briefly, the main reasons for defining water quality criteria that reflect the specific characteristics of effects for urban wet weather discharges are

- The discharges are event-based.
- The discharges are stochastic of nature.
- The time scale effect becomes evident and requires a distinction between effects that are either acute or accumulative.
- Some events, the extreme ones, may result in fatal effects on the ecosystem.

As far as the accumulative effects are concerned, the pollutants originating from the wet weather discharges can be judged parallel with and in addition to the corresponding loads from continuous discharges. As an example, the discharge of nutrients, N and P, from urban wet weather discharges should be considered just a source of input parallel with all other inputs of the nutrients. The relevant time scale for the pollutant load is a season or a year and will therefore include a number of event-based discharges as well as the more or less continuous inflows from the dry weather sources.

Contrary to the accumulative effects, the acute effects from urban wet weather discharges may require attention in terms of specific formulated water quality criteria. Based on the four points mentioned above, a water quality criterion for an event-based acute effect of a specific pollutant must take into account the following:

- The frequency of the effect
- The magnitude of the effect

A criterion that refers to an acute effect must therefore be statistically based and may vary depending on the type of receiving water and the prescribed quality or use of water. It is often stated that CSOs can give rise to acute effects whereas SWR typically should result in accumulative effects. This statement in terms of a clear difference between the impacts of CSOs and SWR is questionable. No doubt, accumulative effects exist but they are very difficult to assess. It is, however, the case that CSOs as well as SWR can result in acute effects observed on the biota of receiving waters (cf. Section 6.5). The principle of a criterion based on assessments and originating from U.S. EPA (1983) is for an acute effect of a heavy metal shown in Figure 11.1. It is readily seen that the two points mentioned above (i.e., the frequency, return period, and the magnitude, concentration, of the effect) are included in the criterion.

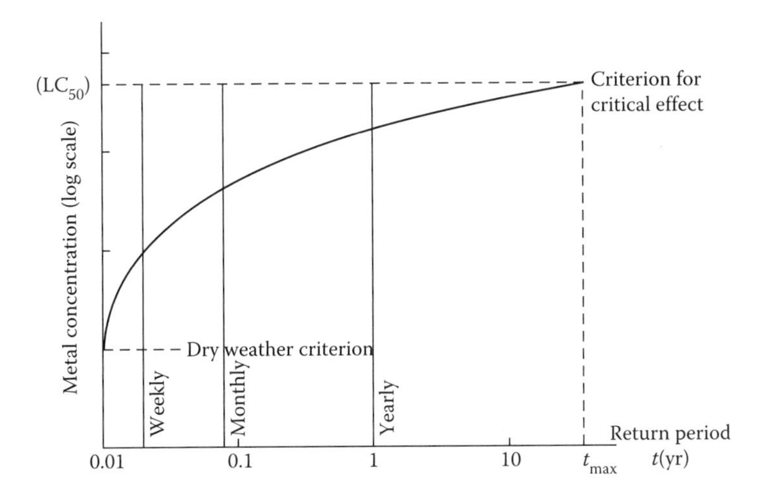

FIGURE 11.1 Principle of a water quality criterion for an acute effect of a pollutant (heavy metal) discharged during a runoff event. Notice that above the return period, t_{max}, for which a critical effect (here defined as LC_{50}) is accepted, the criterion is basically no longer relevant.

Figure 11.1 only expresses a principle and several aspects (e.g., what the definition of an effect or the degree of it), is not discussed in this context. Further details in this respect are dealt with in Chapter 6, particularly in Section 6.3.4. Criteria that follow this principle have been reported in U.S. EPA (1983) and Hvitved-Jacobsen (1986). As previously discussed in this chapter, it is for practical reasons—particularly the big number of discharge points for the wet weather flows and lack of solid knowledge on the effects—not possible to transfer the criterion to a standard with legal consequences (cf. Section 11.4.1).

A criterion requires careful consideration in view of what quality is intended and what is possible in practice. It will therefore depend on a number of local ecosystem characteristics. As an example, the following four types of river systems have been defined for the Ruhr River basin, Germany (Grünebaum et al. 2003):

- Wild river landscape
- Open river plain
- River near residential areas
- Urban river

The names of these different river systems clearly show that the population density (i.e., the degree of urban development) must be considered when assessing quality aspects.

The application of the concept depicted in Figure 11.1 requires that the criterion be compared with data originating from the location in question. To do the statistics in terms of the return period, it is basically required that measurements have been performed during several years. Alternatively, data that substitute measured values might be generated using a model for prediction of the effect. It is, however, clear that this procedure requires careful considerations.

11.4 WATER QUALITY REGULATION FOR WET WEATHER DISCHARGES

11.4.1 Water Quality Standards

A water quality standard is a politically decided and set goal expressing desired uses or states for a particular water. The distinction between a water quality criterion and a water quality standard is subtle but important. A criterion is a statement concerning the limiting levels of concentration or intensity of water quality parameters established with regard to the intended use of the water. On the other hand, a water quality standard is a regulation as to the limiting level of concentration or intensity of a parameter established by a regulatory authority for the purpose of protecting or preparing a water resource for one or more uses. Criteria are therefore typically set by scientifically working institutions whereas standards require a legally based regulatory authority.

The generally decided water quality standards for surface waters, groundwater, and drinking water that are valid in a country will include all types of discharges being from both wet and dry weather sources. However, when it comes to discharge standards that are known from point sources like wastewater treatment plants, it is problematic to recommend standards for urban wet weather discharges. The reason is that standards must be controlled by the regulatory agency. Although concentrations or loads of pollutants theoretically can be measured for discharges from overflow structures and stormwater outlets, the variability within an event and between events makes a control from a practical point of view very difficult if not impossible. Another problem is also what penalty is possible and who should be accused. The sources for the wet weather discharges are diffuse and as a consequence an ultimate solution might require that the regulatory agency close a road to traffic!

Water quality criteria for discharges do not result in such drastic actions as water quality standards might do. Discharge criteria should be considered to help in assessing the impact of potential sources for pollutants and to help in the design of proper systems for discharge for protecting a given receiving water system in terms of the general standards set.

As a consequence, other means than water quality standards are therefore needed by the regulatory agency to manage and control wet weather discharges. Procedures that make use of Best Available Technologies (BATs) and Environmental Impact Assessment (EIA) are means that have been either recommended or required by regulatory agencies (cf. Sections 11.4.2 and 11.4.3).

11.4.2 Best Available Technologies

A BAT can be considered a pragmatic approach for a solution to a problem, often related to treatment and management procedures. A BAT-solution typically originates from a list of methods for recommendation. When dealing with urban and road runoff, a BAT can, as an example, be found among a list of BMP techniques that provide an overview of their characteristics, although such information should generally be recommended for screening purposes only.

A number of BAT-solutions for treatment of urban and road stormwater runoff are dealt with in Chapter 9. The BAT-solutions for reducing the impact of CSOs are included in Chapter 5.

11.4.3 Environmental Impact Assessment

An EIA includes a number of procedures intended to provide information on the environmental impact of a project, often in case of new and large-scale developments. The objective is to evaluate different alternatives to the project with emphasis on their impact onto the environment and requirements of resources, including costs, and to consider methods that observe such goals. An EIA thereby provides information that can be used for decision making among the interested partners of a project (e.g., the developer, the authority responsible for the environment, and the stakeholders.)

The specific contents of an EIA vary but it typically includes a description of the project and the existing environment, the different alternatives of the project considered, the assessment of the proposed project, and its alternatives and a number of possible mitigation measures. An EIA will normally deal with concrete aspects of a project and not assess fundamental alternatives that fall outside the competence of the developer. Furthermore, an EIA is, in principle, limited by the information that is available. There are, because of the different partners, built-in conflicts in an EIA and there is also a debate whether or not EIA should be required by legislation.

Within the area of urban and road drainage the number of solutions to a problem is often high. An EIA can therefore provide valuable information for a final decision among the alternatives assessed.

11.5 INFORMATION AND TECHNICAL SUPPORT

As mentioned in the introduction of this chapter, efficient reduction and control of pollution from wet weather discharges require both participation of the public and skilled technical support. Up-to-date information on subjects related to pollution from urban areas and roads is therefore crucial.

It is not the goal of this text to give detailed guidance in this respect. However, it is important that several public agencies and institutions see it as an obligation, and a challenge, to contribute to a dissemination process. In addition to the information that is given via printed media, like newspapers, brochures, magazines, periodicals, reports, and books, Web sites play a specific and increasing role. Because of the extent and continuous changes, a list of such Web sites is not possible to produce. However, a few addresses where extensive information and support related to knowledge on pollution from nonpoint sources and their control can be found are

- The U.S. EPA: www.epa.gov/nps
- International Stormwater BMP Database Reports and Software (different project sponsors): www.bmpdatabase.org
- A number of Web sites produced in relation to projects financially supported by the European Community

Web sites produced by national and state environmental protection agencies, road authorities, and cities across the world often include both general information and specific recommendations and requirements for management of the wet weather related pollution from urban areas and roads. Reports produced by such agencies and authorities are often disseminated via Web sites.

11.6 FINAL COMMENTS

The politicians define, in terms of legislation, a wide range of conditions for the public and the environment. The decisions they make in terms of legislation and regulation related to the wet weather pollution is, in this respect, no exception. The politicians need sound and relevant information and they are also affected by the feedback they receive. The importance to quantify and to explain in understandable terms the nature and relevance of the wet weather impacts and means to avoid or reduce adverse effects cannot be overlooked. It is therefore gratifying to realize that central formulations in the legislation and regulations of the wet weather pollution, as included in this chapter, have been positively influenced of the scientific and technical facts that have been dealt with in the preceding chapters.

REFERENCES

ASCE. 1998. *Manual and report on engineering practice No. 87.* Reston, VA: American Society of Civil Engineers.

EC. 2000. European commission's directive 2000/60/EC of the European Parliament and of the council establishing a framework for the community action in the field of water policy. http://www.wfd-info.org/

FHWA. 1996. Evaluation and management of highway runoff water quality. U.S. Department of Transportation, Federal Highway Administration (FHWA), Publication No. FHWA-PD-96-032.

Grünebaum, T., G. Morgenschweis, E. A. Nusch, H. Schweder, and M. Weyand. 2003. Measurements for structural improvement with regard to the good status of water bodies: Estimation of expenditure for a river basin in Germany. *Water Science and Technology* 48 (10): 39–46.

Hvitved-Jacobsen, T. 1986. Conventional pollutant impacts on receiving waters, a review paper. In *NATO ASI Series G: Ecological Sciences*, Vol. 10, 345–78, Berlin, Heidelberg, Germany: Springer Verlag.

NRC. 2008. *Urban stormwater management in the United States*. Report from the National Research Council (NRC), U.S. National Academies.

Strecker, E., M. Quigley, B. Urbonas, and J. Jones. 2004. Stormwater management: State-of-the-art in comprehensive approaches to stormwater. *The Water Report* 6: 1–10.

Strecker, E. W., M. M. Quigley, B. R. Urbonas, J. E. Jones, and J. K. Clary. 2001. Determining urban storm water BMP effectiveness. *Journal of Water Resources Planning and Management* 127 (3): 144–49.

U.S. EPA. 1983. Final report of the nationwide urban runoff program (NURP). U.S. Environmental Protection Agency, Water Planning Division, Washington, DC.

WEF. 1998. *Urban runoff quality management,* Manual of Practice No. 23. New York: Water Environment Federation.

WEB SITES

www.epa.gov/watertrain/cwa: U.S. EPA Clean Water Act.

www.epa.gov/npdes/stormwater: U.S. EPA National Pollutant Discharge Elimination System (NPDES).

Appendix 1: Definitions

This appendix includes selected definitions that are generally used in the text. Definitions that mainly refer to a specific chapter where they are further explained and discussed will not be repeated in this list but can be found in the relevant chapter or via the index list. Commonly used terms that are expected to be well known are excluded from the list. A number of terms typically used in urban drainage are explained in Ellis et al. (2004).

Accumulative Effect (Chronic Effect)

A lethal or sublethal effect that occurs over a prolonged period of time, often caused by exposure to relatively low pollutant concentrations or to low doses. In contrast to the acute effect, it is normally difficult to assess an accumulative effect based on reliable tests.

Acute Effect (Short-Term Effect)

A lethal or sublethal effect directly caused by a discharge of substances related to an event, typically a CSO event or a SWR event. An acute effect is typically observed within a period of two to four days and typically assessed against the LC_{50} dose or concentration, often using a 1 to 24 hour exposure test.

Catchment Area

The (impervious) area that contributes (drains) to stormwater runoff and related pollutant load irrespective that the area is served by a separate sewer network, a combined sewer network, or a road surface. The words catchment and watershed are used identically in the text.

Chronic Effect

See *accumulative effect.*

COV

The coefficient of variation, COV = standard deviation divided by mean value.

Combined Sewer Overflow (CSO)

The mixed untreated flow of wastewater and runoff water from impervious urban surfaces and roads that are discharged from an overflow structure in a combined sewer network. The discharge takes place when the capacity of the downstream network (including the treatment plant) is exceeded.

Effective Impervious Area (EIA)

The impervious urban or road area that contributes to the runoff. (Note that EIA is also an acronym for environmental impact assessment.)

Effectiveness of a (Treatment) Method

A measure of how well a (treatment) system in general meets the goals for the inflows to the system.

Efficiency of a (Treatment) Method

A measure of how well a (treatment) method removes pollutants.

First Flush

First flush or first foul flush describes the phenomenon that the first part of the runoff during an event has a relatively higher pollutant concentration or pollutant load than the following part of the runoff. First flush is a phenomenon that can be observed in both combined sewers and storm sewers. (See Section 2.4.1 for further details.)

Highway Runoff

In this text the word is used to identify stormwater runoff from a highway or other types of road surfaces.

Intercepting Sewer

A large sewer in a separate or combined sewer system that conveys the flow of wastewater to a wastewater treatment plant. An intercepting sewer collects the flow from small and trunk sewers and typically has few connections.

LC_{50} Value

A concentration or a dose of a substance (substance with a toxic effect) that leads to a mortality of 50% of a population (cf. acute effect and Section 6.5).

Partitioning

A phenomenon that describes the dynamic behavior of a substance in terms of its transfer and interactions between two phases, either a gaseous phase, a liquid phase, or a solid phase. Partitioning thereby includes a number of different transfer processes like air–water mass transfer, sorption, and ion exchange. The ultimate state of a partitioning reaction is the equilibrium where no net transfer takes place.

Performance of a System

A general measure of how well a system, according to its design, meets the goals in terms of both hydraulics and water quality characteristics. An extended understanding of performance includes costs, regulatory requirements, use of resources, and aspects of risk and safety.

Pollutant

The word pollutant is not well defined and to some extent it can give a wrong signal. In spite of that, it is often used identically with the words "component" and "substance" in the text.

Road Runoff

Stormwater runoff from a road or a highway (cf. *highway runoff*).

Short-Term Effect

See *acute effect*.

Stormwater Runoff (SWR)

The runoff from a catchment that originates from rainfall or snowmelt. Stormwater runoff occurs from urban catchments served by a separate sewer network, roads, and highways; see *highway runoff* and *road runoff*.

Surface Runoff

The runoff of rainwater or snowmelt at an urban (impervious) surface irrespective of the type of sewer network, runoff at a road surface, or runoff at a highway surface.

Trunk Sewer

A sewer that collects wastewater or stormwater from small sewers and conveys the flow to an intercepting sewer.

Watershed

The word in the text is used identically with the word catchment.

REFERENCES

Ellis, B., B. Chocat, S. Fujita, W. Rauch, and J. Marsalek. 2004. *Urban drainage: A multilingual glossary*. London: IWA Publishing.

Appendix 2: Acronyms

This appendix includes selected acronyms used in the text.

ADT	average daily traffic
BAT	best available technology
BMP	best management practice
BOD	biochemical oxygen demand
CEC	cation exchange capacity
COD	chemical oxygen demand
COV	coefficient of variation (standard deviation divided by mean value)
CSO	combined sewer overflow
CWA	Clean Water Act
DO	dissolved oxygen
DWF	dry weather flow
EA	environmental assessment
EC	European Commission/escherichia coli
EIA	effective impervious area/environmental impact assessment
EMC	event mean concentration
EPA	U.S. Environmental Protection Agency
EU	European Union
FC	fecal coliforms
FHWA	federal highway administration
FS	fecal streptococci
IDF	intensity-duration-frequency curve
IE	intestinal enterococci
LID	low impact development
NPDES	U.S. National Pollutant Discharge Elimination System
NURP	U.S. Nationwide Urban Runoff Program
OM	organic matter
PAH	polycyclic aromatic hydrocarbon
RTC	real time control
SMC	site mean concentration
SOD	sediment oxygen demand
SS	suspended solids
SSO	sanitary sewer overflow
SUD	sustainable urban drainage
SUDS	sustainable urban drainage system
SWR	stormwater runoff
TC	total coliforms

TKN	total Kjeldahl nitrogen
TN	total nitrogen
TOC	total organic carbon
TP	total phosphorus
TSS	total suspended solids
U.S. EPA	U.S. Environmental Protection Agency
VSS	volatile suspended solids
WFD	EC water framework directive
WSUD	water sensitive urban design
WWTP	wastewater treatment plant
XOC	xenobiotic organic compound

Index

E

F

Q

R

S

V

W

X

Z